Multimodal Polymers with Supported Catalysts

Alexandra Romina Albunia • Floran Prades
Dusan Jeremic
Editors

Multimodal Polymers with Supported Catalysts

Design and Production

Springer

Editors
Alexandra Romina Albunia
Borealis Polyolefine GmbH
Linz, Austria

Dusan Jeremic
Borealis Polyolefine GmbH
Linz, Austria

Floran Prades
Borealis Polyolefine GmbH
Linz, Austria

ISBN 978-3-030-03474-0 ISBN 978-3-030-03476-4 (eBook)
https://doi.org/10.1007/978-3-030-03476-4

Library of Congress Control Number: 2018966813

This Springer imprint is published by the registered company Springer Nature Switzerland AG
The registered company address is: Gewerbestrasse 11, 6330 Cham, Switzerland

Overview of Polyolefins - Role of Polyolefins in Our Daily Lives

Polyethylene, or, more accurately, low-density polyethylene, LDPE, was the first synthetic polyolefin discovered in the early 1930s by the ICI scientists Gibson and Fawcett [1]. The highly branched thermoplastic material, a pure hydrocarbon, found its application immediately after the discovery. It remains one of the most significant synthetic materials ever used. LDPE is produced under a high-pressure process in bulk ethylene under 2000–3500 bar pressure and at temperatures that are above 200 °C. Polymerisation under these conditions is initiated by free radicals, which are products of thermal decomposition of added peroxides.

The discovery of the catalytic polymerisation processes in the early 1950s, Ziegler catalyst by Ziegler et al. [2] and chromium catalyst by two independent groups at Standard Oil and Phillips Petroleum [3], presented a steep change in the availability of polyethylene. The sheer fact that the catalytic process operates at pressures that are two orders of magnitude lower than the high pressure process needed for manufacturing LDPE simplified the construction and operation of the manufacturing plants.

Fundamental differences between the radical polymerisation and coordinative polymerisation, that is, the mechanism of a catalytic process, are caused by structural differences in the produced polymer. While LDPE molecules are highly branched and cross-linked, material made in a catalytic process is mostly linear, with few branches in the polymer chain that are not deliberatively and carefully introduced. Such a linear material crystallises faster and to a higher degree, consequently increasing the density of the polymer. The catalytic process introduced high-density polyethylene, HDPE.

In the early days of the low-pressure polyethylene, the development moved rather fast by discovering new ways to introduce, use and control additional degrees of freedom for tailoring the primary structure of polymer chains, which influences the mechanical and chemical performance of the material.

The molecular weight of the polymer can be controlled from the stat by regulating the temperature at which the chromium catalyst is being used and by applying and controlling the concentration of a chain transfer agent, usually hydrogen, in the reactor during polymerisation with Ziegler catalysts.

Branching, leading to changing the density of the material, is reintroduced to the linear polyethylene made with Ziegler catalysts by copolymerising α-olefins [4]. Similarly to controlling molecular weight, the concentration of the comonomer during polymerisation directly influences a number of incorporated branches, crystallinity and density of the polymer. The ability to control molecular weight and density of the materials allows the physical performance to be tailored.

Soon after the low-pressure catalytic polymerisation of ethylene was discovered, Natta et al. [4] found out that the same catalyst class, Ziegler catalysts, can also be used for polymerising propylene. This discovery, as well as the ground-breaking body of work done and published by Natta et al., introduced and established another thermoplastic polyolefin that started being industrially manufactured and used in a globally significant amounts.

Pushing the performance boundaries with the significantly higher melting point and stiffness, polypropylene, more precisely syndiotactic polypropylene, rapidly became and has remained one of the most used thermoplastics in the world.

Similarly to the development of polyethylene, polypropylene development further improved the physical and chemical performance of the material and, therefore, the usability for various applications.

One of the principle approaches in modifying the physical and chemical performance of polypropylene is designing the primary structure of the polymer chain. For example, increased stereoregularity, and therefore crystallinity, of the material increases the stiffness of the manufactured objects.

Creating random and block copolymers by combining propylene with ethylene polymerisation in one or more reactors, as well as more sophisticated designs of both matrix and amorphous components of the overall polypropylene material, brings further features such as impact resistance, softness combined with thermal stability, etc.

Polyethylene and polypropylene are two materials that are being manufactured and used in the amount of approximately 160 million tonnes per year globally. PE and PP, together with other, smaller players such as polybutene, are some of the most utilized materials, and represent more than half of the total plastic materials used each year.

Almost a century since their discovery, polyolefins have a large presence in our daily lives. Compared to other synthetic and natural materials, they are simpler to obtain and manufacture, light, with a density lower than 1 g/cm^3, chemically resistant, and not soluble at common ambient temperatures. They are also recyclable and relatively environmentally friendly and durable. In combination with the variety of physical performance features that can be obtained through the grade of the material, it is clear why polyolefins are chosen for material manufacturing for a wide range of products.

Polyolefins can be seen in almost all aspect of our lives. They are most readily found in packaging. The features that make them suitable for the application are predominantly desired and tailorable stiffness/softness, cleanliness, barrier properties, durability, chemical inertia, simplicity in processing, and low cost, among others.

The next largest application of polyolefins is in the construction industry. One of the most demanding application of polyopefins is in manufacturing pipes that will be used for the transport of liquids and gases. Pressure resistance and durability are the required features, so that the pressure pipes used in buildings can last for more than 50 years in service.

Due to its light weight, tailorable stiffness and impact resistance at different temperatures and paintability, polypropylene is increasingly used in the auto industry. Many parts of automobiles are being manufactured from compounds where PP is the principle polymer, often combined with PE, fillers and potentially other materials.

A significant and important application of polyolefins is in the energy transfer and storing. Used for insulation, ultra-clean LDPE can improve the quality of cables to the extent that large amounts of energy can be transmitted over long distances. This sort of energy transfer is essential in implementing and exploiting renewable sources of energy, i.e. wind-powered generators.

The above-mentioned applications are some of the more demanding high-volume applications that are impacting the society. However, by no means they are even close to represent all the ways polyolefins influence our lives. Examples include using UHMWPE for the manufacture of artificial human joints, and as a replacement of PVC in medical applications such as blood bags and tubes. It is also included as corrosion protection for metal objects such as steel pipleines; these are only a few examples of the myriad of applications.

Taking into account the variety of functions that materials based on polyolefins need to fulfil, it is obvious that the above-mentioned ability to tailor physical, mostly mechanical, and chemical performance of the polymer is essential for industry. The primary structure of polymer chains and their composition with similar or different primary structures are some of the principle parameters that are the deciding factor regarding the performance of the material.

So, for example, molecular weight distribution is directly connected with the processability of the material. Broader MWD lowers the energy demand when the polymer is being extruded and vice versa. High molecular weight component in a flexible packaging material contributes to the toughness of the packaging needed to maintain the integrity of the package. Low molecular weight component, on the other hand, enables effective and fast manufacturing, which lowers the overall cost of the package.

The amount and location of the comonomer incorporated in the polymer chain also has an important role in the quality of the polymer. Comonomer-rich long polymer chains create very high impact resistance to film material.

The structure property and processing relationships and correlations are understood at a high level by researchers in both industry and academia.

During the century in which various polyolefins have been developed, several approaches to controlling and tailoring the primary structure have been taken. Most of them can be grouped into three principle categories:

- Creation and the use of sets of multiple polymerisation conditions zones where different polymer chains are generated and combined in situ, often in one particle. This approach is represented mostly by the use of multiple polymerisation reactors in series or parallel.
- The use of specifically designed initiators, catalysts or a combination thereof. This approaches uses multiple types of active sites in a polymerisation catalyst that perform differently and therefore generate the composition of desired polymer chains. The combination of active sites can be a consequence of the inherited catalyst nature; it can be also tailored by combining the catalyst with different initiators before the polymerisation reactor.
- Blending of polymers in the after-reactor process.

Selected contributions in the book describe the role and performance of catalysts in the processes used for manufacturing multimodal polymers. The catalyst construction, recent developments and trends in polymer formation are described in the earlier chapters. Subsequent articles describe the particle growth theory, modelling and consequences of the growth. Finally, the effects of the catalyst performance under various polymerisation conditions on the product performance are discussed.

References

1. TCE Today, http://www.tcetoday.com//media/Documents/TCE/Articles/2011/845/845cewctw.pdf (accessed February 10, 2014).
2. G. Wilke: 50 Jahre Ziegler-Katalysatoren, Angew. Chem. 115 (2003) 5150–5159.
3. H.A. Wittcoff, B.G. Reuben, J.S. Plotkin: Industrial Organic Chemicals, J. Wiley & Sons, New York 2012.
4. D.B.Malpass: Introduction to Industrial Polyethylene, J.Wiley & Sons, New York 2010.

Contents

Chapter 1
Recent Developments in Supported Polyolefin Catalysts: A Review

John R. Severn

1.1 Overview

Over the last 60 years the ability to reduce olefinic refinery gases or liquids into a metastable solid in a controlled manner has created the colossal business of polyolefin materials. Their continued success is thanks to a deep understanding of how to meet and predict a customer's needs in terms of a price/performance package and translate that back through the chain of knowledge (Fig. 1.1). This demand has led to constant evolutions within all areas of the business, punctuated by more than its fair share of revolutionary breakthroughs in the areas of catalyst, polymerization process, and polymer processing technology.

One only has to consider the range of applications where a polyolefin solution finds itself as the preferred option. It is employed in such areas as infrastructure (piping and energy transmission) allowing for the safe consistent supply of water and energy (electricity and gas, etc.) and removal of sewage; advanced packaging, light and reliable packaging that reduces transport emissions (reduced petroleum consumption) and increased shelf life of perishable goods, reducing waste and further reducing transport emissions; automotive application replacing metal with light material allows for lighter automobiles and again further contributes to reduced transport emissions. Examples of the extremely diverse applications of the material can even be found within very niche areas. For example, UHMWPE fibers can be used in securing supertankers, stopping bullets, and surgical sutures.

Polymer synthesis role in this success has come by the ability to control how the macromolecules are put together by sequentially linking α-olefins: the chain length and distribution, skew, and branching present handles by which properties can be

J. R. Severn (✉)
DSM Ahead, Geleen, The Netherlands

SFD Group, Technical University of Eindhoven, Eindhoven, The Netherlands
e-mail: John.severn@dsm.com

A. R. Albunia et al. (eds.), *Multimodal Polymers with Supported Catalysts*,
https://doi.org/10.1007/978-3-030-03476-4_1

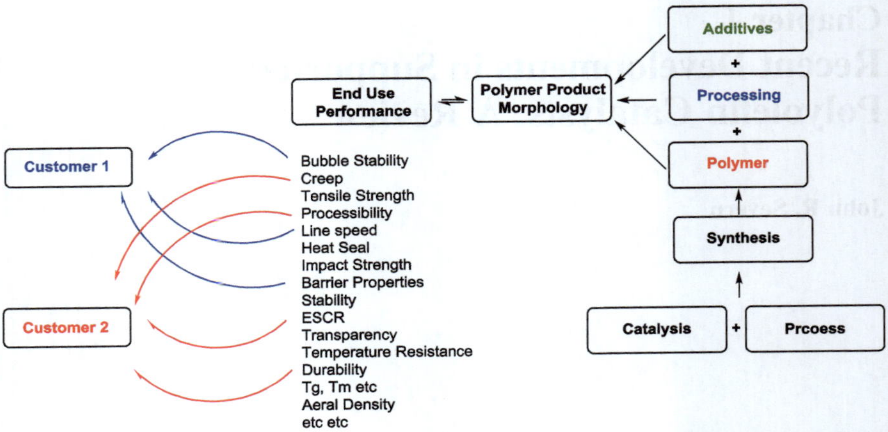

Fig. 1.1 Success through chain of knowledge

tuned. When the monomer is substituted (i.e., an α-olefin), additional opportunities including tacticity (control over stereo- and regio-error type linkages) and comonomer composition distribution become available. In addition, process developments have allowed for the production of advanced materials such as heterophasic alloys. Control over chain length (molecular weight), the population distribution of the chain lengths (molecular weight distribution), and chemical composition distribution (CCD) in and between the chains results in the ability of the chains to form crystallizable, ordered segments. The size and distribution of the crystallites and the ratio of soft amorphous space interspersed between the hard crystallizable segments affect material properties such as melting point, modulus (stiffness), and toughness (resistance to fracture when stressed).

The ability to control the above factors through the use of catalyzed coordination polymerization can be seen throughout the previous and subsequent chapters.

1.1.1 Scope of the Chapter

It is impossible to cover all areas and recent developments within supported polyolefin catalysis utilized in particle-forming process in a single chapter. As a result choices have been made to focus on industrially relevant systems that are at least demonstrated at pilot scale or relevant for the production of multimodal products. As a result there is little academic research discussed in this chapter and will be referred to indirectly via the numerous reviews in this field [1–15]. In most cases the topic will be discussed in terms of illustrative examples of the subject matter.

Finally, the author has focused on patent literature where there is an English-translated version; as a result the author openly acknowledges and apologizes for his ignorance and for not fully covering the tremendous contributions that researchers from companies in Japan, China, Korea, etc. have made to the development in this area.

1.2 Ziegler Catalysts

1.2.1 Ziegler/Natta Polypropylene Catalysis

The history of industrial Ziegler–Natta polypropylene catalysts is generally described in terms of evolutionary generations, corresponding to the chronological order of their development [16–18]. They range from titanium trichloride catalysts, which had their heyday in the late 1950s and the 1960s, to the high-activity magnesium chloride-based catalysts, which have helped fuel the growth and development of these versatile polymers. Advances in process technology have gone hand in hand with the development of polypropylene catalysts, developing from slurry-based processes to the current state-of-the-art, cascade processes which combine bulk or gas-phase technology, allowing for the production of complex copolymers with multiphase structures which have expanded the application range for PP.

$TiCl_3$ Catalysts (First and Second Generation)

The $TiCl_3$ catalysts used in early industrial PP processes were typically prepared by the reduction of $TiCl_4$ with an aluminum alkyl, generating a solid $TiCl_3$ precipitate. $TiCl_3$ exists in four crystalline modifications, α, β, δ, and γ forms. The β-modification has a linear chain-like structure, while α, δ, and γ forms possess layer structures [19, 20]. Typically, the reaction of $TiCl_4$ and $AlEt_3$ (at low Al/Ti ratios) at low temperatures in hydrocarbon solution resulted in the controlled precipitation of catalysts having spheroidal particle morphology, yielding the β-$TiCl_3$ form with cocrystallized $AlCl_3$. This precursor can be converted to the more stereoselective γ-form by heating to 160–200 °C [21]. The catalysts were typically activated by $AlEt_2Cl$ to afford poorly productive systems (*ca.* 1 kg PP/g Cat) which in many cases yielded polypropylene resins that required extractive removal of atactic polymer and removal of catalyst residues (deashing).

An improved (second generation) $TiCl_3$ catalyst developed by Solvay appeared in the 1970s [22]. The catalyst preparation procedure involved the chemical treatment of a $TiCl_3/AlCl_3/AlEtCl_2$ precatalyst, previously produced by the reaction of $TiCl_4$ with $AlEt_2Cl$. The catalytic activity could be greatly improved by extraction of the co-crystallized aluminum chloride with diisoamyl ether, giving a "pure" β-$TiCl_3$. Subsequent treatment with $TiCl_4$ catalyzed the phase transformation from the β- to the δ-form of $TiCl_3$ at a relatively mild temperature (<100 °C) [23]. Lowering the temperature also prevented the growth of the catalyst crystallites. In addition, the extraction of $AlCl_3$ with ether resulted in porous particles with a weakly bonded matrix. The innovation increased the activity and stereoselectivity and resulted in polymerization productivities in the range 5–20 kg/g catalyst in bulk propylene and produced polymers where the removal of the atactic fraction is spared [24]. Although phased out in many locations these catalysts still survive to this day [25, 26].

Third Generation: "Activated" $MgCl_2$

The prehistory of the third generation began in 1960, when Shell patented a catalyst for propylene polymerization that employed $TiCl_4$ supported on $MgCl_2$. The decisive step was achieved however in the late 1960s, when Montecatini and Mitsui independently development "activated" $MgCl_2$ as a support for $TiCl_4$, followed by the incorporation into the catalyst system of electron donors giving higher catalyst stereospecificity [27–29]. Initially, activated $MgCl_2$ was prepared by ball milling a mixture of magnesium chloride and ethyl benzoate, which leads to the formation of very small ($\leq$3 nm thick) primary crystallites of $MgCl_2$ [30].

Third-generation catalysts commonly comprise $MgCl_2$, $TiCl_4$, and an "internal" electron donor and are combined with an aluminum alkyl cocatalyst such as $AlEt_3$ and an "external" electron donor added to the polymerization. The internal donor in third-generation catalysts is typically ethyl benzoate, which is used in combination with a second aromatic ester, such as methyl *p*-toluate or ethyl *p*-ethoxybenzoate (PEEB), as an external donor. An external donor is required due to the fact that, a large proportion of the internal donor is lost as a result of reactions involving the cocatalyst, such as alkylation and/or complexation reactions. The external donor replaces, to a large extent, the internal donor in the solid catalyst, maintaining high catalyst stereospecificity. It has been demonstrated that the most active and stereospecific catalyst systems are those which allow for the highest incorporation of external donors [31], with the effectiveness of a catalyst system depending more on the combination of donors than on an individual internal or external donor. A classic example was disclosed by Scata et al. [32].

Fourth Generation: Phthalate ID/Alkoxysilane ED

The fourth generation of PP catalysts was based on the discovery and development of phthalate/alkoxysilane-based catalyst systems (Fig. 1.2). As was the case for the third-generation catalysts, the function of the external donor is to replace the internal donor lost by alkylation and complexation reactions with the Al alkyl cocatalyst [33, 34]. An additional feature was the development of spherical $MgCl_2$-based catalysts with controlled particle size and porosity, afforded via chemical rather than mechanical activation of magnesium chloride. As a result, catalyst systems with increase stereoselectivity and regioselectivity could be produce, resulting in higher crystallinity polymers, with a considerable increase in catalyst mileage. Phthalate-based catalysts are now regarded as a family of general-purpose systems with which it is possible to cover the majority of product properties and applications. All these

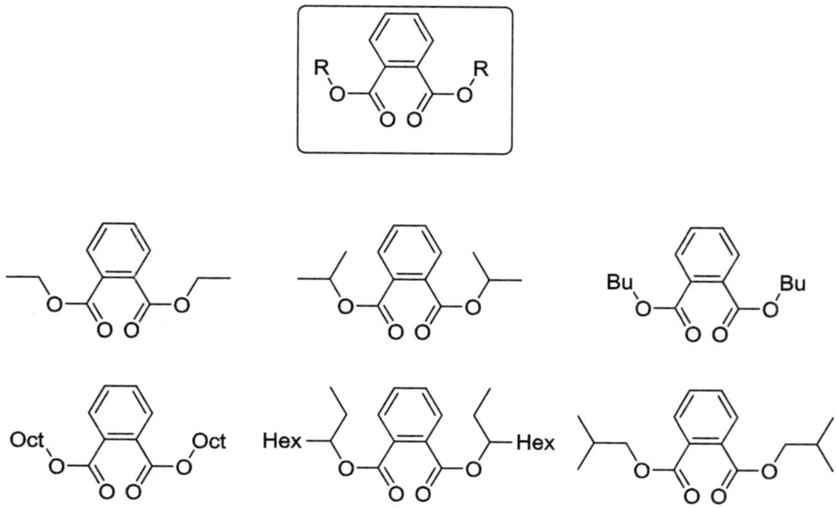

Fig. 1.2 Illustrative example of reported phthalate donors

Fig. 1.3 Illustrative examples of diether internal donors

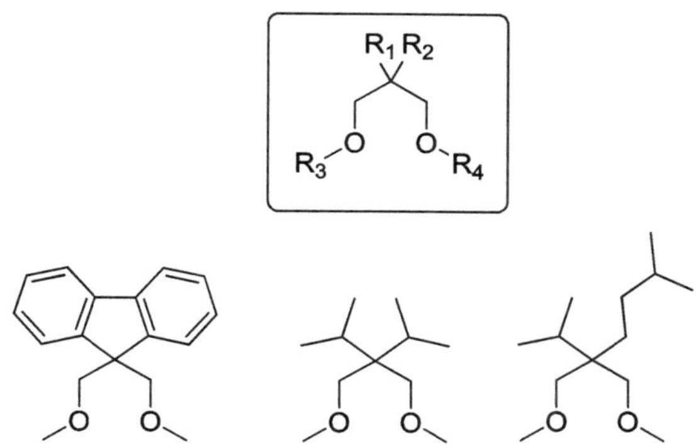

features have contributed to the colossal commercial success of this type of catalyst, and at present they are the most widely used catalyst system in polypropylene production. However, the coming years will see this dominance slowly erode.

Fifth-Generation Catalysts: Diethers, Succinates, and Polyol Esters

The search for further catalyst improvements led to the development of internal donors which were not readily removed from the support on contact with the alkylaluminum cocatalyst. It was found that certain diether compounds, in particular 2,2-disubstituted-1,3-dimethoxypropanes with an oxygen–oxygen distance in the range 2.8–3.2 Å (Fig. 1.3), similar to those of the alkoxysilane external donors are not extracted when the catalyst is brought into contact with the $AlEt_3$ cocatalyst. As a result, high stereospecificity can be obtained even in the absence of external donor [35–39]. Fifth-generation diether catalyst systems show particularly high polymerization activity and good stability, typically giving yields exceeding 100 kg PP/g catalyst. They also give relatively narrow molecular weight distribution and show high sensitivity to hydrogen. Cui et al. recently reported their investigation into the influence of diether donor ratio on the performance of the final catalyst. Experimental data indicates that higher diether loadings cause a decrease in the titanium content of the catalyst, further indicating that the diether is associated with magnesium not titanium. In addition, an optimum ratio was required for activity and stereoselectivity [40].

A seminal work by Chadwick and coworkers has shown that the high hydrogen response of fifth generation, diether-based catalysts arises from chain transfer after the occasional secondary (2,1-) rather than the usual primary (1,2-) monomer insertion [41]. The group rationalized that even the most highly stereospecific active sites are not totally regiospecific. In addition, the broader molecular weight distributions obtained with catalysts containing ester internal donors were attributed to the presence of (some) isospecific active sites having very high regiospecificity and therefore lower hydrogen sensitivity [42].

There is significant evidence from modeling studies that diether donors preferentially adopt a bidentate coordination mode to the (110) lateral cut of $MgCl_2$ [43, 44]. Strong evidence points to the fact that donor molecules reside in the vicinity of stereospecific active species. Therefore, the above reports indicate that the likely location of active titanium is also on the (110) lateral cut. In addition, it has also been shown [45, 46] that the use of a diether as external donor in combination with a phthalate-based catalyst gives active species which are very similar to those present when the diether is used as internal donor, implying that the active species in phthalate-based catalysts are similarly located on the (110) cut.

Lately, new types of internal donor compounds based on aliphatic dicarboxylic esters, such as malonates and glutarates and in particular succinates and polyol esters (Fig. 1.4) have been employed. Researchers at LyondellBasell developed $MgCl_2$-supported catalysts in which the internal donor is a succinate ester [47, 48]. As is the case with phthalate-based catalysts, an alkoxysilane is used as external donor. The essential difference between these catalysts is that the succinate-based systems produce polypropylene having much broader molecular weight distribution. In addition, they also generate ethylene–propylene copolymers having considerably lower glass transition temperature, coupled with better intermolecular and intramolecular comonomer distribution with respect to their conventional counterparts. This

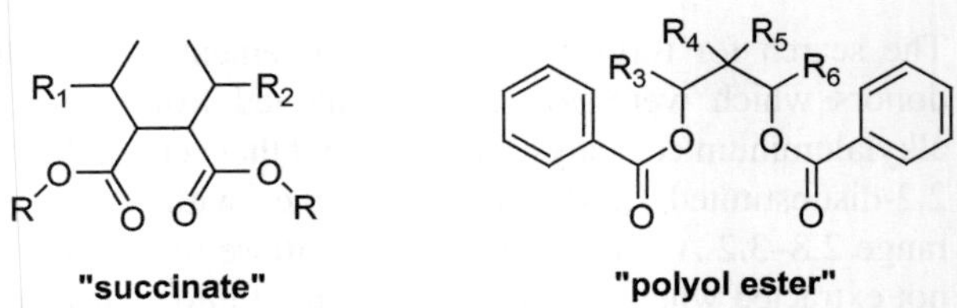

Fig. 1.4 Illustrative examples of succinate and polyol ester internal donors

has allowed for the production of heterophasic copolymers having an improved balance of stiffness and impact strength, taking into account that the incorporation of a rubbery (ethylene–propylene) copolymer phase into a PP homopolymer matrix increases impact strength but leads at the same time to decreased stiffness. Similarly, researchers at BRICI/Sinopec have developed the polyol ester family of internal donors. The catalytic systems resulting from this family of donors have a similar polymerization performance to succinate-based systems. However, unlike the succinates, the polyol ester systems yield high stereoselectivity even in the absence of alkoxysilane external donors [49, 50].

Sixth-Generation Ziegler: Phthalate Replacement

Recent concerns and regulatory restriction (e.g., REACH Regulation (EC) 1907/2006) on the use of phthalates in general has been a source of concern and debate in terms of whether a catalyst is a substance or transported isolated intermediate. From what the author presently understands, dibutyl phthalate (DBP), diisobutyl phthalate (DIBP), and bis(2-ethylhexyl)phthalate (DEHP) are listed on REACH Annex XIV with a sunset date of 21st Feb, 2015. The concentration limit for those three substances requiring an authorization will be 0.3 wt% (3000 mg/kg) and given that PP produced with these catalyst will have at most 1 mg/kg then phthalate/procatalyst-based PP will not be subject to any ban or restriction. All that said, however, offering a customer a totally phthalate-free solution is becoming an attractive offering in terms of responsible care and competitive advantage. As a result, those who treated the debate as an opportunity rather than a threat have reinvigorated research into internal donors, leading to new and interesting donors that not only challenge phthalates. This has led some to define this period as the sixth generation of PP catalysts. Figure 1.5 illustrates some of the numerous disclosures of new donors in recent time [51–70]. In addition to this there has been an increase in mixed donor disclosures, for example blending succinate + diether [71] or succinate + dimethoxytoluene [72].

Undoubtedly it is Dow (now with W.R. Grace) that has been at the vanguard of these developments, with one family appearing to stand out which may well be part of their new non-phthalate offering (Consista Donors). The 1,2-phenylene dibenzoate donors (Fig. 1.6), judging from the recently disclosed large-scale production of PP and a push to protect the donor synthesis routes themselves [73–76], are clearly important as phthalate replacements. It is also clear that these donors may be more complicated in that multiple $TiCl_4$ contact steps and buffering via addition of ethylbenzoate or 2-methoxy ethylbenzonate to the procatalyst and $TiCl_4$, before adding the phenylene dibenzoate, may be required to improve the final morphology and performance of the catalyst [77].

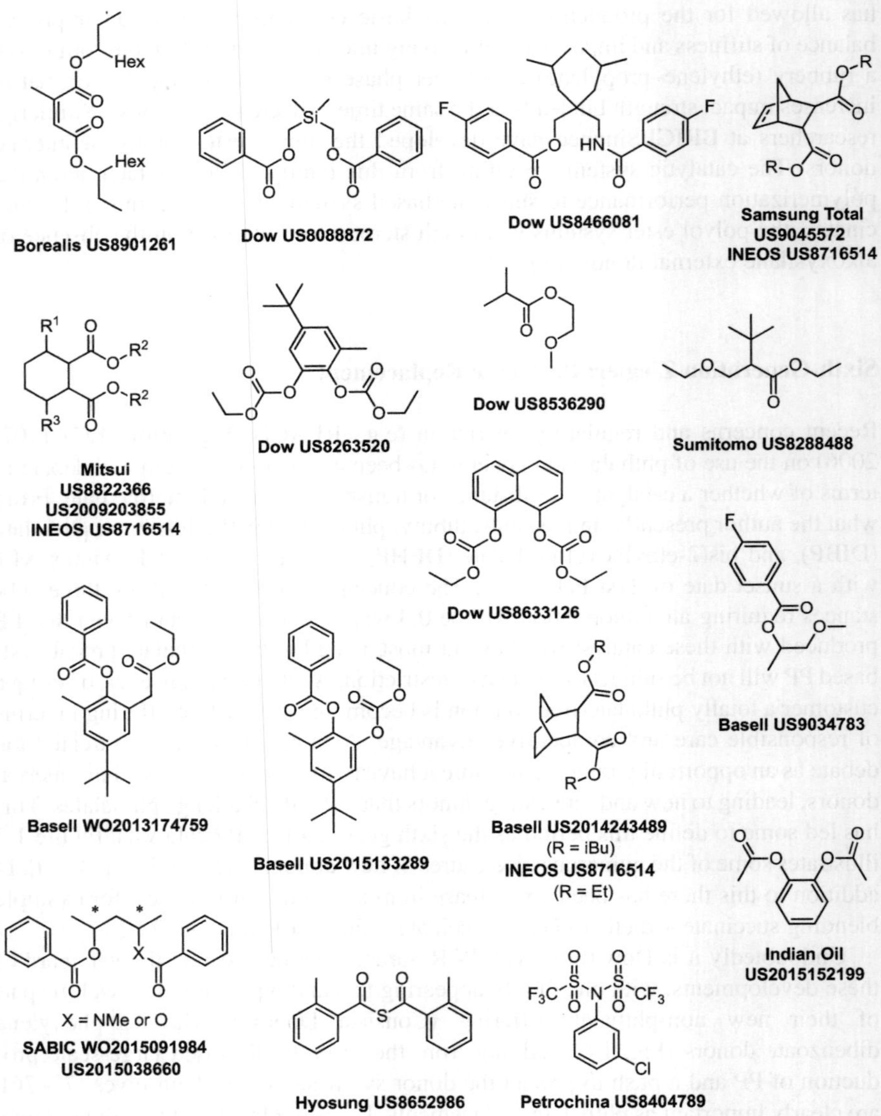

Fig. 1.5 Selection of new donors disclosed in recent time

Catalyst Morphology Control

Several different approaches have been followed for the preparation of catalysts having controlled particle size and morphology. Three main types of reaction have been employed for the controlled preparation of a support material or final catalyst:

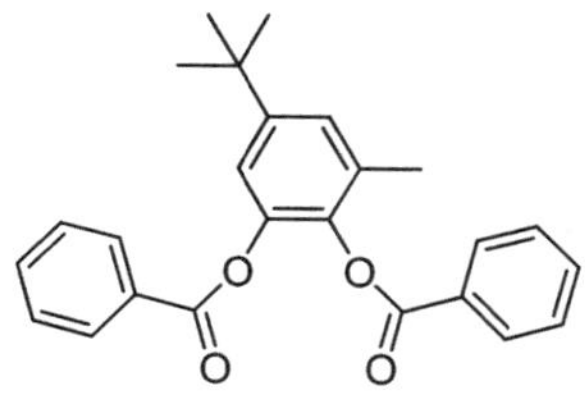

Fig. 1.6 1,2-Phenylene dibenzoate

1. Complexation of $MgCl_2$ with an alcohol, forming a viscous melt at elevated temperature, which is then processed to form a spherical $MgCl_2 \cdot EtOH_n$ support with controllable EtOH content either via spray drying or emulsification/solidification. The support material is treated with the external donor and titanated to form the final catalyst product.
2. Complexation of $MgCl_2$ with an alcohol in a hydrocarbon, followed by controlled precipitation of the reaction products following addition of a chlorinating agent or $TiCl_4$.
3. Controlled precipitation of the reaction products of a magnesium alkyl or alkoxide-containing complex with a chlorinating agent or $TiCl_4$.

Finally followed by the impregnation of a silica carrier with the catalyst reagents.

The latter is the least frequently employed method for the production of Ziegler–Natta PP catalysts, however it is a common technique in Ziegler PE catalysts. Typically, a silica support is treated with a magnesium alkyl complex, followed by chlorination (HCl), addition of $MgCl_2$ and ethanol to form a support material that is converted into the final catalyst via titanation and external donor addition. Representative examples can be found in the work of Novolen/BASF [78–81].

Processes for the preparation of spherical $MgCl_2 \cdot n$EtOH supports are based on spray drying or controlled solidification techniques. The latter is exemplified by the work of LyondellBasell and its previous incarnations. In this preparation molten emulsions of a $MgCl_2 \cdot n$EtOH ($n \approx 2$–3) adduct complex, created in Vaseline and silicone oil, are rapidly cooled in cold hydrocarbon to produce spherical particles with a relatively narrow particle size distribution [82]. As both $MgCl_2$ and ethanol are hydroscopic, care is required in controlling the water content of the final support [83]. This base material has provided the backbone for LyondellBasell developments of Ziegler PP catalyst as well as Zeigler PE catalyst for gas phase polymerization. Recent chemical variation on this base procedure has seen the additions of organoclay such as Cloesite 15A [84] and more recently $Mg(CH_3COO)_2 \cdot 4H_2O$ [85] introduced during the formation of the $MgCl_2 \cdot n$EtOH adduct complex to provide a final catalyst particle with presumably greater strength, as it results in a reduction of the number of "broken" polymer particles. The Cloesite 15A appears to impact the bulk density by possibly increasing the amount of void catalytic space. Similar carrier material is employed in the Sinopec/BRICI DQ-catalyst [86].

$MgCl_2 \cdot n$EtOH (n 2–3) complexes are generally crystalline and consist of a large number of covalent $[MgCl_2]n$ polymeric chains which basically yield the "activated" δ-form of $MgCl_2$ following extraction of ethanol to form the final catalyst.

Relatively recently, Sozzani et al. characterized a number of adducts asserting that $n = 2.8$ or 1.5 represent defined and stable $MgCl_2 \cdot n$EtOH adducts [87]. Auriemma and De Rosa have shown that the high degree of nanoporosity of the derived catalysts may be due to intermediate formation of three-dimensional structures of "amorphous" or highly disordered crystalline $MgCl_2$ polynuclear chain-like species in equilibrium with the "activated" δ-form of $MgCl_2$. The authors postulate that the high degree of "nanoporosity" leads to an increase in the contour length of the lateral faces of $MgCl_2$, which is available for epitactic coordination of titanium and donor species. Further observations point to the fact that control over the $MgCl_2$–EtOH stoichiometry may play a key role in their formation [88]. Experimental observations have indicated that $MgCl_2 \cdot n$EtOH support material in which $n = 3$ is desirable for preparing highly productive and stereoselective catalysts for the homopolymerization of propylene. Reduced alcohol content in the support material either via initial stoichiometry or controlled dealcoholation yields systems with reduced activities and stereoselectivities in homopolymerization but yield catalysts suitable for ethylene–propylene copolymerization [89]. Catalysts derived from support material in which the ethanol contents are lower than 1.5 form the basis of LyondellBasell so called "reactor granule" technology, which when utilized in their proprietary process technology (Catalloy) leads to the formation of heterophasic polyolefin alloys with particularly high rubber contents [48, 90, 91].

LyondellBasell "State-of-the-art" technology allows for high control of catalyst particle dimension, distribution, and surface smoothness (regularity), as illustrated in the scanning electron micrographs of the catalyst particle surface at different levels of magnification in Fig. 1.7a–c.

Spray-dried supports are formed by contacting $MgCl_2$ with ethanol. The adduct is subsequently melted and spray injected into a cooled solvent or gas, which crystallize as spherical particles of $MgCl_2 \cdot n$EtOH ($n \approx 3$) which are screened to extract the required particle size and distribution, the remainder being redissolved and recycled [92–95].

Spherical $MgCl_2 \cdot n$EtOH support materials formed via emulsion or spray drying techniques are converted into the final catalyst via titanation steps. The reaction with excess $TiCl_4$ is undertaken in the presence of the dialkyl phthalate internal donor [17]. Temperatures of at least 80 °C and at least two $TiCl_4$ treatment steps are normally used, in order to obtain high-performance catalysts in which the titanium is mainly present as $TiCl_4$ rather than the $TiCl_3OEt$ generated in the initial reaction with the support. Enhancements in the extraction of superfluous $TiCl_3OR$ from the final catalyst have been reported by Garorff and coworkers at Borealis [96]. The group utilized the addition of dioctyl phthalate as an internal donor "precursor" which undergoes a transesterification reaction to form the final internal donor with the by-product aiding in the extraction of the $TiCl_3OR$.

Catalysts obtained via these routes generally have a BET surface area of around 300 m^2/g and pore volumes in the range 0.3–0.6 cm^3/g [17]. The same supports are now being utilized by Novolen [97] and Süd-Chemie [98], as the basis for some of their PP catalyst offerings.

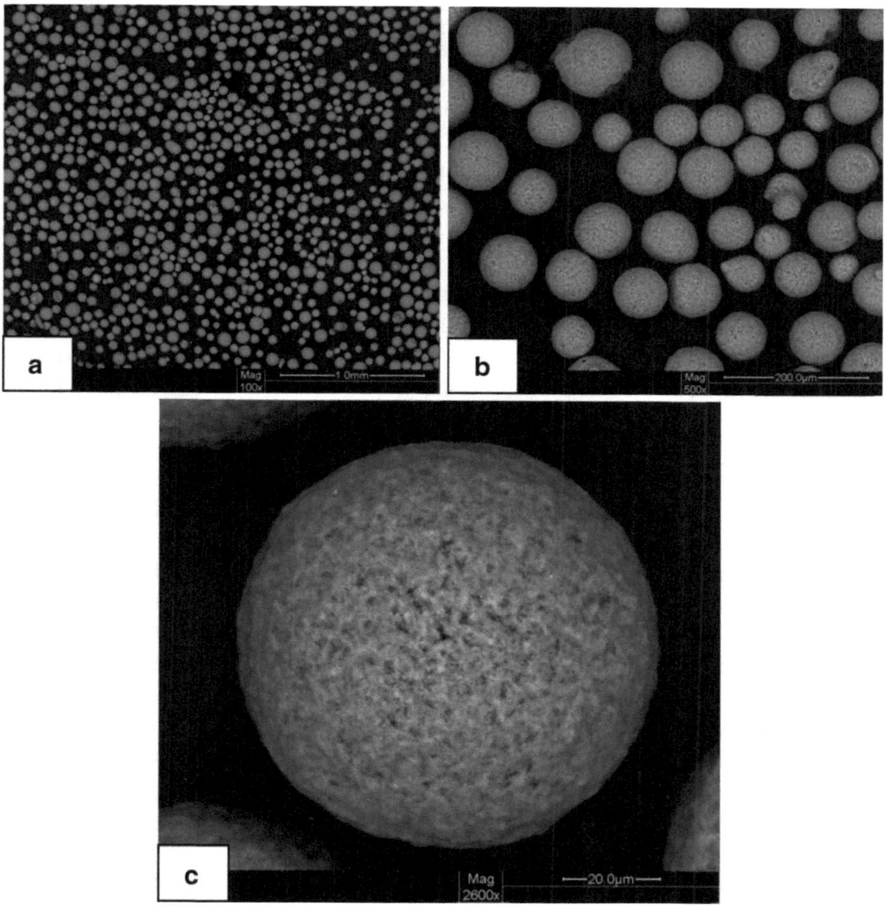

Fig. 1.7 Scanning electron micrographs of catalyst particle surface at different levels of magnification, (**a**), (**b**), and (**c**) (reproduced with permission from LyondellBasell)

Precipitation of a magnesium/titanium precursor with controlled morphology has been a highly successful route and has found significant commercial relevance. The archetypal example of controlled precipitation can be found in the work of Kashiwa and coworkers at Mitsui [99, 100]. In this disclosure $MgCl_2$ is contacted with three equivalents of 2-ethyl hexanol in decane forming a uniform solution. Phthalic anhydride is subsequently added prior to mixing the homogeneous solution with $TiCl_4$. The mixture is allowed to react and precipitate in the presence of an internal donor (diisobutyl phthalate). The titanation step is then repeated on the solid component, prior to washing. These catalysts have been widely applied and have even been claimed to be suitable for challenging film products such as capacitor films where electrical breakdown is directly related to the amount and how finely and homogeneously distributed the catalyst residues are in the product [101].

For the BRICI/Sinopec N-Catalyst and BASF Lynx PP family anhydrous $MgCl_2$ is dissolved in a mixture consisting of an organic epoxide (epichlorohydrin) and tributyl phosphate in toluene to obtain a homogeneous solution. The solution is contacted with phthalic anhydride and treated with an excess of $TiCl_4$. The resultant solid support particles of microspherical form are obtained after heating. The solid catalyst is subsequently treated with the internal donor (diisobutyl phthalate) before isolating the solid, which is further contacted with $TiCl_4$ prior to extensive washing [102, 103]. Factors affecting the particle size of the catalyst produced are discussed in [104, 105].

As stated earlier researchers at Shell were some of the first to note the benefit of coupling titanium to a magnesium surface. This was subsequently followed by the discoveries of Goodall et al. and the development of the Super High Activity Catalyst System (SHAC catalyst) [106–109]. In these basic patents the key breakthrough was the use of $Mg(OEt)_2$ and ethylbenzoate followed by chlorination and titanation. Subsequent iterations of this by Job [110–112] and Kilty and Cuthbert [113], produced morphologically controlled catalyst particles via the creation of a $Mg_3Ti(OEt)_8Cl_2$ precursor from $MgOEt_2$, $Ti(OEt)_4$, $TiCl_4$, ethanol, *o*-cresol, and chlorobenzene. In a subsequent step the catalyst particle is subjected to further titanation in the presence of an internal donor (ethyl benzoate or diisobutyl phthalate) with addition of benzoyl chloride to lower the ethanol content. This produced the so-called BenMag catalyst precursor which had ethyl benzoate in the formulation that could be used as is or easily replaced by another donor, and found commercial relevance in the SHAC 310 (benzoate) and 320 (phthalate) catalyst families. For more recent sixth-generation donor developments the $Mg_3Ti(OEt)_8Cl_2$ precursor has been the starting point without a benzoate-generating step. Similar starting materials have recently been reported by Basell, utilizing $MgCl_2$ and $Mg(OEt)_2$ [114].

$Mg(OEt)_2$ has also been the starting material of choice for the commercially successful Toho Titanium THC family [115–117]. In this case the morphology is set by preparing a spherical $Mg(OEt)_2$ starting point, with chlorination and titanation occurring in the presence of the appropriate donor. Finally $Mg(OEt)_2$ has been converted to a carboxylate via reaction with CO_2 which yields a more soluble Mg starting point, which can either be reacted directly with $TiCl_4$ and $Si(OEt)_4$ to form a solid procatalyst or can be spray-dried with Carbosil TS610 [118]. The former is the basis of the Amoco CD catalyst [119–121]. Catalysts based on mixed Mg(OEt)Cl systems have also been reported by Basell/Akzo Nobel [122, 123] and SABIC [124, 125]. The latter converting a Grignard solution to form controlled morphology Mg(OR)Cl supports. The source of the Grignard appears to be moving from PhMgCl towards an alkyl Grignard such as *n*BuMgCl.

In the past the general belief was that low surface area and compact catalyst particles are connected to low activity due to mass transfer limitations. However, the recent development by Denifl and coworkers at Borealis has shown that a dense low surface area catalyst (ca. 2 m^2/g), prepared via solidification from emulsion, can possess high activity [126].

Borealis emulsion-based catalyst preparation technology is preformed in the absence of an external support. Instead of impregnating a carrier with a solution of

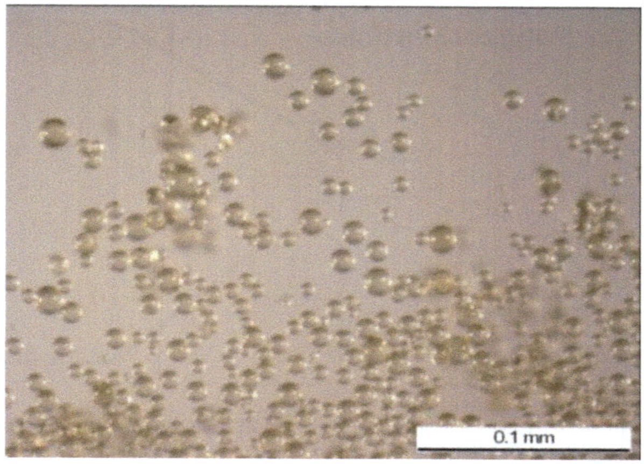

Fig. 1.8 Optical image of "emulsion" derived Ziegler–Natta catalyst (reproduced with permission from Borealis Polymers Oy)

the catalyst components, a two-phase liquid–liquid system is formed in which one phase contains a solution of the catalyst components [81]. The two-phase system itself can be formed either "in situ" by chemical reaction or simply by addition of an inert and immiscible solvent to the catalyst phase. In the second step this liquid–liquid system is then used to make an emulsion, in which the solution of the catalyst components forms the dispersed phase. Finally, the catalyst components of the dispersed phase are solidified and homogeneously distributed in the form of dense transparent spherical particles, of narrow particle size distribution (Fig. 1.8). This technology has the added advantage of reducing the amount of $TiCl_4$ required for titanation. Powder XRD measurements indicated the presence of δ-$MgCl_2$ nanoribbons in the catalyst structure. Scanning electron micrographs have revealed spherical surface species about 40–60 nm in diameter, and chain-like structure within the catalyst particles, with the chains extending from the center of the particle to the surface. Evidently, activation and internal/external donor exchange is not adversely affected by this structure and the monomer does not encounter diffusion problems at the start of the polymerization. Such catalysts undergo rapid and complete fragmentation with controlled particle growth during polymerization, yielding compact polymer particles with high bulk densities (Fig. 1.9) [127]. The fine homogeneous distribution of catalyst residues coupled with the activity of these system leads to potential use in challenging applications such as capacitor films [128]. In addition, the compact nature of the particles does not restrict the ability to produce heterophasic polymers.

The emulsion-based ZN PP catalyst has an inherent ability to insert the ethylene molecules evenly between the propylene molecules, due to the homogeneous nature and distribution of its active sites. The good randomness with these catalyst reduces the problem with sticky ethylene-rich short chains, giving about 5 °C lower melting point (when compared with a standard reference catalyst having the same internal donor producing a random PP with similar ethylene content). Furthermore, good randomness leads to resins with excellent optical properties, such as transparency.

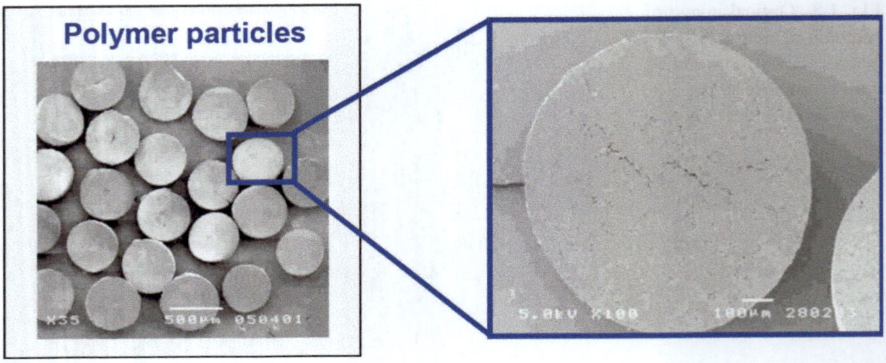

Fig. 1.9 SEM microtomed cross section images of polymer particles resulting from "emulsion"-- based Ziegler–Natta catalyst (reproduced with permission from Borealis)

The amount of oligomers, <C15, in the final product is also significantly lower when using this catalyst compared to the abovementioned conventional ZN PP catalysts. This results in additional advantages in the end product with regard to organoleptic properties, migration, blooming, and fuming [129].

External Donors

As stated earlier, an external donor is often required to buffer the extraction of certain internal donors via the AlR_3 cocatalyst. The most effective alkoxysilane external donors for high catalyst stereospecificity are short chain (Me or Et) alkoxysilanes containing relatively bulky groups alpha to the silicon atom (Fig. 1.10) [130–141]. The relationships between the structure of the silane compounds and their polymerization behavior have been discussed in detail by Härkönen and Seppala [133–136]. It has been recognized that silanes containing hydrocarbon substituents with the appropriate size and oxygen atoms with the appropriate electron density are required to produce highly isotactic polypropylenes.

The external donor not only decreases "less-tactics" formation but can also increase the degree of steric control at isospecific sites [139]. Sacchi et al. investigated the effect of silane donors on the highly sensitive first stereoregular insertion of propylene into a Ti-Et bond formed using a phthalate-based catalyst activated with ^{13}C-enriched $AlEt_3$ [140]. The mole fraction of *erythro* (isotactic) placement in the isotactic polymer fraction was 0.67 with no external donor, 0.82 with $MeSi(OEt)_3$ and 0.92 with $PhSi(OEt)_3$. It was therefore concluded that the alkoxysilane external donor was present in the environment of at least part of the isospecific centers.

Typically, an industrial silane donor contains at least one secondary or tertiary carbon linked directly to the silicon atom. This bulky hydrocarbyl group is believed to protect the silane against removal from the catalyst surface via complexation with aluminum alkyl [142]. Examples of such donors can be seen in Fig. 1.17, with

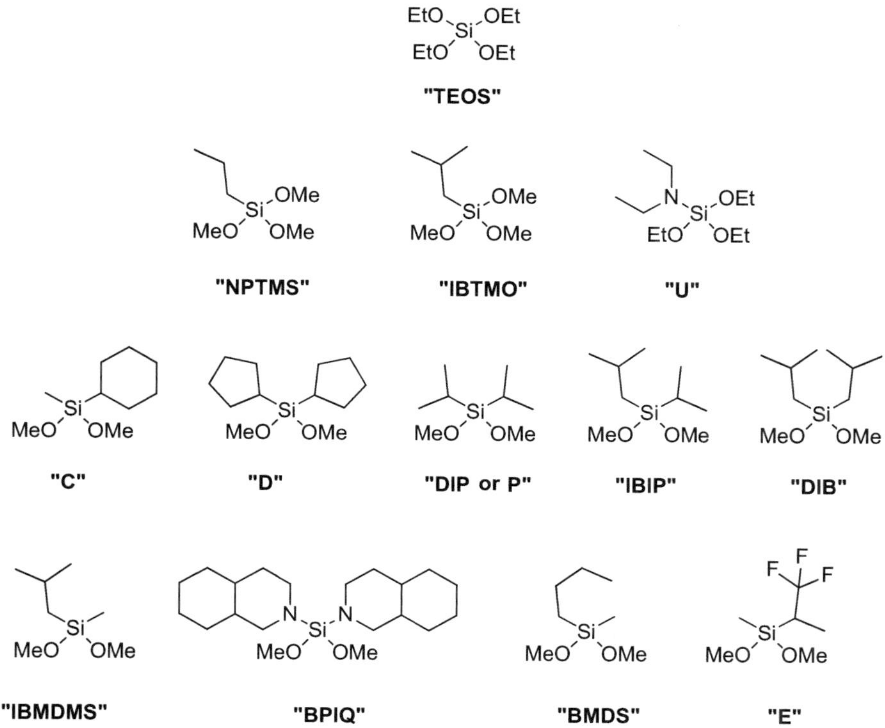

Fig. 1.10 Examples of alkoxysilane external donors

cyclohexyl(methyl)dimethoxysilane ("C" donor) and dicyclopentyldimethoxysilane ("D" donor) being the most commonly employed [142]. When comparing the polymerization performance of cyclohexyl(methyl)dimethoxysilane and dicyclopentyldimethoxysilane, the latter gives particularly high stereospecificity [143] and broader molecular weight distribution [144]. To illustrate the point Streeky et al. compared different silane donors for their impact on activity, microtacticity, melting behavior, and film blowing processability and performance [124].

The desired molecular weight distribution (MWD) of a polypropylene depends on the end-use application of the polymer. A narrow MWD, as well as relatively low molecular weight, is advantageous in fiber spinning applications. In contrast, extrusion of pipes and thick sheets requires high melt strength, and therefore relatively high molecular weight and broad molecular weight distribution. A broad molecular weight distribution, along with high isotactic stereoregularity, is also beneficial for high crystallinity and therefore high rigidity.

High PP stereoregularity and broad molecular weight distribution has also been obtained by employing dimethoxysilanes containing polycyclic amino groups, as exemplified by di(perhydroisoquinolinyl)dimethoxysilanes (*BPIQ* donor Fig. 1.10) [145]. Cyclic amino substituted silane donors first appeared in a Himont patent application [146], however it is the work of scientists in Ube Industries that further developed the potential of these donors [147, 148]. Ikeuchi and coworkers reported that dipiperidinyldimethoxysilane, which possesses a similar geometrical structure to dicyclopentyldimethoxysilane, yields more isospecific polypropylene and rationalized this increase via interaction of the nitrogen atom of the piperidinyl moiety with the active titanium center, increasing the steric influence at the metal center [148].

The increased demand for melt processible TPO, for applications in vehicles and household electrical appliances, has seen the need for a homo-PP matrix that has not only a high stereoregularity but also a high melt flow rate (low Mw). In a multimodal process this places demands on the operation of the homopolymerization phase. Industrially the molecular weight is controlled by hydrogen. In these instances, a larger quantity of hydrogen is required to achieve this goal, but a high hydrogen concentration a bulk phase process may lead to the formation of bubbles (cavitation on loop pumps), which can lead to reactor shutdown. In a gas phase process the amount of hydrogen required may exceed the safety limits of the reactor in terms of pressure. In this case the partial pressure of monomer needs to be reduced to achieve the desired $C_3^{=}/H_2$ ratio, resulting in a decrease in productivity of the catalyst. One way of solving such an issue is to increase the hydrogen response of the catalyst with a suitable external donor. To this end amino trialkoxy silane donors, such as diethylaminotriethoxylsilane (*U* donor Fig. 1.10), have been employed to yield catalytic systems with an increased hydrogen response [145, 149]. Further developments have been made by Toho Titanium with the use of mixed amino-alkoxysilanes [150].

Mixed Donors

An alkoxysilane external donor, as illustrated, can be used to control the stereoregularity and molecular weight distribution of a polymer resin, as well as the H_2 response and ethylene copolymerization capability of the catalyst system. However, it is often the case that an external donor that yields desirable control over one polymer property is ineffective or detrimental with respect to additional properties of the polymer. To combat this shortcoming, combinations of external donors have been employed in order to adjust polymer properties either according to an expected average or through the use of multiple reactors in the hope of achieving beneficial effects over an individual external donor [151–155].

Employing combinations of external donors does have some disadvantages. It is not uncommon for one of the external donors to be dominant, so as to exclude the effect of the other(s). As a result, polymerization in single/multi stage processes often produces resins whose properties are determined essentially by the dominating external donor. Miro and coworkers at Exxon have exploited the dominant behavior of certain external donors [154]. When a combination of dicyclopentyldimethoxysilane ("D" donor) and tetraethoxysilane (TEOS) is used, "D" donor acts as the dominant donor, forming resins similar to those produced by "D" donor alone, that is, high stereoregularity and low melt flow rate. However, the group discovered that they could take advantage of this dominant effect by using a two-stage polymerization process. The first stage utilized TEOS alone, forming polymer with a high melt flow rate, "D" donor being added at the second stage. The combination of mixed donor and a two-stage process achieved a final polymer with high stereoregularity, a relatively high melt flow rate and moderately broad molecular weight distribution. Recently, Chen and Nemzek extensively studied, mixed "D" donor and TEOS systems [155]. The group discovered that at lower total molar ratio of donor to titanium (E.D./Ti = 1–10) it is still possible for the weaker donor TEOS to participate in the reaction, specifically with respect to its effect on the melt flow of the resultant polypropylene resins and hence hydrogen response. The dominating influence of the "D" donor is unsurprisingly related to the total molar ratio of donor to titanium (E.D./Ti) and therefore the total amount of dominant donor. In addition to the increased hydrogen sensitivity, the mixed donor system showed higher incorporation of ethylene in continuous copolymerization experiments with ethylene for a given feed rate [155]. More recently Exxon has shown that a combination of a LyondellBasell succinate-based catalyst and a combination of NPTMS and "D" donor (Fig. 1.10) can produce rheologically interesting homo-PP [156]. The mixed donors addition points can be split across different reactors ("D"-donor first loop; TEOS second loop) as with SINOPEC Split donor technology to also produce high melt strength polymers [157]. An additional example of mixed donors can be seen in the combination of NPTMS and BPIQ [158].

The main advantage of the fourth-generation catalysts over the third-generation systems is their greater stability during polymerization. The ethyl benzoate-based catalysts exhibit very high initial activity but then decay rapidly, losing around 90% of their activity during the course of a polymerization. This limits their productivity to around 30 kg PP/g catalyst under typical polymerization conditions (bulk liquid monomer, approx. 70 °C). In contrast, productivities up to around 70 kg PP/g catalyst are achievable with the fourth-generation, phthalate-based systems. The increased stability also allowed for the further development of cascaded process technologies, which has allowed for further tailoring of technical polymer resins. However, this increase in activity and stability may lead to operability problems, particularly in gas phase processes. For example some of Dow's SHAC™ catalyst systems operating in the Unipol gas-phase process require catalyst kill systems to be installed in order to kill the catalyst in the event of a pressure or fluidization loss [159]. This problem has led to the development of the "self-extinguishing/limiting" catalysts concept. In the initial disclosures a mixture of external donors (alkyl benzoates (PEEB) and alkoxysilanes) are employed. It is claimed that this results in catalyst systems that are very active at the standard operating conditions, but rapidly lose their activity at higher temperatures and therefore prevent the formation of sheets or large polymer chunks in the fluidized bed reactor. Therefore, if a disruption does occur, the catalyst will deactivate quickly, leaving an intact particle bed. Upon recovery, the bed can be refluidized and reactivated with a minimum of downtime and effort [159–161]. Further development has seen the preferred use of aliphatic esters such as isopropyl myristate or di-*n*-butyl sebacate (Fig. 1.11) in combination with NPTMS (Fig. 1.10) [162] and the extension to "self-limiting" polyethylene catalysts systems.

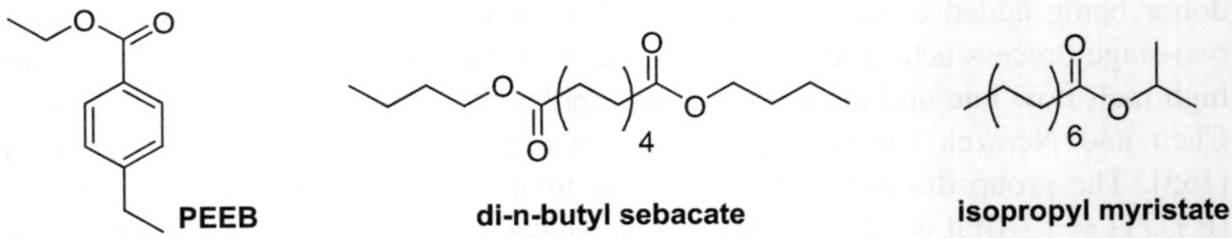

Fig. 1.11 Examples of "self-limiting" donors

1.2.2 Ziegler Catalysts: Polyethylene

It is harder to classify Ziegler PE catalysts in terms of generations, when compared to PP. However, there have been common development areas that are focused on making a catalyst more suited to the process to make an added-value product range. For example, operators of Ziegler catalysts in single phase reactors (GPR and Slurry) sought the need to increase the molecular weight distribution (MWD) of their products to improve polymer processability, and the ability to alter the comonomer composition distribution (CCD) to reduce hexane-solubles, for example. The MWD and CCD of Ziegler were limiting when compared to Cr-based catalyst systems, which still are alive and well in production of MDPE-HDPE grades. This in turn ultimately led to the development of bimodal Ziegler processes, due to restricted availability to license Cr technology in certain areas of the world. The development of bimodal technology has called for improved CCD and slightly narrower MWD for certain applications, but mainly an improvement in hydrogen response. High hydrogen response is required to maintain high catalyst activity (in contrast to propylene polymerization, addition of hydrogen leads to decreased ethylene polymerization activity) and to avoid loss of ethylene due to ethane formation. In addition, control over the particle size and, equally importantly, the particle size distribution of the catalyst, has become extremely important for bimodal operation. Smaller polymer particles allow for better mixing of the interparticle heterogeneity in terms of split (ratio of LMW to HMW) on the polymer particle as a consequence of operating in a multistage continuous process with residence time distributions. This reduces imperfections in the final product such as "white-spots" and gels. However, it is paramount that the generation of fines be kept to an acceptable level for smooth and safe operation, and hence the smaller the polymer particle targeted, the narrower the particle size distribution required [163–168].

In the following section we will concentrate on commercially relevant catalyst developments in these areas.

$MgCl_2$–Titanium Catalysts on Silica

Silica-based catalysts have found wide use, especially in gas-phase processes where a controlled particle size and distribution is needed in order to keep the bed in position. The silica support acts as a carrier material, imparting morphology control to the catalyst. In theory, it is not considered as a part of the active center, but it offers a large pore volume and surface area. A wide variety of catalysts can also be achieved by altering the physical properties of the silica support material (particle size and distribution, pore volume, surface area, etc.). The resulting catalyst carrier may consist of up to 50 wt% of the procatalyst. In addition, the calcination conditions for the support can be altered, thereby altering the proportions of isolated and hydrogen-bonded silanol and siloxane bridge groups and also affecting the friability of the support. The activities of these catalysts depend to some degree on the amount of titanium that is chemically bound and the amount of "free" $TiCl_4$ (the more "free" titanium, the higher the activity).

The silica is either physically impregnated with a solution of $MgCl_2/TiCl_4$ in THF or chemically impregnated with the catalyst components, which can either be added together or in differing orders. Common practice is to pacify the surface with an alkyl-aluminum, magnesium complex or an organosilane [169–188]. Chemical and physical impregnation procedures are exemplified in the work of Mobil [189] and UCC [190, 191], respectively. These catalysts, when combined with $AlMe_3$ as a cocatalyst, were the first to produce so called "super-hexene" grade LLDPE, with improved dart impart strength and hexane extractables [192, 193].

The development of impregnated silica catalysts has more recently been led by the work of NOVA Chemical, resulting in their NOVACAT range. In NOVACAT type catalysts, silica is first pacified by treatment with an aluminum alkyl. Magnesium chloride is then deposited on the treated silica by chlorinating a magnesium alkyl compound. To this is then added an electron donor, a titanium compound and a second aluminum compound (so-called split Al addition); the original goal of the development was to produce polymers with high MFR (broader molecular weight distributions) [194]. Figure 1.12 depicts the general synthetic route, however, Kelly et al. also elegantly showed that by varying the order of addition of the components following the magnesium alkyl addition, one could produce a range of polymers (density, M.I. and hexane extractables) under the same polymerization conditions. The group even succeeded in making "super-hexene" grade LLDPE without the need for $AlMe_3$ as a cocatalyst ($AlEt_3$) [195]. Cocatalyst concentration is also crucial in defining the mechanical performance of the final product in film applications, for example dart impact strength [196]. Deviation from THF as an electron donor has also been claimed to influence the hydrogen response of these systems [197].

With respect to bimodal processes, the silica-based catalysts of Borealis for Borstar and CX processes have excelled [181, 198, 199].

Spray-Dried $MgCl_2$–Titanium Catalysts

Spray-drying techniques are commonly used to prepare $MgCl_2$-based Ziegler PE systems, and is core to the ubiquitous Union Carbide UCAT family of catalysts [191, 200, 201] that were primarily designed to operate in a UNIPOL process, but have been extended to bimodal hexane slurry processes [202]. In spray drying, droplets containing a solution or slurry of the catalyst components are sprayed into

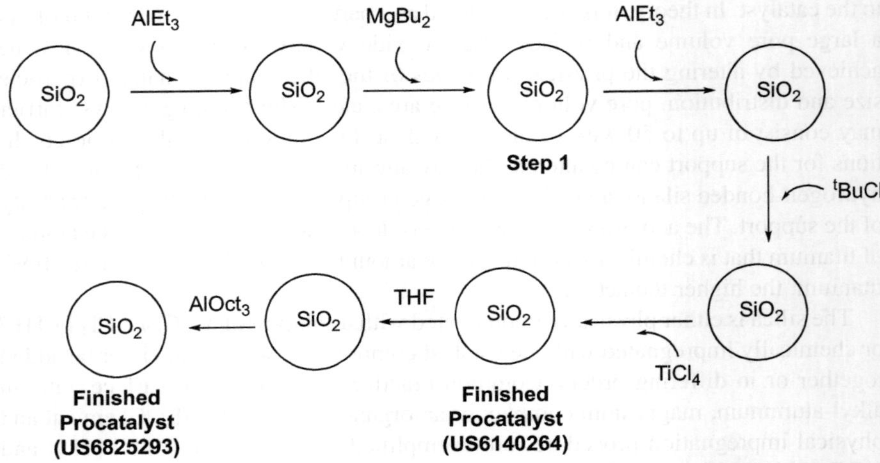

Fig. 1.12 Generic example of a NOVACAT Ziegler procatalyst synthesis

a chamber under drying conditions to remove the solvent or slurry diluent, leaving behind a solid spheroidal catalyst.

The characteristics of the droplets formed can be used to tailor the particle size of the final catalyst. Structural reorganization of the particle can be influenced by volume and size changes. In addition, spray-drying conditions can be used to form substantially spheroidal catalyst particles (large or small) or aggregated particles, with a homogeneous composition or porosity. Typically, spray-drying techniques employ a "filler" component to aid control of the shape and composition of the catalyst. Common fillers are hydrophobic fumed silicas, which impart a relatively high viscosity to the slurry to be spray-dried and also improve the mechanical strength of the final catalyst (Fig. 1.13). Further modifications of this recipe include the addition of EtOH to increase the solubility of $MgCl_2$ and hence the maximum achievable concentration in THF [203], improving the final morphology of the product, and the addition of a combination of one or more group 4 halides ($ZrCl_4$ or $HfCl_4$) [204]. The filler which makes up to 30 wt% of the catalyst may also generate increased porosity in the final catalyst [205–208].

The catalyst systems are then partially activated with two different alkyl aluminum-containing compounds (Fig. 1.14) before full activation in the final reactor with TEA for example. The proportions and nature of the alkyl can affect the MFR and hexane extractables in LLDPE production. For example, if $Al(n\text{Hex})_3$ and DEAC are replaced with $AlMe_3$ and $AlMe_2Cl$, respectively, and $AlMe_3$ is used in the reactor as the final cocatalyst, it is possible to produce so-called "Super-hexene" grades [193].

More recently the application of self-limiting donors, previously discussed for polypropylene, has been extended to these types of catalysts [209, 210].

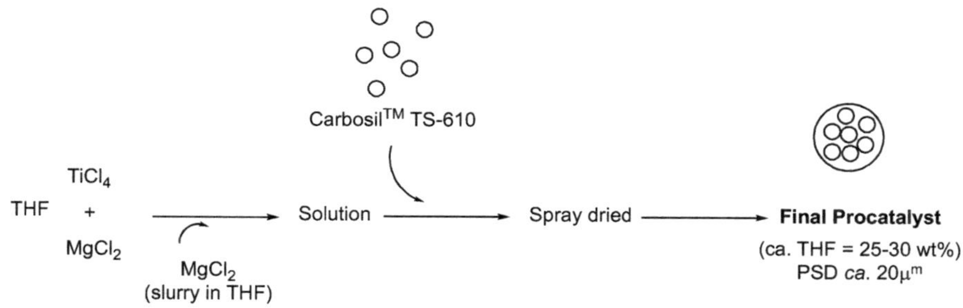

Fig. 1.13 Generic example of a UCAT Ziegler procatalyst synthesis

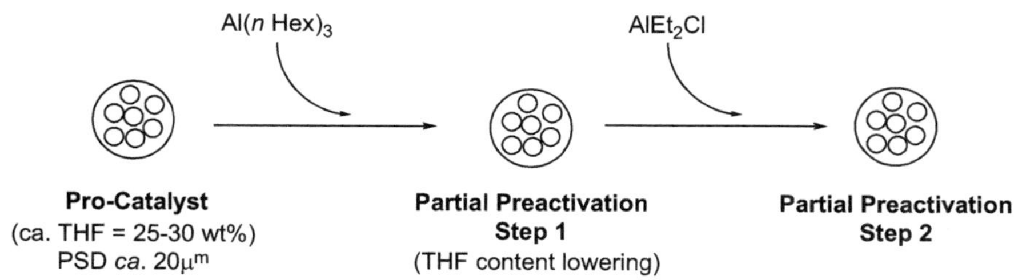

Fig. 1.14 Partial activation of the spray-dried procatalyst

MgCl2–Titanium Catalysts Based on $Mg(OR)_2$

Ziegler catalysts are prepared by first bringing $Mg(OR)_2$ into a dissolved/dispersed state or preparing the alkoxide in situ via reaction of MgR^1R^2 with an alcohol. $MgCl_2$ is then precipitated as an amorphous material, imparting high catalytic activity to the system. Usually, this is done by the addition of a suitable chlorination agent. Titanium can be added either as a soluble complex or used as one of the chlorination agents. By using different "tricks" during precipitation, and by adding suitable donor compounds, particle size and particle size distribution can be directed to a certain extent. Catalysts based on $Mg(OR)_2$ are by far the most utilized systems in bimodal slurry processes.

Researchers at Solvay developed the original system, which was based on making a solution of $Mg(OEt)_2$ and $Ti(OBu)_4$ and then chlorinating via an aluminum chloroalkyl containing species [211–213]. Further evolution of this system can be seen in the work of researchers at INEOS [214, 215] (Figs. 1.15 and 1.16) and SABIC [216, 217]. Figure 1.16 illustrates the example of an ethylbenzoate-modified catalyst; as one can imagine, the final procatalyst may also have a variety of alkylbenzoate donors produced by the reaction of ethylbenzoate and the other components. In the disclosure of Friederichs et al. we can also see that the addition of 1-allyl-3,4-dimethoxybenzene as an electron donor results in a significant reduction of unwanted ethane formation in the LMW phase of a bimodal composition when compared to unmodified and THF- or ethyl benzoate-modified formulations [216].

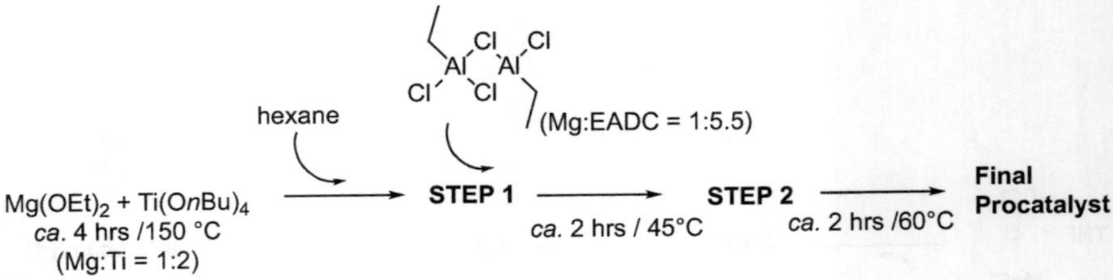

Fig. 1.15 Example of $Mg(OEt)_2$-based procatalyst

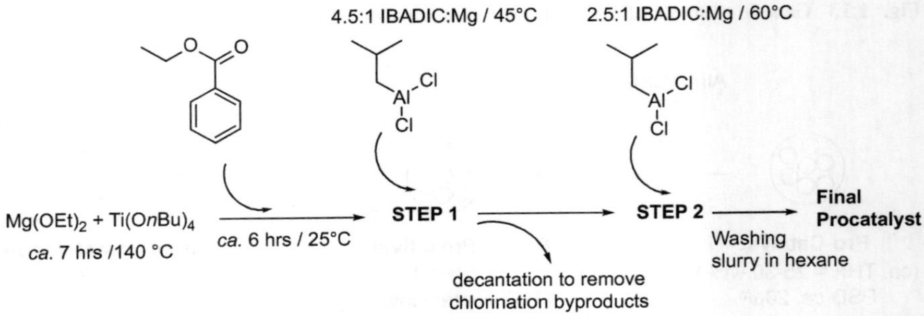

Fig. 1.16 Example of donor modified $Mg(OEt)_2$-based procatalyst

The $Mg(OR)_2$ may not require to be 100% dissolved, depending on its particle size. Lecht, Blöm, Berthoud, and others from Hoechst utilized finely dispersion Mg $(OEt)_2$ (Fig. 1.17) [218] and $Mg(OMe)_2$ (Fig. 1.18) [219, 220] in the preparation of catalyst systems. The two synthetic pathways described below differ in the strength of the reducing agent employed and one could imagine that $Mg(OEt)_2$ is the preferred support. These $Mg(OR)_2$ supported catalysts technology are current the backbone of recent developments in the multimodal Hostalen process to produce multimodal pipe, blown film, and blow molding [221–224]. In the case of blow molding the group has also shown the possibility to form LCB via a Ziegler catalyst [221].

Total and its previous incarnation, on the other hand, have utilized a path that includes creating a soluble $Mg(OR)_2$ via reaction of $MgEtBu/AlEt_3$ (1:0.03) with 2-ethylhexanol [225, 226]. The compound is treated with successively stronger chlorinating titanium reagents $TiCl(OiPr)_3 \rightarrow 2\ TiCl_4/Ti(OnBu)_4 \rightarrow TiCl_4$ to form the final catalyst. Vizzini et al. have also disclosed how the catalyst can be tuned via altering the reaction conditions and reagent ratios [227, 228]. This system produces a catalyst well suited for operation in bimodal slurry processes due its narrow particle size distribution, high activity, good hydrogen response and the relatively narrow Mw/Mn of the resulting polymer [229, 230]. Similar long-chain $Mg(OR)_2$ starting points have very recently been disclosed by DSM Stamicarbon [231], Borealis [232, 233], and SINOPEC [234]. The latter converted soluble $Mg(OR)_2$ to a final procatalyst via mixed chlorination steps (step 1: BCl_3 or $SiCl_4$ etc.; step 2: $TiCl_4$) in the presence of a SBS triblock copolymer (Kraton FG1901X) to form morphologically controlled catalysts [234].

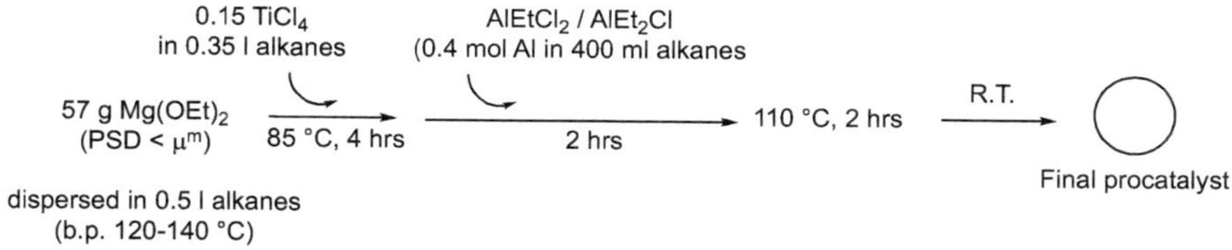

Fig. 1.17 Synthetic route from particulate $Mg(OEt)_2$

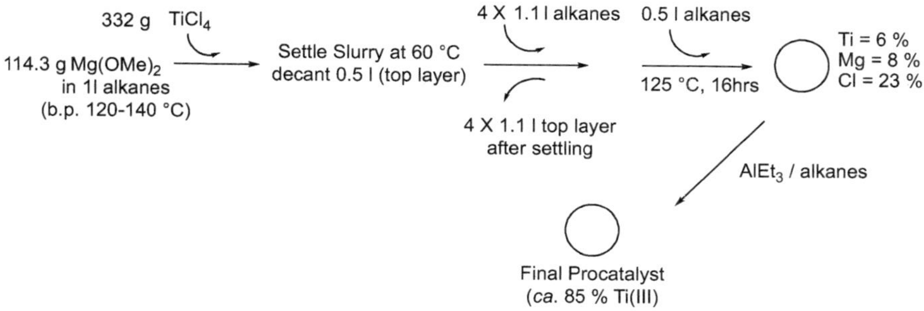

Fig. 1.18 Synthetic route from $Mg(OMe)_2$

$MgCl_2$–Titanium Catalysts Based on $MgCl_2$ alcohol adducts

There are two distinct synthetic routes with respect to $MgCl_2$–Titanium catalysts based on $MgCl_2$ alcohol adducts, as we have seen for PP catalysts. Controlled formation of a solid spherical $MgCl_2$ adduct support that can be titanated in a separate step, or controlled precipitation using a solubilized $MgCl_2$ adduct. For the latter, illustrative routes to catalysts relevant for multimodal operation will be discussed, based on the use of 2-ethylhexanol to form a soluble adduct with $MgCl_2$. The genesis of this is the work of the morphology master of Mitsui, who have taken this technology from PP to PE and to the point of a fine art. For example, the system can be controlled to almost submicron levels with retention of a narrow PSD, which has allowed for the production of fine UHMWPE particle grades known as the Mipelon product line3 [235] (Fig. 1.19). In addition, the support can be combined with AlR_3 and a phenoxyimine complex to form an excellent MAO-free support [236]. Examples of the synthetic routes can be found below in Fig. 1.20 [237, 238]. The use of different combinations of alcohol and donors and their effect has also been disclosed [239].

The second illustrative example of a relevant multimodal catalyst, based on in excess of 100 patent citations as being the preferred embodiment of titanium/magnesium catalyst systems, is that of Invetep [240] and subsequent similar disclosures from BASF [241, 242]. The basic difference between these systems appears in the composition of the mixed aluminum alkoxide–siloxide additive that is added to the $MgCl_2$ adduct. In addition, Chang et al. discovered that the addition of $Si(OEt)_4$

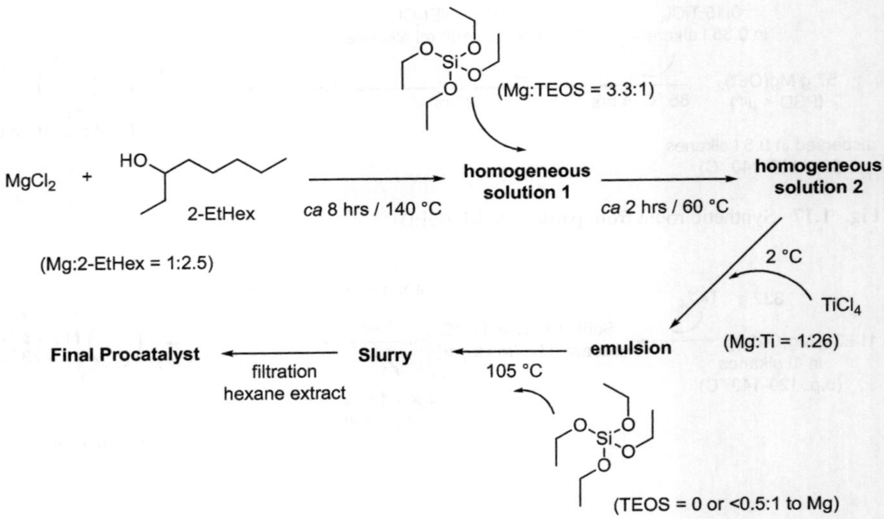

Fig. 1.19 Synthetic route to fine catalyst particles

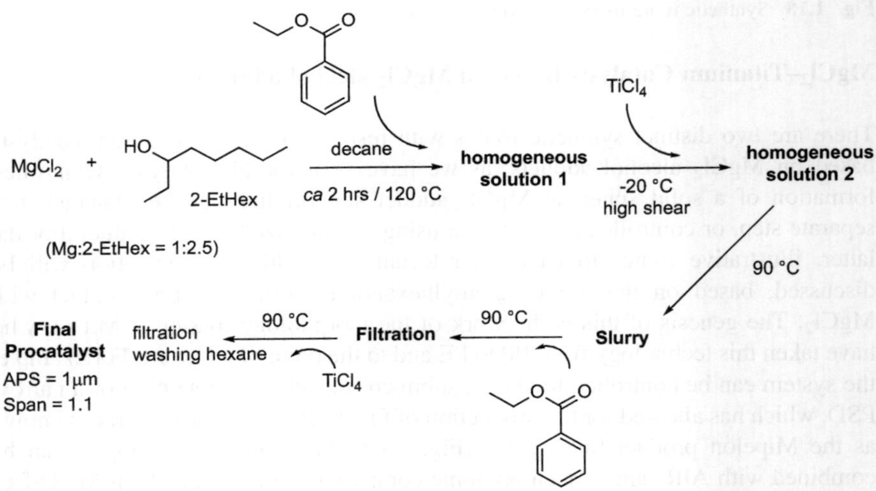

Fig. 1.20 Example of Mitsui $MgCl_2{\cdot}n$ROH adduct-based procatalyst synthesis

and ethylbenzoate yielded a catalyst that had high activity, hydrogen response, and reduced fines [242], while Garoff et al. utilized malonate donors to reduce the molecular weight distribution (Fig. 1.21) [241].

Finally, the ubiquitous $MgCl_2{\cdot}n$EtOH adducts [82, 83], have found extensive application in gas phase PE, in particular the Spherilene S and C processes [243, 244]. In this case the support particles are usually larger than for the PP disclosures and are often employed at lower ethanol contents. As with the PP system,

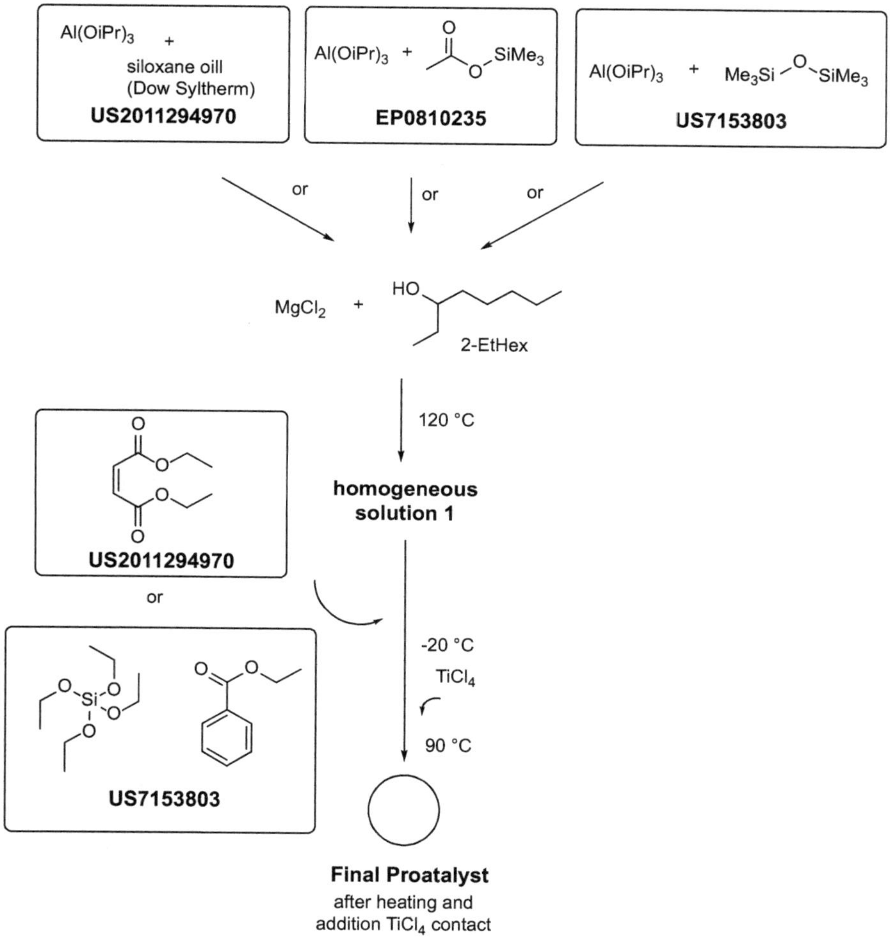

Fig. 1.21 Example of catalyst synthesis with malonate donors

the removal of the residual ethanol is done by $TiCl_4$. A donor, most commonly ethyl acetate, can be added during this phase. The catalyst is prepolymerized with propylene. Recent reported developments in donors are 1,2-dimethoxy-benzene derivatives [245, 246], diethers [247], methyl *tert*-butyl ether [246], silanes [248], and the addition of $GeCl_4$ as a mileage enhancer [249].

MgCl2–Titanium Catalysts Based on MgRCl

Numerous catalysts have been synthesized utilizing a step involving a Grignard reagent, including the formation of $Mg(OEt)_2$. The following section will focus on

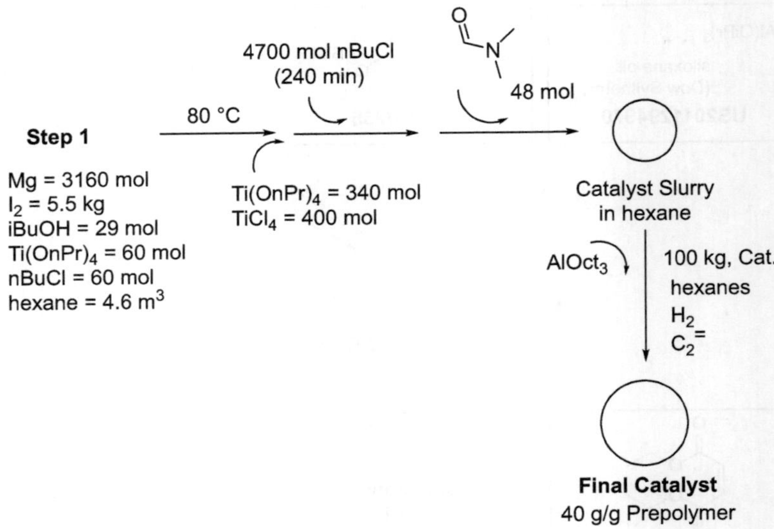

Fig. 1.22 Example of an industrial scale recipe for an Innovene G catalyst

two illustrative systems developed for unimodal gas phase and bimodal slurry processes. For the former, one of the catalysts for the Innovene G gas phase process has been selected, mainly due to the sheer scale of the disclosed preparation and further in-reactor modification [250, 251] (Fig. 1.22).

The choice of cocatalyst in the polymerization, as we have seen above, is crucial in achieving added value LLDPE and plant operability. Berardi et al. clearly illustrated that in the case of this catalyst and process $AlEt_2Cl$ is preferred over Al-Me_2Cl [251]. The catalyst is also utilized in Westlake Energx Technology [252, 253], which is used to make Mxsten grades slate of LLDPE, which are unsurpassed in terms of the performance of Ziegler film grades. The Grignard-based catalyst system in Fig. 1.23 [254, 255] was developed for the Mauzren bimodal slurry process [256–258]. The system has also utilized mixed V/Ti systems [259]. Similar systems have also been disclosed by Sumitomo [260].

1.3 Supported Molecular Catalysts

1.3.1 Polypropylene Complex Development

In today's world, classic heterogeneous Ziegler–Natta catalysts are beyond comparison in performance and cost. Nevertheless, the quest for understanding the polymerization mechanism and the promise of greater control over the synthetic macromolecular process to produce new materials still fuels the development of supported molecular PP catalysts. Stereoregular polypropylene and copolymers

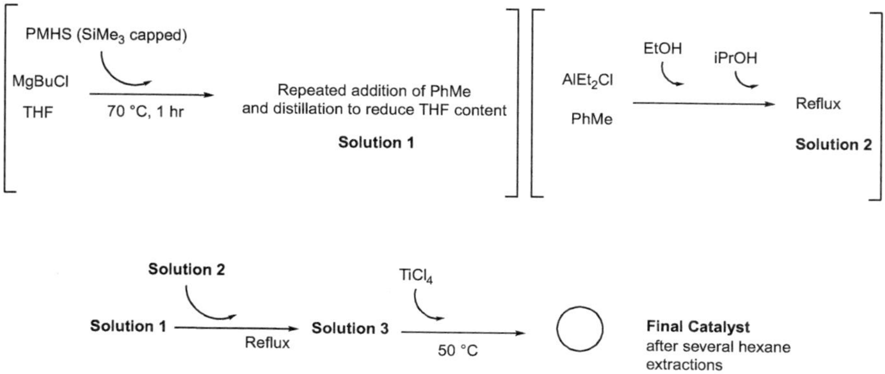

Fig. 1.23 Example of a procatalyst for the Mauzren bimodal slurry process

made with single site catalysts have been extremely slow in penetrating into the market, due in part to the amount of chemistry that can be done on PP in an extruder, for example LCB for high melt strength (HMS)PP and controlled rheology. However, the author's organometallic blood finds it hard to resist the briefest discussion of the structural elegance of homogeneous catalyst complexes for PP and the ingenuity of the researchers to find holes in apparently seamless patents. It is also a reminder of the majesty of catalysis in how seemingly minor changes in substitution patterns affect relative activation energy barriers, often resulting in major impact on resultant polymer properties.

The areas of focus in terms of complex development from the matriarch Spaleck complex (*rac*-$Me_2Si(2\text{-}Me,4\text{-}PhInd)_2ZrCl_2$) [261], have been in the area of ligand substitution to overcome the limitations compared to Ziegler–Natta catalysts and to exploit a definite advantage. Substitution of the standard ligand framework has sought to increase the melting point of the homo-PP through better regio- and stereo-selectivity and to increase the molecular weight of propylene/ethylene copolymers. The area in which metallocenes outperform Ziegler–Natta catalysts, and the source of present interest, is in the area of copolymerization of propylene with longer-chain olefins (1-butene, 1-hexene, 1-octene, etc.), giving access to new crystalline morphologies. Figure 1.24 illustrates recent complexes that have sought to tackle the above limitations and opportunities [262–276].

1.3.2 *Polyethylene Complex Development*

Supported molecular catalyst use for polyethylene, unlike polypropylene, is still seeing reasonable year-on-year growth and is becoming somewhat commoditized. Reviews of the development of molecular catalysts are numerous [1, 3, 4–7, 14, 15]. The illustrative examples of metallocenes in Fig. 1.25 are currently either the workhorses for current mPE products or are well on the road, predominantly on

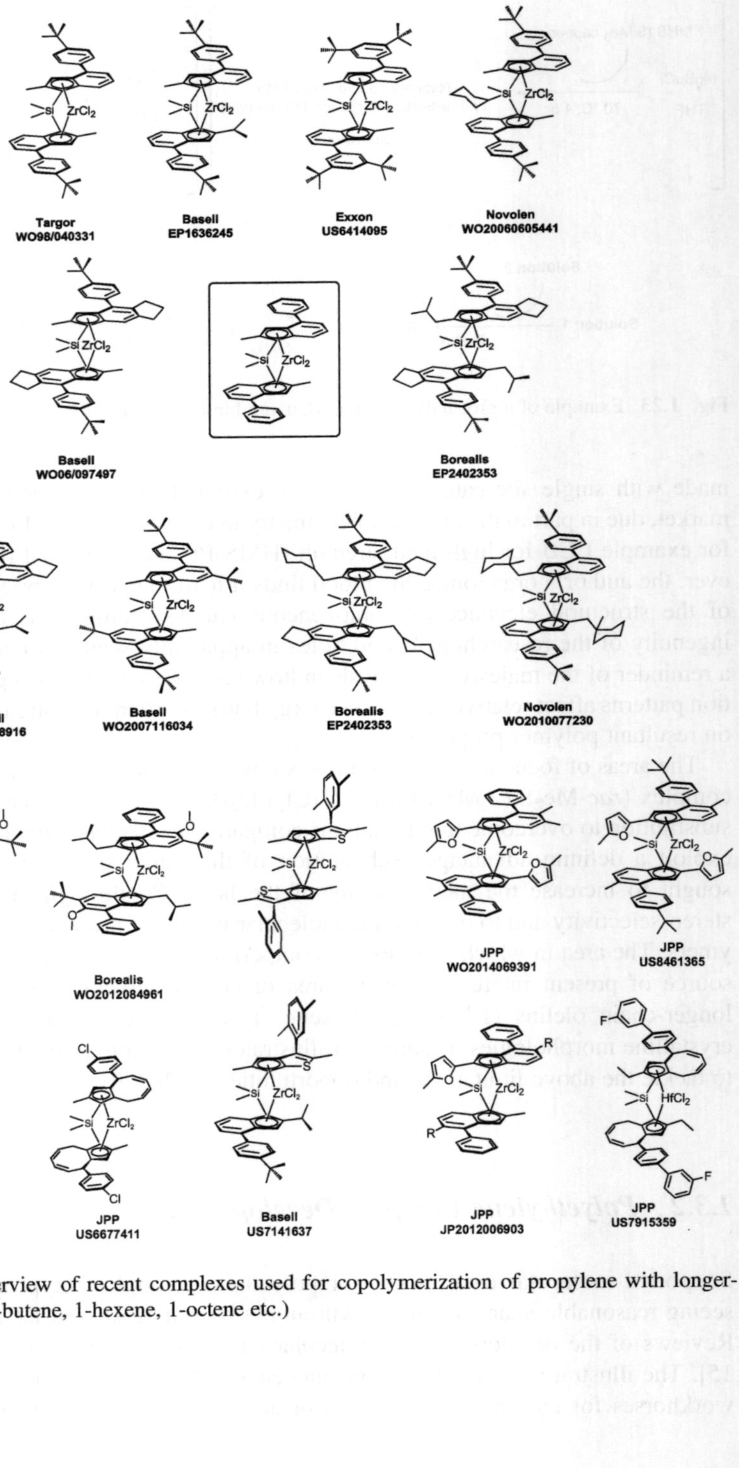

Fig. 1.24 Overview of recent complexes used for copolymerization of propylene with longer-chain olefins (1-butene, 1-hexene, 1-octene etc.)

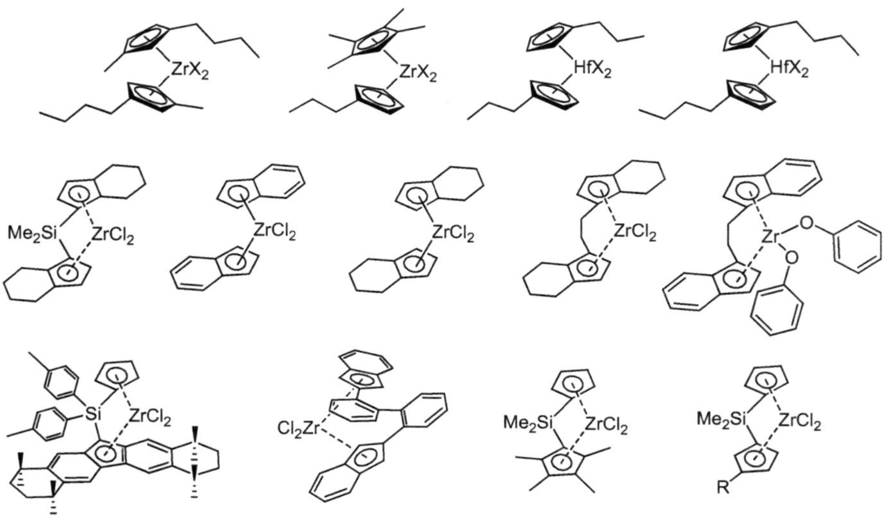

Fig. 1.25 Metallocene complexes used for mPE production

silica/MAO carriers. For example, the current Univation catalyst family [277–282] has been commercially successful for a number of years, particularly in single phase reactors. Simple unbridged metallocenes have also been employed in the Mitsui Evolue bimodal gas phase process [283, 284] and the Borealis Borstar loop/gas phase process [285–287].

Bridged complexes such as *rac*-bis(tetrahydroindenyl)zirconium dichloride have been utilized by Total in their bimodal double-loop processes and have been shown to be capable of making a wide range of advanced grades, including high pressure PE100+ pipe material based on 1-hexene comonomer [288–291]. The challenge of producing the next generation pipe and blown molding material has also seen Mitsui develop the Evolue H grades based on a bridged Cp-Fluorenyl catalyst system [292, 293].

In terms of post-metallocenes, in single component systems the constrained geometry catalysts (CGC, Fig. 1.26) based on borate-supported activators developed by researchers at Dow [294–296], have now been commercialized in supported systems by Asahi who cite reference [294] (HDPE slurry) and by INEOS [297] (LLDPE gas phase), Dow's former collaboration partners for the utilization of CGC in particle-forming processes until Dow's merger with UCC. At Asahi the system has formed the basis of their Creolex product range, from reasonably low molecular weight HDPE to their Sunfine UHMWPE powder. The latter is interesting in that the process needs to run with hydrogen to avoid fouling but then needs Tebbe reagent to be constantly added to the reaction to help the catalyst achieve UHMWPE ranges. The response to hydrogen in terms of activity with the CGC-7 complex is also an added advantage when compared to some complexes and the system as a result has also found use by Dow in a bimodal process to once again produce outstanding PE100+ material [298].

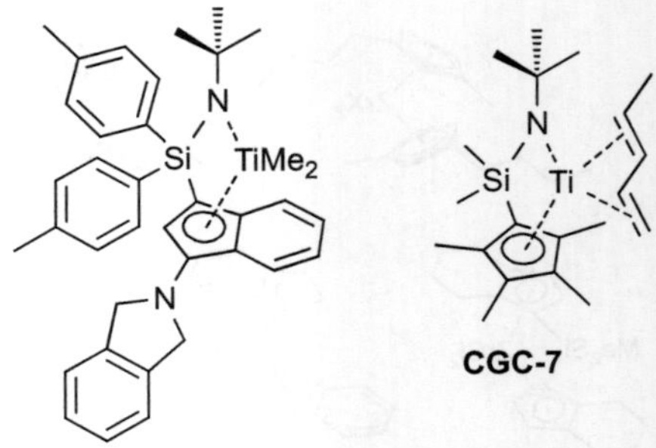

Fig. 1.26 Examples of supported constrained geometry catalysts (CGC)

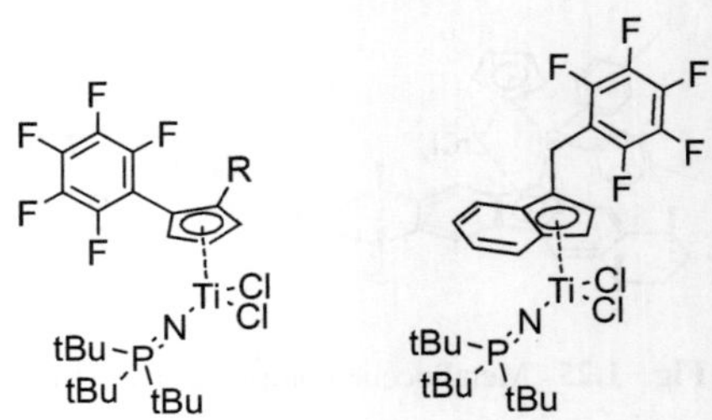

Fig. 1.27 Illustrative examples of supported phosphinimide catalysts

Nova Chemicals appears to be quite far down the road with a supported version of its phosphinimide family of catalysts (Fig. 1.27), previously developed for its bimodal solution process [299–303]. Adaption to the original bonding pattern via addition of the pentafluorophenyl moiety has brought in a better hydrogen response. This, coupled to the indenyl ligand which had previously been shown to be relatively fluxional, has led to the development of the Ind–CH_2–C_6F_5 ligand. These developments have led to systems that can be tuned via the addition of an antistatic agent to produce materials that are heterogeneous in terms of comonomer distribution (reverse) [303]. In the case of Ind–CH_2–C_6F_5 the system has a broader Mw/Mn due to the indenyl moiety; the incorporation of the antistatic agent significantly increase the heterogeneity in terms of comonomer distribution and a significant increase in Mz.

The combination of CGC catalyst and phoshinimides has also been employed as a hybrid catalyst for bimodal polyethylene in a single gas phase reactor, a concept that will be dealt with in more detail in the subsequent section [303].

The newest kids on the block are the bis(hydroxylarylaryloxy) "O_4" based systems (Fig. 1.28). This catalyst system, with the extensively varied structure modification points, has been a veritable gold mine for innovation, and as a result these complexes have been employed in numerous processes and product families [304–309]. The complexes have also been supported on silica/MAO or the tethered borate system by Symyx/Ticona [310, 311] and Dow, respectively [312]. The former demonstrated the prodigious molecular weight capabilities that these catalysts possess and also the effect that scavenger selection can have on the activity and molecular weight capability.

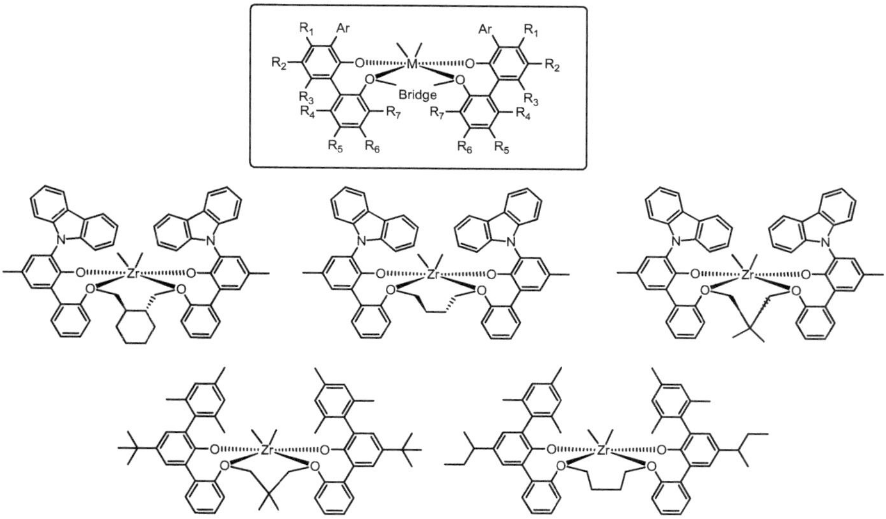

Fig. 1.28 Illustrative examples of supported bis(hydroxylarylaryloxy)-based catalysts

1.3.3 Supported Activator Development

There have been numerous disclosures on how to support a molecular catalyst and on what sort of support. The extent of these efforts can be seen in the reviews of Hlatky [313], Severn et al. [314], and Severn and Chadwick [315]. The following section concentrates on the latest commercial developments in this area.

Silica-Supported Methylaluminoxane or Boron-Based Cocatalysts

Predominantly, combinations of silica and MAO have been used to support a molecular catalyst, with extensive variation being disclosed and optimized to match a molecular catalyst with a certain process. An overview of what has been disclosed can be seen in the review of Severn [315]. Within this area current developments appear to be on doping either the MAO or the silica to improve the performance.

Researchers at Albemarle have shown a near doubling in activity for a SiO_2/MAO system via addition of $AlMe_2F$ after the MAO has been contacted with the silica. The group links this boost in activity to an increase in the production of $[AlMe_2]^+$ cations that form part of the activation step [316].

The addition of SO_4^{2-}, NO_3^{2-}, F^-, and SiF_6^{2-} containing salts to the silica prior to calcination has also been shown to provide significant improvements in the performance of the final catalyst. Nova Chemicals for example have introduced NaF [317], $[(NH_4)_2SiF_6]$ [318], and more recently $Zr(SO_4)_2$ [299, 319]. While

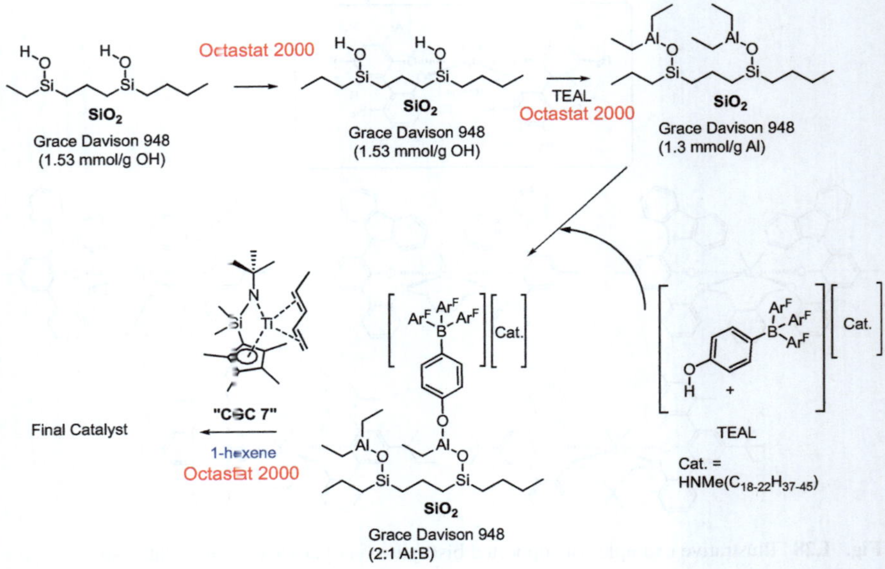

Fig. 1.29 Synthetic route to INEOS HP-LL catalyst

Exxon have focused more on the addition of $[(NH_4)_2SiF_6]$ to silica prior to contact with either MAO/molecular catalyst [320] solution or MAO [321].

As mentioned earlier, the supported borate activators developed by researchers at Dow [294–296] have now been commercialized in supported systems by Asahi (HDPE slurry) and INEOS [297] (LLDPE gas phase) and are appearing in recent disclosures of Dow. Key to this development was the synthesis of a soluble borate starting material and numerous treatments with antistatic agents to reduce the static loading of the fresh catalyst that is injected into the reactor (Fig. 1.29). The advantages of addition of an antistatic agent to the catalyst have also been seen for the metallocene catalysts discussed earlier.

Supported Methylaluminoxane System Free of a Support

Solidifying MAO with controlled morphology has several attractions, one of which is the reduction in the "dead weight" that the carrier brings to the catalyst. Two companies have developed systems to form solidified MAO support where the morphology of the final catalyst can be controlled. The first system is the Borealis Sirius catalyst system developed by Denifl et al., in which the catalyst is activated in a homogeneous phase before the activated solution is mixed with a perfluorocarbon containing a perfluorocarbon-containing surfactant [322–325]. An emulsion is formed at low temperature with a perfluorocarbon and the solidification is induced via dropping this solution in a reactor containing excess perfluorocarbon at elevated

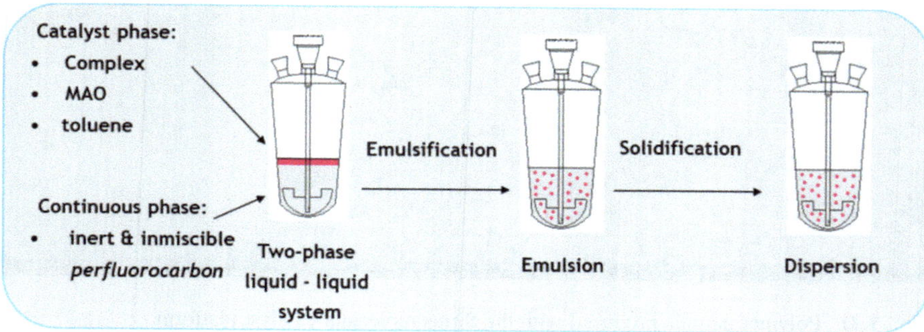

Fig. 1.30 Schematic representation of the Sirius molecular catalyst platform

temperature. Solidification occurs by virtue of the toluene in the MAO/Complex phase being extracted into the perfluorocarbon phase (Fig. 1.30). The catalyst is filtered hot, and the catalyst requires prepolymerization before use [326]. Continuous production of this catalyst system has also been disclosed.

Modifications to this method have been the introduction of an additional boron-containing cocatalyst [327, 328] or the utilization of clathrated MAO, previously described by Albemarle [329, 330].

The preferred clathrating agent until recently appeared to be octamethyl-disiloxane which forms a two-phase system with a MAO-lean top phase that contains free TMA and a viscous bottom layer that increases in viscosity on washing with PhMe [331]. A way was then found to solidify clathrated MAO via non-perfluorocarbon routes that produced spherical compact catalyst particles with high activity [332–334]. An example of polyethylene particles formed by these systems can be seen in Fig. 1.31, in which the image on the right hand side is a microtomed cross section of an individual polymer particle illustrating the dense polymer particles formed with these catalysts.

More recently, Tosoh Fine Chemicals have produced a solid MAO system that can be precontacted with a metallocene to form a support-free catalyst system [335, 336].

Supported System Free of Methylaluminoxane

Several groups have made good use of MAO-free solutions to produce commercially relevant supported molecular catalysts. CPChem has gained a tremendous knowledge on doping SiO_2, SiO_2/Al_2O_3, and Al_2O_3 supports with salts as part of their chromium-based catalyst developments. They have found that doped supported systems can activate molecular catalysts even in the absence of methylaluminoxane. The development of these systems has been reviewed by McDaniels, one of the lead developers of this technology [337]. Looking at the tremendous amount of patent literature on this topic, it appears that two technologies have been developed,

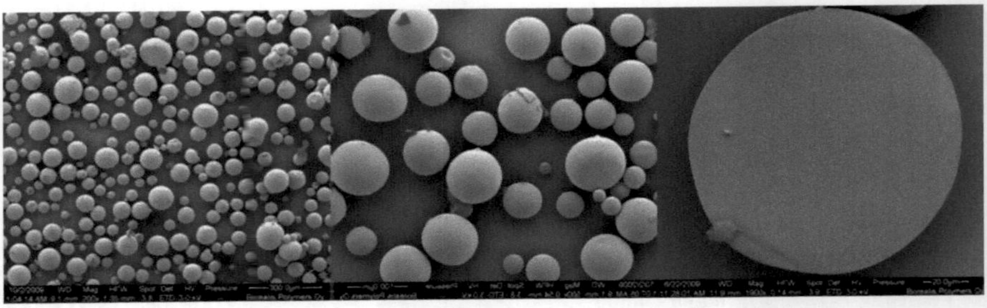

Fig. 1.31 Polymer particle produced with the Sirius molecular catalyst platform

depending on the ease of activation of the metallocene complex used. For example, relatively easy to activate metallocenes such as $^{n}Cp_2ZrCl_2$ and Ind_2ZrCl_2 are employed in combination with SiO_2/Al_2O_3 (ca. 13 wt% Al_2O_3) supports that have been precontacted with $[NH_4][HF_2]$ prior to an optimized calcination step. This in turn is treated with a "precontact solution" consisting of the metallocene, $AlEt_3$ and hexene to form the final active species. The precontact step leads to significant activity improvements via formation of a cyclic dialkylaluminum complex and has been disclosed with an example at full commercial scale [338]. The second technology for more challenging to activate complexes is the use of Al_2O_3 (Bohemite) with $[(NH_4)_2(SO_4)]$; as with the above, the ratio of reagents and the thermal history of the calcination step are crucial in gaining the most from these systems [339].

Fluorophenols have been used to form an active heterogeneous molecular catalyst system. Albemarle utilized C_6F_5OH in the scene below to produce a Brönsted acidic activator (Fig. 1.32) [340, 341]. Ineos have also recently been looking into these systems [342].

Researchers at Targor [343] and Basell [344, 345] have also produced MAO-free activators based on simple alkylaluminum and $B(C_6F_5)_2OH$ (Fig. 1.33) and $B(C_6F_5)(OH)_2$, (Fig. 1.34) respectively.

Additional use of organofluoro modifiers can be seen in the work of researchers at Exxon and Sumitomo. Holtcamp and coworkers at Exxon were able to produce heterogeneous cocatalyst packages via contacting 4,5,6,7-tetrafluoroindole with $AlEt_3$ or Al_2OEt_4 silica [346–348]. Sumitomo on the other hand have demonstrated the formation of a zinc-based supported cocatalyst system with a trifluoro-phenol complex (Fig. 1.35) [349].

Last but certainly not least is the work of Japan Polypropylene and the supported cocatalyst for their Wintec and Welnex product lines [274, 349–352]. In this concept the clay particles are treated with H_2SO_4 to form a larger controlled morphology particle, with trioctylaluminum (TOAL) added to finish the product (Fig. 1.36).

Another notable clay-based development can be seen in W.R. Grace IOLA-A [353] and IOLA-B [354] supports produced by spray-drying hydrogels with clay and MgO.

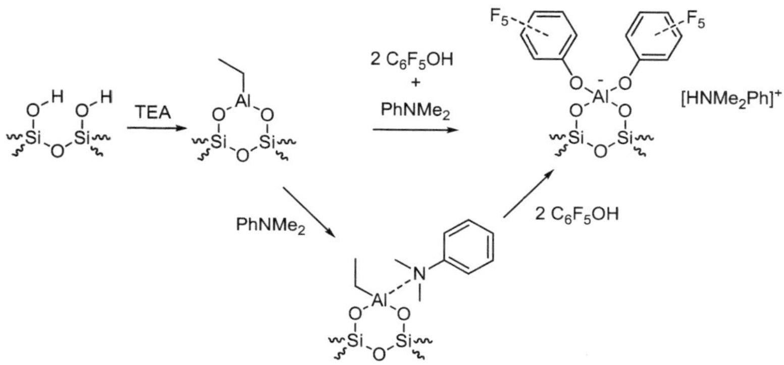

Fig. 1.32 Example of Albemarle MAO-free supported activator

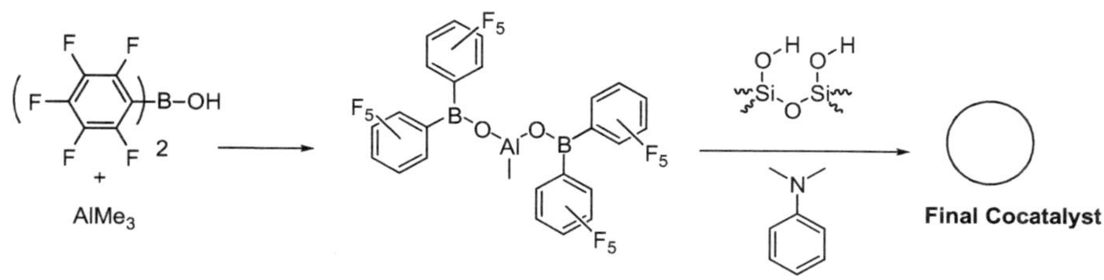

Fig. 1.33 MAO-free supported activator

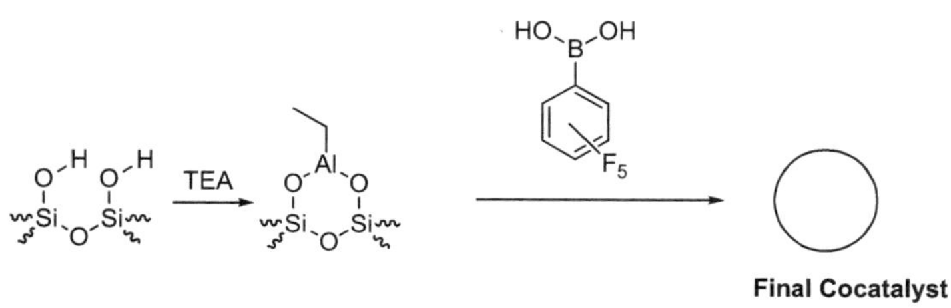

Fig. 1.34 MAO-free supported activator

1.3.4 Supported Hybrid Catalyst for Multimodal Polyethylene

The success of single gas-phase processes (UNIPOL, Innovene-G, and Spherilene-S) and slurry loop processes (Phillips Process and Innovene-S) has seen a significant amount of capacity installed in licensors' and licensees' manufacturing sites. For economic reasons and in view of the wider availability of such reactors, efforts have been made for some time to prepare polymers having broader (multimodal) and

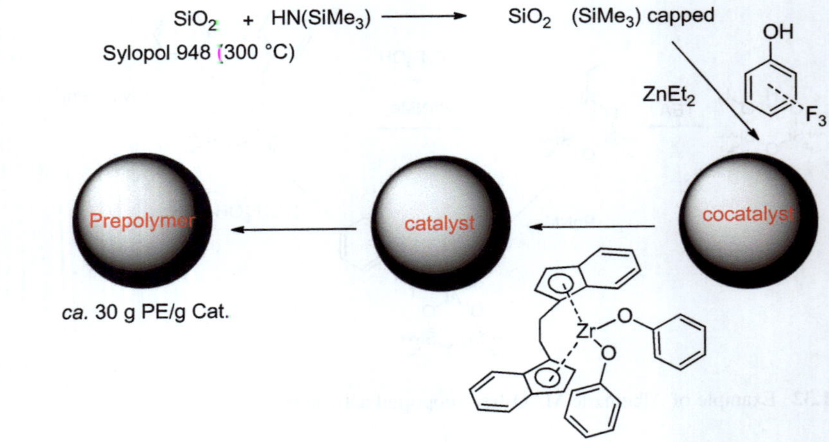

Fig. 1.35 Example of Sumitomo zinc-based supported activator

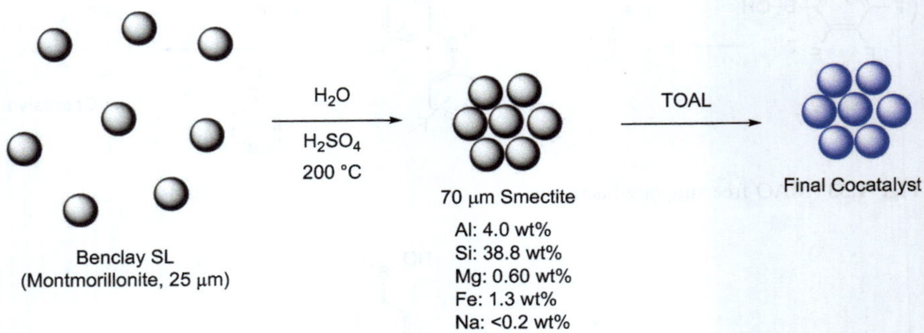

Fig. 1.36 Example of Japan Polypropylene clay-based activators

tailored comonomer compositions in continuous single reactor processes. To achieve this objective, catalyst mixtures or catalysts having different active centers on the same carrier particle (Hybrid Catalysts) have frequently been described and have been discussed by Friederichs et al. [355].

Hybrid catalysts can comprise the same or different classes of catalyst.

- Ziegler or chromium catalyst + molecular catalyst
- molecular catalyst 1 + molecular catalyst 2, etc.

In the following section we will only focus on the combination of molecular catalysts for polyethylene and refer the reader to the illustrative examples of Mink et al. [356, 357] and the excellent overview of Friederichs et al. for discussions on the former and polypropylene systems.

Challenges of Single Supported Systems

Before discussing hybrid systems it is worthwhile recapping on some of the challenges that impact even a single component system. It has already been shown above and in the literature that a molecular catalysts is very sensitive to its supporting protocol, carrier material and activator [315]. This combination impacts the final catalyst in terms of kinetics, plant operability and powder properties. Unfortunately there is no one size fits all procedure. It has also been shown, and this is the core of the hybrid concept, that there is a wide range of response to hydrogen and comonomer, for example [358]. So different catalyst complexes can make different products under the same polymerization conditions. Unfortunately however, this asymmetric response to process variables does not stop at just hydrogen, ethylene, and comonomer. For example, asymmetric responses can occur with respect to process poisons, scavenger, continuity/antistatic agents, and operation in condensed mode.

A little discussed fact outside of industry is the asymmetric ability for a molecular catalyst to generate hydrogen, especially during copolymerization (Figs. 1.37 and 1.38).

This impacts the upper molecular weight capability of the system. The ability to generate hydrogen also does not translate to response to hydrogen [359, 360]. Solutions to increase the upper molecular weight capability have involved the use of Tebbe reagent as a hydrogen scavenger [361].

Challenges of Hybrid Supported Systems

In the continuous polymerization of olefins in a single reactor with hybrid catalysts, there is the problem that the properties of the polymers obtained depend not only on the polymerization conditions but also significantly on the ratio of the individual active sites present. Fluctuations in the compositions of different hybrid catalyst batches used can thus lead to different proportions of the polymer components formed by the individual catalyst components. The shift in active site ratio can also be impacted for example by ageing, that is, if one of the components is more sensitive than the other to storage or shipment, this can also result in different products even when the same batch is used [362]. Finally, fluctuations of the polymerization conditions can also influence the activity of the individual sites in a disproportionate way. Therefore a significant amount of the academic work published on such systems in a batch process is somewhat irrelevant from an

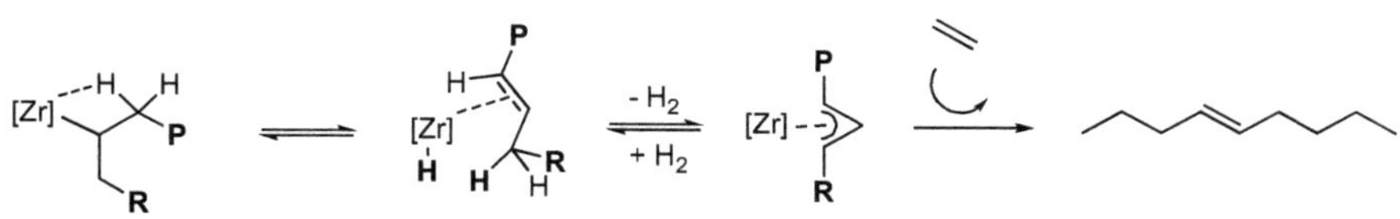

Fig. 1.37 Schematic representation of the proposed mechanism for hydrogen formation

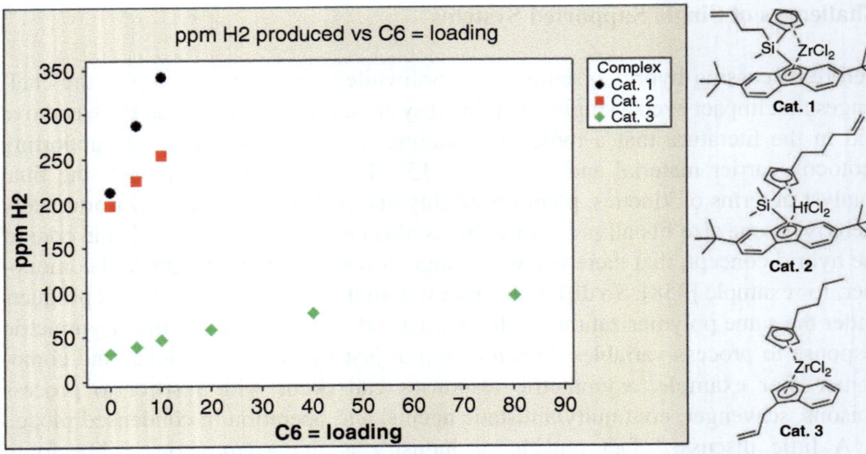

Fig. 1.38 Hydrogen generation as a function of 1-hexene concentration for different complexes

industrial perspective. For the abovementioned issues of composition control in continuous single reactor trains, there is a great need for finding ways of controlling this with hybrid catalysts. That said however, one cannot help being impressed by the scientific competence and sheer ingenuity of the industrial scientists who have risen to this challenge.

Hybrid Systems

A combination of metallocene complexes in the same process is not a new concept with, Ewen and Welborn being the first to combine different metallocenes in a solution process [363, 364]. The following section, however, is focused on particle-forming processes.

Possibly the best known hybrid systems are the Prodigy BMC catalyst series developed by Univation [358, 365]. An illustrative example is the BMC-200 system in which unbridged metallocene (Cp)(^{Me4}Cp)$ZrCl_2$ is used to make low molecular weight comonomer-lean material, while the dibenzyl McConville complexes produce the high molecular weight comonomer-rich fraction (Fig. 1.39) [366]. The catalyst is made via spray-drying the two complexes together with MAO and Carbosil TS-610, in a ratio of *ca.* 3:1 McConville to metallocene. As mentioned earlier there is no one size fits all and within this technology of supportation difference can be seen in the preparation method. For example, the spray drying temperature can impact the activity and active site ratio of the final catalyst [367].

As we saw earlier there needs to be a control mechanism in the process to compensate for variation in split and so on. To this end Univation developed its

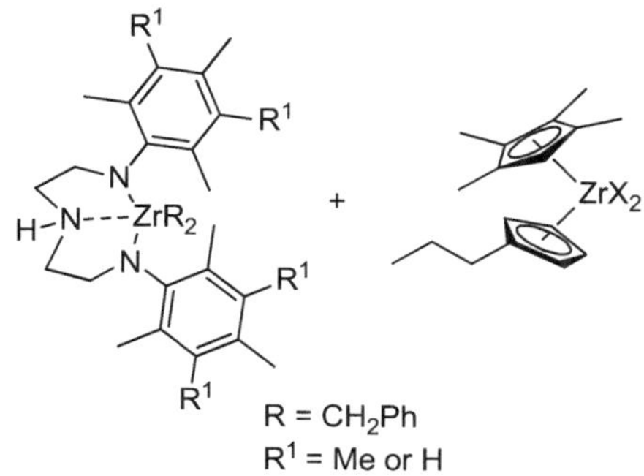

Fig. 1.39 Example of combination of different metallocene complexes

own means to solve the problem. The spray-dried catalyst (complexes, MAO, Carbosil TS-610) system is injected into the reactor in a mineral oil. Additional single component catalyst is added as a liquid into this mineral oil suspension on the way to the reactor, and is supported/activated onto the particles of the binary system. This trim system often uses the alkylated version of the LMW metallocene component to fine tune the molecular weight and split control of the polymer being made [367, 368]. Fine tuning it appears can also be done via an asymmetric response to condense mode operation [369, 370]. The system, when utilized in a single reactor, can make impressive PE100+ pipe material. However, somewhat of a drawback of these systems is that production of several grades with varying degrees of split between the molecular weights is challenging. To overcome this, more recent developments have seen the use of a component (LMW or HMW complex) that can be considered unresponsive to hydrogen. In case of the HMW component the "O_4" complexes described above have been employed (Fig. 1.40) [371, 372], while for the LMW a modified phenoxyimine complex has been utilized (Fig. 1.41) [373].

These catalysts all form extremely pronounced bimodal material which make excellent pipe products, but for some applications like blow molding a very broad distribution is required that would make the split too pronounced, leading to such issues as melt fracture. In this concept there needs to be an additional mode to fill the valley, so to say. So, proving that three is not a crowd, three-component hybrid catalysts have been disclosed to tackle this issue (Fig. 1.42) [373, 374].

In terms of disclosure the main rival to the Univation hybrid crown is CPChem., which is also the substantial licensor of the unimodal Phillips slurry loop technology. They have concentrated mainly on the combinations of two or more metallocenes on their proprietary supported activators previously discussed [375–379]. More recent disclosures have appeared to settle on one main system based on a unbridged and a bridged metallocene for the LMW phase and HMW phase, respectively [380]. As above, the addition of a third component really does assist in broadening the Mw distribution (Fig. 1.43).

As for split control it appears that the addition of an alcohol such as isopropyl alcohol can be used to selectively poison the LMW site, shifting the split to high

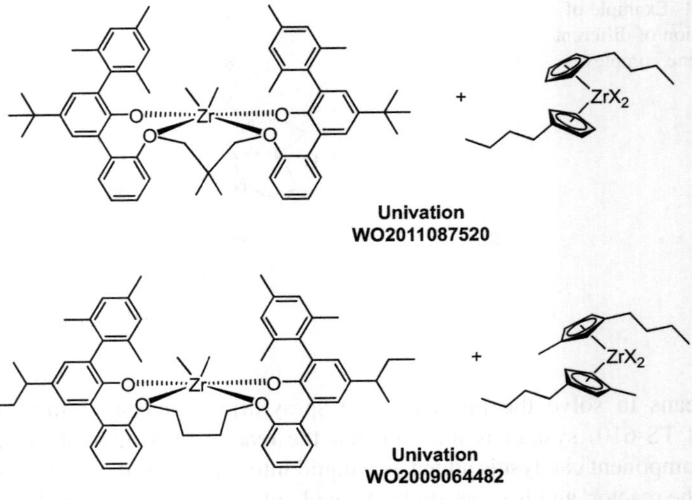

Fig. 1.40 Example of bicomponent hybrid catalyst with poor hydrogen response on the HMW component

Fig. 1.41 Example of bicomponent hybrid catalyst with poor hydrogen response on the LMW component

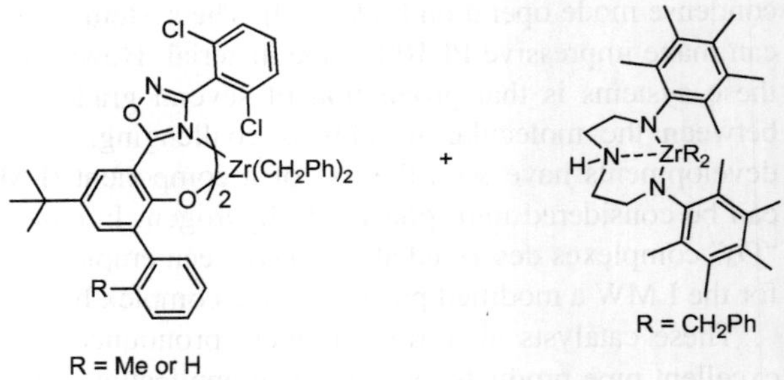

HMW with relatively little drop in the productivity [380]. Recent patenting has seen the company disclose multiple loop technology [381], opening another route to multimodal polymers.

Japan Polyethylene have also started to look at combinations of bridged metallocenes to form similar multimodal applications [382–384].

The iron-based Brookhart–Gibson catalyst has been utilized and finally commercialized by Mihan and coworkers at LyondellBasell (Fig. 1.44). In the numerous disclosures there are several combinations, but judging by the scale of the disclosed examples and also the application for patents protecting transitions between catalysts, one would expect that the combination outlined above is the leading combination [385–389]. Similar combinations were also reported by Repsol [390, 391].

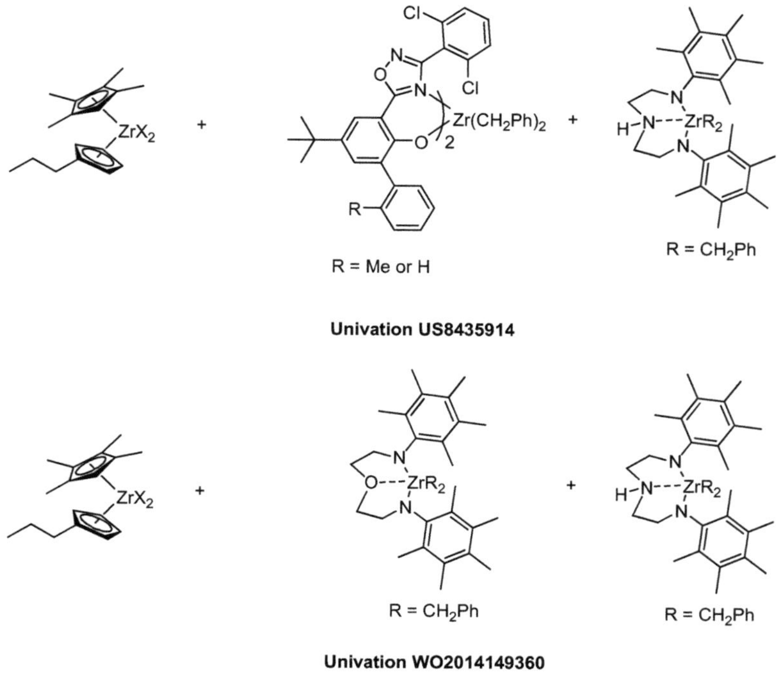

Fig. 1.42 Examples of tri-component hybrid catalysts

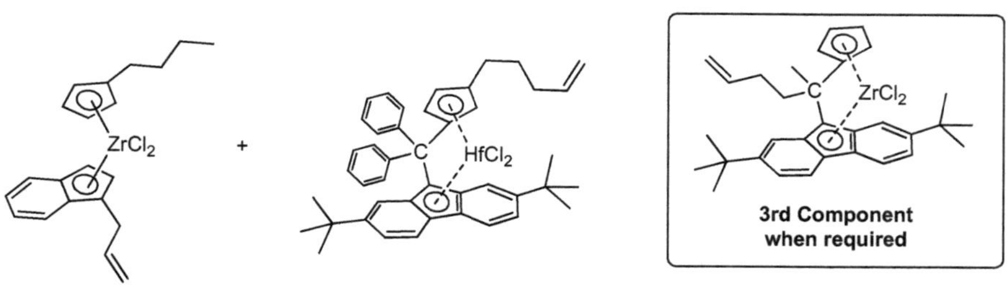

Fig. 1.43 Example of CPChem bi- and tri-component hybrid catalyst

Fig. 1.44 Mixed metallocene and Brookhart/Gibson hybrid catalyst

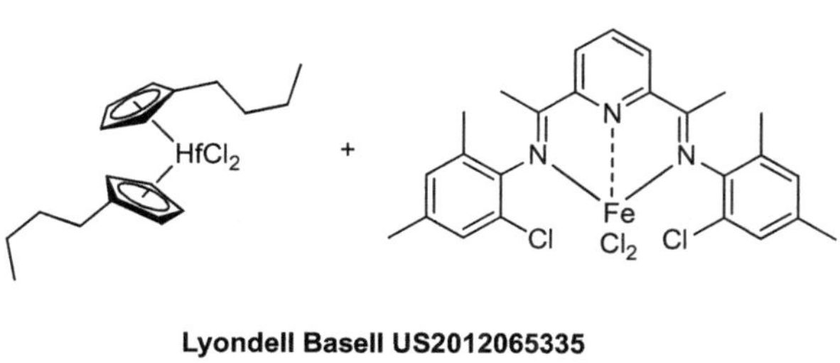

1.4 Conclusion

In this chapter the author tries to cover the latest industrial developments in the area of supported catalysts. From the vast number of references one can see that this area is still seeing significant developments. The reader should also be aware that what is missing from this review is the tremendous work being carried out towards multi-modal products via solution processes. Taking the above into account, one can see the tremendous opportunities that catalysis continues to bring to the polyolefin industry and why it remains a healthy area for research, at least within an industrial environment.

Acknowledgements The author is grateful to DSM and his family for giving him the time to write this chapter and to Ines Mingozzi (LyondellBasell), Nicolaas Friederichs (SABIC), John F. Walzer (ExxonMobil), and John C. Chadwick for providing advice and support during the preparation of this chapter.

References

1. Resconi, L., Chadwick, J. C., & Cavallo, L. (2007). In D. M. P. Mingos & R. Crabtree (Eds.), *Comprehensive, organometallic chemistry III*. New York: Elsevier.
2. Kaminsky, W. (Ed.). (2013). *Polyolefins: 50 years after Ziegler and Natta I: Polyethylene and Polypropylene*. Advances in Polymer Science (Vol. 257). Berlin: Springer.
3. Kaminsky, W. (Ed.). (2013). *Polyolefins: 50 years after Ziegler and Natta II: Polyolefins by metallocenes and other single-site catalysts*. Advances in Polymer Science (Vol. 257). Berlin: Springer.
4. Baier, M. C., Zuideveld, M. A., & Mecking, S. (2014). *Angewandte Chemie International Edition, 53*, 9722.
5. Scheirs, J., & Kaminsky, W. (Eds.). (2000). *Metallocene-based polyolefins, preparation, properties and techniques*. New York: Wiley.
6. Benedikt, G. M. (Ed.). (1998). *Metallocene technology and modern methods in commercial applications catalyzed polymers*. New York: William Andrew Publishing.
7. Reiger, B., & Saunders, L. (2003). In S. Bauch, S. Kacker, S. Striegler (Eds.), *Late transition metal polymerization catalysts*. Weinheim: Wiley.
8. Soga, K., & Terano, M. (Eds.). (1994). *Catalyst design for Taylor – Made polyolefins*. Studies in surface science and catalysis (Vol. 89, p. 277). Amsterdam: Elsevier.
9. Patil, A. O., & Hlatky, G. G. (Eds.). (2004). *Beyond Metallocenes: Next-generation polymerization catalysts*. ACS Symposium Series 857. Oxford University Press.
10. Hlatky, G. G. (1999). Metallocene catalysts for olefin polymerization: Annual review for 1996. *Coordination Chemistry Reviews, 181*, 243.
11. Shapiro, P. J. (2002). *Coordination Chemistry Reviews, 231*, 67.
12. Bochmann, M. (2004). *Journal of Organometallic Chemistry, 689*, 3982.
13. Pédeutour, J.-N., Radhakrisknan, K., Cramail, H., & Deffieux, A. (2001). *Macromolecular Rapid Communications, 22*, 1095.
14. Gibson, V. C., & Spitzmesser, S. K. (2003). *Chemical Reviews, 103*, 283–315.
15. Nomura, K., Liu, J., Padmanabhan, S., & Kitiyanan, B. (2007). *Journal of Molecular Catalysis A: Chemical, 267*, 1.
16. Porri, L. (2004). *Macromolecular Symposia, 213*, 1.

17. Albizzati, E., Cecchin, G., Chadwick, J. C., Collina, G., Giannini, U., Morini, G., et al. (2005). In N. Pasquini (Ed.), *Polypropylene handbook* (2nd ed., p. 15). Munich: Hanser Publishers.
18. Busico, V. (2013). In W. Kaminsky (Ed.), *Polyolefins: 50 years after Ziegler and Natta I: Polyethylene and Polypropylene*. Advances in Polymer Science (Vol. 257, p. 37). Berlin: Springer.
19. Natta, G., Corradini, P., & Allegra, G. (1961). *Journal of Polymer Science, 51*, 399.
20. Boor, J. (1979). *Ziegler-Natta catalysts and polymerizations*. New York: Academic Press.
21. Giannini, U. (1981). *Die Makromolekulare Chemie, 5*, 216.
22. Hermans, J. P., & Henrioulle, P. (1972). *US Patent No. 4,210,738.*
23. Goodall, B. L. (1990). In van der S. Ven (Ed.), *Polypropylene and other polyolefins. polymerization and characterization* (p. 1). Amsterdam: Elsevier.
24. Bernard, A., & Fiasse, P. (1990). In T. Keii & K. Soga (Eds.), *Catalytic olefin polymerization* (p. 405). Amsterdam: Elsevier.
25. De Smet, E., Hendrickx, L., Verwimp, W., Verheijen, M., Allemeersch, P., Vanzeir, E., et al. (2014). *US Patent Appl. 2014/0171606.*
26. Tsunori, R., Otsuki, Y., Onodera, T., Ikeda, S., & Shinohara, M. (2010). *US Patent No. 7,732,532.*
27. Kashiwa, N. (2004). *Journal of Polymer Science, Part A: Polymer Chemistry, 42*, 1.
28. Kuran, W. (2001). *Principles of coordination polymerisation* (p. 43). New York: John Wiley and Sons.
29. Galli, P., Luciani, L., & Cecchin, G. (1981). *Angewandte Makromolekulare Chemie, 94*, 63.
30. Galli, P., Barbè, P. C., Guidetti, G. P., Zannetti, R., Martorana, A., Marigo, A., et al. (1983). *European Polymer Journal, 19*, 19.
31. Sacchi, M. C., Tritto, I., Shan, C., Mendichi, R., & Noristi, L. (1991). *Macromolecules, 24*, 6823.
32. Scata, U., Luciani, L., & Barbè, P. C. (1982). *US Patent No. 4,315,835.*
33. Noristi, L., Barbè, P. C., & Baruzzi, G. (1991). *Die Makromolekulare Chemie, 192*, 1115.
34. Parodi, S., Nocci, R., Giannini, U., Barbè, P. C., & Scatà, U. (1981). *European Patent No. 45977.*
35. Albizzati, E., Giannini, U., Morini, G., Smith, C. A., & Ziegler, R. (1995). In G. Fink, R. Mülhaupt, & H.-H. Brintzinger (Eds.), *Ziegler catalysts. Recent scientific innovations and technological improvements* (p. 413). Berlin: Springer.
36. Albizzati, E., Barbè, P. C., Noristi, L., Scordamaglia, R., Barino, L., Giannini, U., et al. (1989). *European Patent No. 0,361,494.*
37. Morini, G., & Cristofori, A. (1996). *European Patent No. 0,728,724.*
38. Albizzati, E., Giannini, U., Morini, G., Galimberti, M., Barino, L., & Scordamaglia, R. (1995). *Macromolecular Symposia, 89*, 73.
39. Barino, L., & Scordamaglia, R. (1995) *Macromolecular Symposia, 89*, 101.
40. Cui, N., Ke, Y., Li, H., Zhang, Z., & Guo, Z. (2006). *Journal of Applied Polymer Science, 99*, 1399.
41. Chadwick, J. C., Morini, G., Albizzati, E., Balbontin, G., Mingozzi, I., Cristofori, A., et al. (1996). *Macromolecular Chemistry and Physics, 197*, 2501.
42. Chadwick, J. C., van der Burgt, F. P. T. J., Rastogi, S., Busico, V., Cipullo, R., Talarico, G., et al. (2004). *Macromolecules, 37*, 9722.
43. Scordamaglia, R., & Barino, L. (1998). *Macromolecular Theory and Simulations, 7*, 399.
44. Toto, M., Morini, G., Guerra, G., Corradini, P., & Cavallo, L. (2000). *Macromolecules, 33*, 1134.
45. Sacchi, M. C., Forlini, F., Tritto, I., Locatelli, P., Morini, G., Noristi, L., et al. (1996). *Macromolecules, 29*, 3341.
46. Chadwick, J. C., Morini, G., Balbontin, G., Mingozzi, I., & Albizzati, E. (1997). *Macromolecular Chemistry and Physics, 198*, 1181.
47. Morini, G., Balbontin, G., Gulevich, Y., Duijghuisen, H., Kelder, R., Klusener, P. A., et al. (2004). *US Patent No. 6,818,583.*

48. Cecchin, G., Morini, G., Pelliconi, A. (2001). *Macromolecular Symposia, 173*, 195.
49. Gao, M., Lui, H., Li, Z., Wang, J., Yang, J., Li, T., et al. (2008). *US Patent No. 7,388,061.*
50. Gao, M., Liu, H., Wang, J., Li, C., Ma, J., & Wie, G. (2004). *Polymer, 45*, 2175.
51. Haikarainen, A., Denifl, P., & Leinonen, T. (2014). *US Patent No. 8,901,261.*
52. Chen, L., Leung, T. W., & Tao, T. (2012). *US Patent No. 8,088,872.*
53. Chen, L., Leung, T. W., Tao, T., Gao, K. & Huang, X. (2013). *US Patent No. 8,466,081.*
54. Kim, S. Y., Lee, J. W., Kim, E. I., & Park, J. R. (2015). *US Patent No. 9,045,572.*
55. Ernst, A. B., Streeky, J. A., & Oliver, W. L. (2014). *US Patent No. 8,716,514.*
56. Coalter, J. N. III, Leung, T. W., Tao, T., & Gao, K. (2012). *US Patent No. 8,263,520.*
57. Chen, L., Leung, T. W., & Tao, T. (2013). *US Patent No. 8,536,290.*
58. Coalter, J. N. III, Leung, T. W., Tao, T., & Gao, K. (2012). *US Patent No. 8,633,126.*
59. Matsunaga, K., Tsuru, K., Kawakita, K., Jinnai, T., & Shinozaki, T. (2014). *US Patent No. 8,822,366.*
60. Matsunaga, K., Tsuru, K., & Shinozaki, T. (2009). *US Patent Appl. 2009/0203855.*
61. Fujiwara, Y., Hirahata, W., & Hamaki, H. (2012). *US Patent No. 8,288,488.*
62. Mignogna, A., Guidotti, S., Morini, G., & Pater, J. T. M. (2015). *US Patent Appl. 2015/0158957.*
63. Mignogna, A., Guidotti, S., Morini, G., & Pater, J. T. M. (2015). *US Patent Appl. 2015/133289.*
64. Mignogna, A., Guidotti, S., Morini, G., Pater, J. T. M., Balboni, D., & Cristofori, A. (2015). *US Patent No. 9,034,783.*
65. Nifant'ev, I., Mignogna, A., Bagrov, V., Esposito, S., Guidotti, S., Morini, G., et al. (2014). *US Patent Appl. 2014/0243489.*
66. Siddiqui, I. H., Al-Bahiyl, K. A., & Al-Gahtani, K. M. (2015). PCT Int. Appl. 2015/091984.
67. Taftaf, M. I., Ghalit, N., Bukatov, G. D., Sergeev, S. A., Zakharov, V. A., Sainani, J. B., et al. (2015). *US Patent Appl. 2015/0038660.*
68. Son, K. C., Koh, H. L., Ahn, J. K., & Lee, S. H. (2014). *US Patent No. 8,652,986.*
69. Yi, J., Cui, C., Li, H., Yin, B., Zhang, Y., Lang, X., et al. (2013). *US Patent No. 8,404,789.*
70. Bantu, B., Singh, G., Kaur, S., Kumar, N., Kapur, G. S., Kant, S., et al. (2015). *US Patent Appl. 2015/0152199.*
71. Guidotti, S., Morini, G., Esposito, S., Mignogna, A., Pater, J. T. M., Fabrizio Piemontesi, F., et al. (2014). *US Patent Appl. 2014/0046010.*
72. Guidotti, S., Piemontesi, F., Pater, J. T. M., & Morini, G. (2014). *US Patent No. 8,829,126.*
73. Chen, L., Leung, T. W., & Tao, T. (2012). *US Patent No. 8,288,585.*
74. van Egmond, J. W. (2015). PCT. Int. Pat. Appl. 2015/081254.
75. Chen, L., Leung, T. W., Roth, G. A., Tao, T., & Gao, K. (2013). *US Patent No. 8,507,717.*
76. Gullo, M. F., Roth, G. A., Leung, T. W., & Williams, C. C. (2013). *US Patent Appl. 2013/0053525.*
77. Coalter, J. N. III, Chen, L., & Williams, C. C. (2013). *US Patent Appl. 2013/0338321.*
78. Spaether, W., Lynch, J., Rösch, J., & Hemmerich, R. (2007). *US Patent No. 7,232,785.*
79. Spaether, W., & Calderone, A. T. (2004). PCT Int. Appl. WO2004/016660.
80. Kerth, J., Zolk, R., & Hemmerich, R. (1992). *US Patent No. 5,162,465.*
81. Zolk, R., Kerth, J., & Hemmerich, R. (1991). *US Patent No. 5,006,620.*
82. Ferraris, M., Rosati, F., Parodi, S., Giannetti, E., Motroni, G., & Albizzati, E. (1983). *US Patent No. 4,399,054.*
83. Evangelisti, D., & Collina G. (2006). *US Patent No. 7,060,763.*
84. Collina, G., Evangelisti, D., Morini, G., & Ferrara, G. (2009). PCT Int. Appl. WO2009080568.
85. Gaddi, B., Collina, G., & Evangelisti, D. (2014). PCT Int. Appl. WO2014/095523.
86. Mao, B., Yang, X., Li, Z., & Yang, A. (1994). Chin. Pat. CN 1091748C.
87. Sozzani, P., Bracco, S., Comotti, A., Simonutti, R., & Camurati, I. (2003). *Journal of the American Chemical Society, 125*, 12881.
88. Auriemma, F., & De Rosa, C. (2007). *Chemistry of Materials, 173*, 195.

89. Sacchetti, M., Govoni, G., & Clarrocci, A. (1993). *US Patent No. 5,221,651*.
90. Govoni, G., Clarrocci, A., & Sacchetti, M. (1993). *US Patent No. 5,231,119*.
91. Cecchin, G., Guglielmi, F., Pellicani, A., & Burgin, E. (1994). *US Patent No. 5,286,564*.
92. Invernizzi, R., & Ligorati, F. (1985). *US Patent No. 4,506,027*.
93. Iiskola, E., & Koskinen, J. (1989). *US Patent No. 4,829,034*.
94. Karbasi, A. K., Leinonen, T., & Sormunen, P. (1994). *European Patent Appl. 0,627,449 A1*.
95. Koskinen, J., & Jokinen, P. (2013). P.C.T. Appl. 93/19100.
96. Garoff, T., Leinonen, T., & Iiskola, E. (1997). *European Patent No. 0,586,390*.
97. Klendworth, D. D., Johnson, K. W., Winter, A., & Langhauser, F. (2010). *US Patent Appl. 2010/0069586*.
98. Zhu, B., Jia, J., Zhao, X., Kersting, M., Gückel, C., Wei, C., et al. (2012). *European Patent No. 2,194,070*.
99. Kioka, M., & Kashiwa, N. (1990). *US Patent No. 4,952,649*.
100. Kusumoto, T., Sugimoto, Y., & Fujisawa, M. (2015). *US Patent No. 8,975,354*.
101. Tan, N., & Tamura, S. (2012). *US Patent Appl. 2012/0010318*.
102. Mao, B., Yang, A., Zheng, Y., Yang, J., & Li, Z. (1992). *European Patent No. 0,258,485*.
103. Marin, V. P., Hintolay, A., & Spencer, M. D. (2014). *US Patent Appl. 2014/0128556*.
104. Spencer, M. D., & O'Reilly, N. (2014). *US Patent No. 8,685,879*.
105. Spencer, M. D. (2013). *US Patent No. 8,344,079*.
106. Goodall, B. L., van der Nat, A. A., & Sjardijn, W. (1982). *US Patent No. 4,329,253*.
107. Goodall, B. L., van der Nat, A. A., & Sjardijn, W. (1983). *US Patent No. 4,393,182*.
108. Goodall, B. L., van der Nat, A. A., & Sjardijn, W. (1983). *US Patent No. 4,400,302*.
109. Goodall, B. L., van der Nat, A. A., & Sjardijn, W. (1983). *US Patent No. 4,414,132*.
110. Job, R. C. (1991). *US Patent No. 5,077,357*.
111. Job, R. C. (1992). *US Patent No. 5,082,907*.
112. Job, R. C. (1992). *US Patent No. 5,124,298*.
113. Kilty, P. A., & Cuthbert, T. R. (2004). *US Patent No. 6,825,146*.
114. Morini, G., Cristofori, A., Gaddi, B., Liguori, D., Pater, J. T. M., & Vitale, G. (2011). *US Patent No. 8,062,989*.
115. Terano, M., Murai, A., Inoue, M., & Miyoshi, K. (1989). *US Patent No. 4,816,433*.
116. Terano, M., Soga, H., & Kimura, K. (1989). *US Patent No. 4,829,037*.
117. Murai, A., Terano, M., Kimura, K., & Inoue, M. (1989). *US Patent No. 4,839,321*.
118. Wagner, B. E., Zilker, D. P., & Jorgensen, R. J. (1997). *US Patent No. 5,604,172*.
119. Arzoumanidis, G. G., Karayannis, N. M., Khelghatian, H. M., Lee, S. S., & Johnson, B. V. (1989). *US Patent No. 4,866,022*.
120. Arzoumanidis, G. G., Karayannis, N. M., Khelghatian, H. M., Lee, S. S., & Johnson, B. V. (1991). *US Patent No. 4,988,656*.
121. Streeky, J. A., Bersted, B. H., Blake, J. W., Feng, D., Hoppin, C. R., & Tovrog, B. S. (2001). *US Patent No. 6,201,079*.
122. Smith, G. M., Amata, R. J., Tirendi, C. F., & Band, E. I. (1993). *US Patent No. 5,262,573*.
123. Epstein, R. A., & Wallack, W. T. (2009). *US Patent No. 7,504,352*.
124. Ramjoie, Y. J. E., Sergeev, S. A., Vlaar, M., Zakharov, V. A., & Bukatov, G. D. (2011). *US Patent No. 7,947,788*.
125. Zuideveld, M. A., Sainani, J. B., & Vimalkumar, M. P. (2015). *European Patent Appl. 2,837,634*.
126. Leinonen, T., & Denifl, P. (2003). *European Patent No. 1,273,595*.
127. Abboud, M., Denifl, P., & Reichert, K.-H. (2005). *Macromolecular Materials and Engineering, 290*, 1220.
128. Denifl, P., Jääskeläinen, P., Leinonen, T., Malm, B., Nymark, A. E., & Vestberg, T. (2015). *European Patent No. 2,565,221*.
129. Vestberg, T. (2007). *ACS advances in polyolefins*. Santa Rosa.
130. Härkönen, M., Seppälä, J. V., & Väänänen, T. (1990). In T. Keii & K. Soga (Eds.), *Catalytic olefin polymerization* (p. 87). Amsterdam: Elsevier.

131. Proto, A., Oliva, L., Pellecchia, C., Sivak, A. J., & Cullo, L. A. (1990). *Macromolecules, 23*, 2904.
132. Okano, T., Chida, K., Furuhashi, H., Nakano, A., & Ukei, S. (1990). In T. Keii & K. Soga (Eds.), *Catalytic olefin polymerization* (p. 177). Amsterdam: Elsevier.
133. Härkönen, M., & Seppala, J. V. (1989). *Macromolecular Chemistry, 160*, 2535.
134. Härkönen, M., & Seppala, J. V. (1990). *Studies in Surface Science, 56*, 87.
135. Härkönen, M., & Seppala, J. V. (1991). *Macromolecular Chemistry, 192*, 721.
136. Härkönen, M., & Seppala, J. V. (1992). *Macromolecular Chemistry, 193*, 1413.
137. Härkönen, M., Seppala, J. V., & Salminen, H. (1995). *Polymer Journal, 27*, 256.
138. Härkönen, M., Seppala, J. V., Chûjô, R., & Kogure, Y. (1995). *Polymer, 36*, 1499.
139. Kakugo, M., Miyatake, T., Naito, Y., & Mizunuma, K. (1988). *Macromolecules, 21*, 314.
140. Sacchi, M. C., Forlini, F., Tritto, I., Mendichi, R., Zannoni, G., & Noristi, L. (1992). *Macromolecules, 25*, 5914.
141. Taniike, T., & Terano, M. (2013). In W. Kaminsky (Ed.), *Polyolefins: 50 years after Ziegler and Natta I: Polyethylene and Polypropylene*. Advances in Polymer Science (Vol. 257, p. 81). Berlin: Springer-Verlag.
142. Ishimaru, Kioka, M., & Toyota, A. (1989). *European Patent No. 0,350,170*.
143. Chadwick, J. C., van Kessel, G. M. M., & Sudmeijer, O. (1995) *Macromolecular Chemistry and Physics, 196*, 1431.
144. Chadwick, J. C. (2001). *Macromolecular Symposia, 173*, 21.
145. Yoshikiyo, M., Fukunaga, T., Sato, H., Machida, T., Ikeuchi, H., Yano, T., et al. (2007). *US Patent No. 7,238,758*.
146. Stewart, C. A. (1991). *European Patent Appl. 0,410,443*.
147. Yao, S., & Tanaka, Y. (2001). *Macromolecular Theory and Simulations, 10*, 850.
148. Ikeuchi, H., Yano, T., Ikai, S., Sato, H., & Yamashita, J. (2003). *Journal of Molecular Catalysis A: Chemical, 193*, 207.
149. Tanaka, Y., Sato, H., & Fujita, H. (2006). *Studies in Surface Science and Catalysis, 172*, 527.
150. Yano, T., Hosaka, M., Sato, M., & Kimura, K. (2012). *US Patent No. 8,247,504*.
151. Ishimaru, N., Kioka, M., & Toyota, A. (1990). *European Patent No. 0,385,765*.
152. Miro, N. D., & Ohkura, M. (1999). *European Patent No. 1,080,122*.
153. Shamshoum, E., Rauscher, D., & Burmaster, D. (2001). *European Patent No. 0,676,419*.
154. Miro, N. D., Georgellis, G. B., & Swei, H. (1998). *European Patent No. 0,743,960*.
155. Chen, L., & Nemzek, T. L. (2006). *US Patent No. 7,141,635*.
156. Meka, P., Abubakar, S. M., Chen, S. Y., & Edwards, T. S. (2014). PCT Int. Appl. 2014/070386.
157. Song, W., Guo, M., Zhang, S., Wei, W., Huang, H., Yu, L., et al. (2015). *US Patent No. 9,068,030*.
158. Li, R. T., Lawson, K. W., Mehta, A. K., & Meka, P. (2007). *US Patent No. 7,183,234*.
159. Peil, K. P., Neithamer, D. R., Patrick, D. W., Wilson, B. E., & Tucker, C. J. (2004). *Macromolecular Rapid Communications, 25*, 119.
160. Chen, L. (2008). *US Patent No. 7,420,021*.
161. Campbell, R. E., & Chen, L. (2008). *US Patent No. 7,381,779*.
162. Cai, P., Van Egmond, J. W., Fedec, M. J., Goad, J. D., Brady, R. C. III, & Chen, L. (2013). *US Patent Appl. 2013/0005923*.
163. Alt, F. P., Böhm, L. L., Enderle, H. F., & Berthold, J. (2001). *Macromolecular Symposia, 163*, 135.
164. Nowlin, T. E., Mink, R. I., & Kissin, Y. V. (2010). In R. Hoff & R. T. Mathers (Eds.), *Handbook of transition metal polymerization catalysts* (p. 131). New York: John Wiley and Sons.
165. Wu, L., & Wanke, S. E. (2010). In R. Hoff & R. T. Mathers (Eds.), *Handbook of transition metal polymerization catalysts* (p. 231). New York: John Wiley and Sons.
166. Krentsel, B. A., Kissin, Y. V., Kleiner, V. J., & Stotskaya, L. L. (1997). *Polymers and copolymers of higher α-Olefins*. Munich: Carl Hanser Verlag.

167. Böhm, L. L. (2003). *Angewandte Chemie International Edition, 42*, 5010.
168. Böhm, L. L. (2013). In W. Kaminsky (Ed.), *Polyolefins: 50 years after Ziegler and Natta I: Polyethylene and Polypropylene*. Advances in Polymer Science (Vol. 257, p. 37). Berlin: Springer.
169. Karol, F. J., Goeke, G. L., Wagner, B. E., Fraser, W. A., Jorgensen, R. J., & Friis, N. (1981). *US Patent No. 4,302,566*.
170. Hartshorn, A. J., & Jones, E. (1982). *US Patent No. 4,324,691*.
171. Hagerty, R. O., Mohring, R. O., & Allen, L. M. (1995). *European Patent No. 0,231,102*.
172. Nowlin, T. E., & Mink, R. I. (1997). *European Patent No. 0,612,327*.
173. Nowlin, T. E., & Mink, R. I. (1999). *European Patent No. 0,729,478*.
174. Nowlin, T. E., & Mink, R. I. (1999). *European Patent No. 0,701,575*.
175. Nowlin, T. E., & Mink, R. I. (1995). PCT Int. Appl. 9513873.
176. Apecetche, M. A., Cao, P. A., Awe, M. D., Schoed-Wolters, A. D., & Impleman, R. W. (2008). *US Patent No. 7,381,780*.
177. Spencer, L., & Springs, M. C. (1997). *US Patent No. 5,633,419*.
178. Mavridis, H., Reinking, M. K., Shroff, R. N., Mutchler, J. A., Holland, C. S., Lindstrom, K. M., et al. (2001). *US Patent No. 6,171,993*.
179. Reinking, M. K. (2006). *US Patent No. 7,151,145*.
180. Sillantaka, L., Plamqvist, U., Iiskola, E., & Koavujäri, S. (2000). *US Patent No. 6,043,326*.
181. Ala-Huikku, S., & Lommi, M. (1996). *European Patent No. 0,573,633*.
182. Dombro, R. A. (1982). *US Patent No. 4,335,016*.
183. Pullukat, T. J., & Hoff, R. E. (1983). *US Patent No. 4,374,753*.
184. Brun, C., Cheux, A., & Barthel, E. (1992). *European Patent No. 0,296,021*.
185. Lalange-Magne, C., & Royer-Mladenov, C. (2006). *European Patent No. 1,490,415*.
186. BP Lavera SNC. (2005). *European Patent Appl. 1,502,924*.
187. Kioka, M., Kawakita, K., & Toyota, A. (1989). *European Patent No. 0,494,084*.
188. Kioka, M., Kawakita, K., & Toyota, A. (1991). *European Patent Appl. 408,750*.
189. Hagerty, R. O., Petsche, I. B., & Schurzky, K. G. (1985). *US Patent No. 4,562,169*.
190. Karol, F. J., Levine, I. J., & George, F. C. (1987). *European Patent No. 0,120,503*.
191. Levine, I. J., & Karol, F. J. (1988). *US Patent No. 4,719,193*.
192. Allen, L. M., Hagerty, R. O., & Mohring, R. O. (1988). *US Patent No. 4,732,882*.
193. Cook, P. J., Hagerty, R. O., Husby, P. K., & Nowlin, T. E. (1992). *US Patent No. 5,139,986*.
194. Kelly, M., Jeremic, D., Ker, V., & Russell, C. (2000). *US Patent No. 6,140,264*.
195. Kelly, M., Goyal, S. K., Ker, V., Montyn de Wit, P., Kimberley, B. S., & Hoang, P. P. M. (2007). *US Patent No. 7,211,535*.
196. Goyal, S. K., Wiwchar, T. W., Ker, V., & Kelly, M. (2004). *US Patent No. 6,825,293*.
197. Kelly, M., & Kimberley, B. S. (2010). *US Patent No. 7,671,149*.
198. Garoff, T., Johansson, S., Palmqvist, U., Lindgren, D., Sutela, M., Waldvogel, P., et al. (2000). *US Patent No. 6,034,026*.
199. Lindroos, J., Johansson, S., & Waldvogel, P. (2005). *US Patent No. 6,924,343*.
200. Hamer, A. D., & Karol, F. J. (1981). *US Patent No. 4,293,673*.
201. Karol, F. J., Goeke, G. L., Wagner, E. B., Frazer, A., & Jorgensen, R. J. (1981). *US Patent No. 4,302,566*.
202. Schouterden, P. J. C., Nicasy, R. A. J., Munjal, S., Jorgensen, R. J., & Wagner, B. E. (2014). *US Patent No. 8,916,649*.
203. Wagner, B. E., & Jorgensen, R. J. (2006). *US Patent No. 6,982,237*.
204. Jorgensen, R. J., Kapur, M., Michie, W. J., & Wagner, B. E. (2009). PCT Int. Appl. 2009/085922.
205. Zoeckler, M. T., Wagner, B. E., & Kao, S.-C. (2008). *US Patent No. 7,348,383*.
206. Spriggs, T. E., Turner, M. D., Wagner, B. E., Levandovsky, A., & Lacks, D. J. (2006). PCT Int. Appl. 2006/020623.
207. Jorgensen, R. J., Upham, S. M., Madden, J. D., & Michie, W. J. (2001). *US Patent No. 6,187,866*.

208. Wagner, B. E., Job, R., Schoeb-Wolters, A., & Jorgensen, R. J. (2010). PCT Int. Appl. 2010/017393.
209. Beigzadeh, D., Campbell, R. E., Ewart, S. W., Froese, R. D., Jorgensen, R. J., & Margl, P. M. (2015). *US Patent No. 8,993,692*.
210. Campbell, R. E., Chen, L., Painter, R. B., Reib, R. N., & Tilston, M. W. (2008). *US Patent No. 7,393,910*.
211. Berger, E., & Derroitte, J.-L. (1975). *US Patent No. 3,901,863*.
212. Bienfait, C. (1986). *US Patent No. 4,617,360*.
213. Bian, J. (2003). *US Patent No. 6,545,106*.
214. Cermelli, I., Dheur, L. M. G., & Siberdt, F. (2013). PCT Int. Appl. 2013/178673.
215. Ameye, T. F., Frederich, A., & Jan, D. (2013). *US Patent No. 8,445,619*.
216. Batinas-Geurts, A. A., Friederichs, N. H., Schoffelen, T., Zuidema, E., & Garg, P. (2013). PCT Int. Appl. 2013/087167.
217. Friederichs, N. H. (2010). PCT Int. Appl. 2010/006756.
218. Böhm, L. (1991). PCT Int. Appl. 1991/018934.
219. Lecht, R. (1994). *European Patent No. 0,401,776*.
220. Berthold, J., Diedrich, B., Franke, R., Hartlapp, J., Schäfer, W., & Strobel, W. (1985). *European Patent No. 0,068,257*.
221. Dotsch, D., Marczinke, B. L., Meier, G., & Salinas, M. S. (2014). *US Patent No. 8,802,768*.
222. Berthold, J., Heinicke, L.-G., & Meier, G. (2014). *US Patent No. 8,673,437*.
223. Berthold, J., Nitz, H., Rothhöft, W., & Vogt, H. (2015). *US Patent No. 9,051,458*.
224. Berthold, J., Nitz, H., Rothhoeft, W., Schulte, U., & Vogt, H. (2008). PCT Int. Appl. 2008/049551.
225. Chen, H., Coffy, T. J., & Shamshoum, E. S. (2001). *US Patent No. 6,174,971*.
226. Knoeppel, D. W., Coffy, T. J., Enriquez, H., & Gray, S. D. (2005). *US Patent No. 6,864,207*.
227. Vizzini, K., Knoeppel, D., Gray, S., Rauscher, D., Coffy, T., & Enriquez, H. (2009). *US Patent No. 7,473,664*.
228. Enriquez, H., Vizzini, K., & Gray, S. (2010). *US Patent No. 7,655,590*.
229. Vantomme, A., Siraux, D., Sinoy, A. V., & Gielens, J.-L. (2013). *US Patent No. 8,609,792*.
230. Coffy, T. J., Guenther, G., & Gray, S. D. (2012). *US Patent No. 8,138,264*.
231. van den Berg, C. E. P. V. (1979). *US Patent No. 4,178,300*.
232. Denifl, P., Leinonen, T., & Kipiani, G. (2103). PCT Int. Appl. 2013/098138.
233. Garoff, T., & Waldvogel, P. (2012). *US Patent No. 8,143,184*.
234. Wang, S., Liu, D., Zhou, J., Lü, X., Zhang, L., Mao, B., et al. (2015). *US Patent No. 9,068,025*.
235. Yasuda, K., Matsumoto, T., & Mizumoto, K. (2010). *US Patent Appl. 2010/0196711*.
236. Nakayama, Y., Saito, J., Bando, H., & Fujita, T. (2006). *Chemistry: A European Journal*, 7546.
237. Tsutsui, T., Yashiki, T., & Funabara, Y. (2004). *US Patent No. 6,716,924*.
238. Yashiki, T., & Minami, S. (2004). *US Patent No. 6,806,222*.
239. Matsunaga, K., Yamamoto, H., Yamamoto, K., Shinozaki, T., & Takahashi, K. (2013). *US Patent No. 8,383,541*.
240. Agrifoglio, G., Diaz-Barrios, A., Liscano, J., Matos, J. O., & Trujillo, M. (2004). *European Patent No. 0,810,235*.
241. Chang, M., & Garoff, T. (2011). *US Patent Appl. 2011/0294970*.
242. Zhu, Z., Chang, M., & Aarons, C. J. (2006). *US Patent No. 7,153,803*.
243. Cuffiani, I., Pennini, G., & Sacchetti, M. (1996). *US Patent No. 5,578,541*.
244. Morini, G., Dall'Occo, T., Piemontesi, F., Spoto, R., Vincenzi, P., & Vitale, G. (2009). *US Patent No. 7,592,286*.
245. Fushimi, M., Liguori, D., Dall'occo, T., Morini, G., Pater, J. T. M., & Vitale, G. (2011). PCT Int. Appl. 2011/015553.
246. Fushimi, M., & Schneider, M. (2009). PCT Int. Appl. 2009/027269.
247. Fushimi, M., Schneider, M., & Morini, G. (2009). PCT Int. Appl. 2009/098177.
248. Fushimi, M., Schneider, M., & Morini, G. (2009). PCT Int. Appl. 2009/027266.

249. Fushimi, M. (2009). PCT Int. Appl. 2009/098198.
250. Gelus, E. (2002). *European Patent No. 0,703,246*.
251. Berardi, A., Frederich, A., Jan, D., Kuhlburger, J.-J., Sgard, A., & Van Daele, A. (2012). *US Patent No. 8,293,857*.
252. Ford, R. R., & Stuart, R. K. (2001). *US Patent No. 6,291,613*.
253. Ford, R. R., & Stuart, R. K. (2001). *US Patent No. 7,652,113*.
254. Tsubaki, K., Morinaga, H., Iwabuchi, T., & Kawahara, M. (1980). *US Patent No. 4,223,118*.
255. Iizuka, T., Yamamoto, S., Kawahara, M., Takemori, T., Iida, M., & Nishitani, K. (2004). *US Patent No. 6,743,863*.
256. Tsubaki, K., Morinaga, H., Matsuo, Y., & Iwabuchi, T. (1982). *US Patent No. 4,357,448*.
257. Mavridis, H., Mehta, S. D., Mack, M. P., Garrison, P. J., & Lynch, M. W. (2007). *US Patent No. 7,230,054*.
258. Mehta, S. D., Reinking, M. K., Joseph, S., Garrison, P. J., Lewis, E. O., Schwab, T. J., et al. (2009). *US Patent Appl. 2009/0304966*.
259. Kamiyama, S., Kawahara, M., Inamatsu, K., Kase, K., Mizokami, K., & Matsumoto, T. (1995). *US Patent No. 5,422,400*.
260. Okada, M., & Hirahata, W. (2012). *US Patent No. 8,105,970*.
261. Spaleck, W., Antberg, M., Rohrmann, J., Winter, A., Bachmann, B., Kiprof, P., et al. (1992). *Angewandte Chemie International Edition, 31*, 1347.
262. Okumura, Y., Seidel, N., & Kölling, L. (2009). *European Patent No. 1,636,245*.
263. Burkhardt, T. J., Haygood, W. T., Li, R. T., Vizzini, J. C., Kuchta, M. C., Stehling, U. D., et al. (2002). *US Patent No. 6,414,095*.
264. Paczkowski, N., Winter, A., & Langhauser, F. (2006). PCT Int. Appl. 2006/060544.
265. Okumura, Y., Nifant'ev, E. I., Michael Elder, M., Ivchenko, P. V., & Bagrov, V. (2005). PCT Int Appl. 2005/058916.
266. Nifant'ev, E. I., Ivchenko, P. V., Okumura, Y., Ciaccia, E., & Resconi, L. (2006). PCT Int. Appl. 2006/097497.
267. Resconi, L., Castro, P., Maaranen, J., Voskoboynikov, A. V., Asachenko, A. F., Tsarev, A. A., et al. (2012). *European Patent Appl. 2402353*.
268. Resconi, L., Focante, F., Balboni, D., Nifant'ev, E. I., Ivchenko, P. V., & Bagrov, V. (2007). PCT Int. Appl. 2007/116034.
269. Sell, T., Winter, A., Thorn, M. G., Dimeska, A., & Langhauser, F. (2010). PCT Int. Appl. 2010/077230.
270. Castro, P., Izmer, V. V., Konovich, D. S., Resconi, L., & Voskoboynikov, A. Z. (2012). PCT Int. Appl. 2012/084961.
271. Kashimoto, M., Takahashi, T., Naoshi Iwama, N., Nakano, M., & Toshinori Suzuki, T. (2014). PCT Int. Appl. 2014/069391.
272. Nakano, M., Uchino, H., Iwama, N., Kashimoto, M., & Kato, T. (2013). *US Patent No. 8,461,365*.
273. Hafner, N., Castro, P., Kulyabin, P. V., Izmer, V., Voskoboynikov, A., Resconi, L., et al. (2013). PCT Int. Appl. 2013/007650.
274. Uchino, H., Nakano, H., Toriu, S., Tayano, T., Niwa, H., Ishihama, Y., & Sugano, T. (2004). *US Patent No. 6,677,411*.
275. Japan Polypropylene. (2012). *Japanese Patent No. 5,639,813*.
276. Kitade, S., Uchino, H., Shinozaki, J., Takahashi, K., & Masuda, K. (2011). *US Patent No. 7,915,359*.
277. Bamberger, R. L., German, P. M., Locke, L. K., & Malpass, G. D. (1998) *European Patent No. 0,699,219*.
278. Lue, C.-T., Merrill, N. A., Muhle, M. E., & Vaughan, G. A. (2001). *US Patent No. 6,255,426*.
279. Agapiou, A. K., & Russell, K. A. (2005). *US Patent No. 6,936,226*.
280. Szul, J. F., Farley, J. M., McCullough, L. G., & Impelman, R. W. (2007). *US Patent No. 7,179,876*.
281. Szul, J. F., & Farley, J. M. (2007). *US Patent No. 7,157,531*.

282. McCullough, L. G. (2005). *US Patent No. 6,884,748.*
283. Takahashi, M., Todo, A., Ikeyama, S., Tsutsui, T., Matsunaga, S., & Kaneshige, N. (2005). *US Patent No. 6,894,120.*
284. Tohi, Y., Yoshitsugu, K., Akiyama, N., Fujita, T., & Chinaka, M. (2013). *US Patent No. 8,445,609.*
285. Kokko, E., Pakkanen, A., Vahteri, M., Palmlof, M., & Oderkerk, J. (2011). *US Patent Appl. 2011/0262670.*
286. Helland, I., & Skar, M. (2012). *US Patent No. 8,314,187.*
287. Pannell, R. B. (2014). *US Patent No. 2014/143526.*
288. Miserque, O., Michel, J., Dupire, M., Siberdt, F., Costa, J.-L., Bettonville, S., et al. (2005). *US Patent No. 6,946,521.*
289. Slawinski, M. (2013). *US Patent No. 8,445,607.*
290. Michel, J., Slawinski, M., & Debras, G. (2012). *US Patent No. 8,153,734.*
291. Belloir, P., & Bertrand, C. (2014). *US Patent No. 8,691,354.*
292. Tasaki, T., Akashi, T., Matsubara, S., & Okamoto, M. (2009). *US Patent Appl. 2009/0018299.*
293. Fukushi, K., Iwamasa, K., & Okamoto, M. (2009). *US Patent Appl. 2009/0036584.*
294. Jacobsen, G. B., Matsushita, F., Spencer, L., & Wauteraerts, P. L. (2003). *US Patent No. 6,506,866.*
295. Jacobsen, G. B., Loix, P. H. H., & Stevens, T. J. P. (2001). *US Patent No. 6,271,165.*
296. Arriola, D. J., Timmers, F. J., Devore, D. D., & Redwine, O. D. (2007). *US Patent No. 7,193,024.*
297. Chai, C. K. (2011). *US Patent No. 7,968,659.*
298. Van Dun, J. J., Schouterden, P. J. C., Sehanobish, K., van den Berghen, P. F., Jivraj, N., Vanvoorden, J., et al. (2012). *US Patent No. 8,338,538.*
299. Lam, P., Ker, V., Carter, C. A. G., Shaw, B. M., Baar, C. R., Kazakov, A., et al. (2012). *US Patent No. 8,829,137.*
300. McKay, I., Jeremic, D., Jacobsen, G. B., & Mastroianni, S. (2011). *US Patent No. 7,863,213.*
301. Ker, V., Lam, P., Jiang, Y., Hoang, P. P. M., Carter, C. A. G., & Morrison, D. J. (2014). PCT Int. Appl. 2014/08967.
302. Hoang, P. P. M., Lam, P., Ker, V., Baar, C. R., Carter, C. A. G., Jiang, Y., et al. (2015). *US Patent Appl. 2015/099856.*
303. Jacobsen, G. B., Jeremic, D., Mastroianni, S., & McKay, I. D. (2013). *US Patent No. 8,536,081.*
304. Hermel-Davidock, T. J., Demirors, M., Hayne, S. H., & Cong, R. (2014). *US Patent No. 8,729,200.*
305. Karjala, T. P., Ewart, S. W., Eddy, C. R., Vigil, A. E., Demirors, M., Munjal, S., et al. (2014). *US Patent No. 8,722,817.*
306. Konze, W. V., Stevens, J. C., & VanderLende, D. D. (2012). *US Patent No. 8,202,953.*
307. Ewart, S. W., Munjal, S., Vigil, A. E., Karjala, T. P., & Demirors, M. (2015). *US Patent No. 9,045,628.*
308. Boone, H. W., Iverson, C. N., Konze, W. V., & Vanderlende, D. D. (2007). PCT Int. Appl. 2007/136494.
309. Hustad, P. D., Szuromi, E., Timmers, F. J., Carnahan, E. M., Clark, T. P., Roof, G. R., et al. (2014). *US Patent No. 8,907,034.*
310. Diamond, G. M., Leclerc, M. K., & Zhu, G. (2014). *US Patent No. 8,637,618.*
311. Robert, D. R., Hufen, J., Lüdtke, K., & Ehlers, J. (2015). *US Patent No. 9,034,999.*
312. Carnahan, E. M., Devore, D., Godziela, G., Vosejpka, P., Wagner, B., & Coalter, J. (2014). *European Patent No. 1778738.*
313. Hlatky, G. G. (2000). *Chemical Reviews, 100*, 1347.
314. Severn, J. R., Chadwick, J. C., Duchateau, R., & Friederichs, N. (2005). *Chemical Reviews, 105*, 4073.
315. Severn, J. R., & Chadwick, J. C. (Eds.). (2008). *Tailor-made polymers. Via immobilization of alpha-olefin polymerization catalysts*. Weinheim: Wiley-VCH Verlag GmbH.

316. Luo, L., Sangokoya, S. A., Wu, X., Diefenbach, S. P., & Kneale, B. (2013). *US Patent No. 8,354,485.*
317. Gao, X., Chisholm, P. S., Donaldson, R. D., & McKay, I. (2004). *US Patent No. 6,710,143.*
318. Gao, X., Chisholm, P. S., Kowalchuk, M. G., & Donaldson, R. D. (2004). *US Patent No. 6,734,266.*
319. Gao, X., Santos, B. G., Minh Hoang, P. P. M., Jones, A. M., Shaw, B. M., & Jobe, I. R. (2012). *US Patent No. 8,298,978.*
320. Speca, A. N. (2002). *US Patent No. 6,368,999.*
321. Farley, J. M., Adetunji, P. A., Mirams, S. J., & Beckton, G. L. (2014). PCT Int. Appl. 2014/144397.
322. Denifl, P., Van Praet, E., Bartke, M., Oksman, M., Mustonen, M., Garoff, T., et al. (2003). PCT Int. Appl. 2003/051934.
323. Bartke, M., Oksman, M., Marja Mustonen, M., & Denifl, P. (2005). *Macromolecular Materials and Engineering, 290*, 250.
324. Tynys, A., Saarinen, T., Bartke, M., & Löfgren, B. (2007). *Polymer, 48*, 1893.
325. Heiskanen, H., Denifl, P., Hurme, M., & Pitkänen, P. (2010). *Chemical Engineering and Technology, 33*, 682.
326. Aumo, J., Matikainenn, P., Bartke, M., Elovirta, T., Vijay, S., Lylykangas, M., et al. (2013). *US Patent No. 8,501,881.*
327. Kallio, K., Mustonen, M., Huhtanen, L., Severn, J., Castro, P., Virkkunen, V., et al. (2014) *European Patent Appl. 2,722,346.*
328. Kallio, K., Mustonen, M., Huhtanen, L., Severn, J., Castro, P., Virkkunen, V., et al. (2014). *European Patent Appl. 2,722,345.*
329. Sangokoya, S. A. (1997). *US Patent No. 5,670,682.*
330. Sangokoya, S. A. (1999). *US Patent No. 5,922,631.*
331. Elo, P., Severn, J., Denifl, P., Rautio, S., Mustonen, M., & Hongell, A.-L. (2013). *US Patent No. 8,420,562.*
332. Kallio, K., Mustonen, M., Elo, P., Severn, J., & Denifl, P. (2014). *US Patent No. 8,828,901.*
333. Valonen, J., & Mustonen, M. (2014). *US Patent No. 8,822,365.*
334. Reznichenko, A., Ajellal, N., Castro, P., & Saeed, I. (2015). PCT Int. Appl. 2015/062936.
335. Kaji, E., & Yoshioka, E. (2013). *US Patent No. 8,404,880.*
336. Kaji, E., & Yoshioka, E. (2015). *US Patent Appl. 2015/0057418.*
337. McDaniel, M. P. (2008). In J. R. Severn & J. C. Chadwick (Eds.), *Tailor-made polymers. Via immobilization of alpha-olefin polymerization catalysts*. Weinheim: Wiley-VCH Verlag GmbH.
338. Jensen, M. D., Hawley, G. R., McDaniel, M. P., Crain, T., Benham, E. A., Martin, J. L., et al. (2007). *US Patent No. 7,294,599.*
339. Benham, E. A., & McDaniel, M. P. (2010). PCT Int. Appl. 2010/151537.
340. Luo, L., Lee, J. Y., Diefenbach, S. P., Sangokoya, S. A., & Bauch, C. G. (2007). PCT Int. Appl. 2007/005676.
341. Luo, L., Wu, K., & Diefenbach, S. P. (2011). *US Patent No. 7,928,172.*
342. Clarembeau, M., Pannier, G., & Paye, S. (2012). PCT Int Appl. 2012/080314.
343. Bohnen, H., & Fritze, C. (2002). *US Patent No. 6,482,902.*
344. Richter, B., & Heike, G. (2011). *US Patent No. 8,076,259.*
345. Seidel, N., Richter, B., & Kratzer, R. (2008). *US Patent No. 7,442,667.*
346. Holtcamp, M. W., & Cano, D. A. (2004). *US Patent No. 6,703,338.*
347. Holtcamp, M. W., & Cano, D. A. (2005). *US Patent No. 6,858,689.*
348. Casty, G. L., & Piland, E. J. (2010). *US Patent No. 7,741,417.*
349. Goto, T., & Onodera, Y. (2012). *US Patent No. 8,101,537.*
350. Isobe, E., Maruyama, Y., Shimizu, F., Suga, Y., & Suzuki, T. (1994). 5,308,811.
351. Suga, Y., Uehara, Y., Maruyama, Y., Isobe, E., Ishihama, Y., & Sagae, T. (1999). *US Patent No. 5,928,982.*

352. Takahashi, K., Ishihama, Y., Akashige, E., Ikehata, F., Uchida, H., & Kawase, M. (2003). *US Patent No. 6,632,911*.
353. Shih, K.-Y., Carney, M. J., & Denton, D. A. (2002). *US Patent No. 6,399,535*.
354. Shih, K.-Y. (2004). *US Patent No. 6,686,306*.
355. Friederichs, F., Ghalit, N., & Xu, W. (2008). In J. R. Severn & J. C. Chadwick (Eds.), *Tailor-made polymers. Via immobilization of alpha-olefin polymerization catalysts*. Weinheim: Wiley-VCH Verlag GmbH.
356. Mink, R. I., Schurzky, K. G., Shirodkar, P. P., & Santana, R. L. (2006). *US Patent No. 6,995,109*.
357. Mink, R. I., Nowlin, T. E., Shirodkar, P. P., Diamond, G. M., Barry, D. B., Wang, C., et al. (2006). *US Patent No. 7,129,302*.
358. Vaughan, G. A., Szul, J. F., McKee, M. G., Farley, J. M., Lue, C.-T., & Kao, S.-C. (2006). *US Patent No. 7,141,632*.
359. Burkhardt, T. J., Canich, J. A. M., Poirot, E. E., Sagar, V. R., Walzer, J. F., & Welborn, H. C. (1996). PCT Int. Filing 1996/000246.
360. Masino, A. P., Murray, R. E., Yang, Q., Secora, S. J., Jayaratne, K. C., Beaulieu, W. B., et al. (2013). *US Patent No. 8,450,436*.
361. Jensen, M. D., Elder, M. J., Singleton, A. G., Schmidt, S. R., Kerwin, P. J., Hain, J. H., et al. (2015). *US Patent No. 9,045,569*.
362. Muruganandam, N., Abichandani, J., Terry, K. A., Patel, H. G., & Rodriguez, G. (2013). *US Patent No. 8,563,458*.
363. Ewen, J. A., & Welborn, H. C. (1985). *US Patent No. 4,530,914*.
364. Ewen, J. A., & Welborn, H. C. (1990). *US Patent No. 4,937,299*.
365. Lue, C.-T., & Crowther, D. J. (2002). *US Patent No. 6,492,472*.
366. Liu, H.-T., & Mure, C. R. (2013). *US Patent No. 8,378,029*.
367. Lynn, T. R., Hussein, F. D., Pequeno, R. E., Zilker, D. P., Savatsky, B. J., & Awe, M. D. (2014). PCT Int. Appl. 2014/109832.
368. Mawson, S., Kao, S.-C., Kwalk, T. H., Lynn, T. R., McConville, D. H., McKee, M. G., et al. (2004). *US Patent No. 6,689,847*.
369. Savatsky, B. J., Oskam, J. H., Blood, M. W., Davis, M. B., Jackson, D. H., Lynn, T. R., et al. (2012). *US Patent No. 8,318,872*.
370. Pequeno, R. E., Hagerty, R. O., & Savatsky, B. J. (2010). *US Patent No. 7,754,834*.
371. Rix, F. C., Kao, S.-C., Kolb, R., Li, D., & Garcia-Franco, C. A. (2014). *US Patent No. 8,835,577*.
372. Kolb, R., Li, D., Rix, F. C., & Garcia-Franco, C. A. (2012). *US Patent No. 8,088,704*.
373. Kao, S.-C., Rix, F. C., Li, D., Harlan, C. J., & Khokhani, P. A. (2013). *US Patent No. 8,435,914*.
374. Wagner, J. B., Giesbrecht, G. R., Kao, S.-C., & Jaker, S. P. (2014). PCT Int. Appl. 2014/149360.
375. Jensen, M. D., McDaniel, M. P., Martin, J. L., Benham, E. A., Muninger, R., Jerdee, G., et al. (2006). *US Patent No. 7,041,617*.
376. Martin, J. L., Benham, E. A., Kertok, M. E., Jensen, M. D., McDaniel, M. P., Hawley, G. R., et al. (2007). *US Patent No. 7,199,073*.
377. Jayaratne, K. C., Jensen, M. D., & Yang, Q. (2007). *US Patent No. 7,226,886*.
378. Martin, J. L., Thorn, M. G., McDaniel, M. P., Jensen, M. D., Yang, Q., DesLauriers, P. J., et al. (2007). *US Patent No. 7,312,283*.
379. Yang, Q., Jayaratne, K. C., Jensen, M. D., McDaniel, M. P., Martin, J. L., Thorn, M. G., et al. (2012). *US Patent No. 8,268,944*.
380. Yang, Q., McDaniel, M. P., Crain, T. R., Masino, A. P., Cymbaluk, T. H., & Stewart, J. D. (2015). *US Patent No. 8,957,168*.
381. Bhandarkar, M. B., Benham, E. A., Gonzales, R. A., Kufeld, S. E., Mutchler, J. A., Gill, G. M., et al. (2014) PCT Int. Appl. 2014/093082.
382. Satoh, Y., & Harada, Y. (2013). *US Patent No. 2015/0018491*.

383. Bando, H., Satoh, Y., Yukita, T., Harada, Y., Sonobe, Y., Tohi, Y., et al. (2014). *US Patent No. 8,785,574.*
384. Ishihama, Y., Asakawa, R., Sakuragi, T., Fukuda, T., Sakata, K., Aoki, M., et al. (2014). *US Patent Appl. 2014/0194277.*
385. Mihan, S., Karer, R., Schmitz, H., & Lilge, D. (2011). *US Patent No. 8,003,740.*
386. Mihan, S., Fraaije, V., & Schmitz, H. (2014). *US Patent No. 8,859,451.*
387. Fantinel, F., Mannebach, G., Mihan, S., Meier, G., & Vittorias, I. (2015). *US Patent No. 8,957,158.*
388. Vogt, H., Mihan, S., Mannebach, G., Richter-Lukesova, L., Meierhöfer, M., & Brüning, H. (2015). PCT Int. Appl. 2015/055392.
389. Schmitz, H., & Mihan, S. (2013). *US Patent No. 8,435,911.*
390. Vega, W. M., & Munoz-Escalona, A. (2003). *US Patent No. 6,605,676.*
391. Lopez, R. M., Martin, M. C., Prieto, A. O., Sancho, R. J., Campora, P. J., Pilar Palma, R. P., et al. (2008). *European Patent No. 2,003,166.*

383. Bando, H., Satoh, N., Mikata, T., Higuchi, Y., Sonoda, K., Tada, Y. [illegible] (20[illegible]). *US Patent No. 8,785,374*.
384. Ushikubo, Y., Akiyama, R., Sakamoto, T., Takada, T., Shimizu, A. [illegible] (201[illegible]). *US Patent Appl. 2014/0[illegible]127*.
385. Mihan, S., Karch, R., Schmitz, H., Liegel, P. (2014). *[illegible] Patent No. [illegible]*.
386. Mihan, S., Fraaije, V., & Schmitz, H. (2014). *US Patent No. [illegible]*.
387. Fantinel, F., Mannebach, G., Mihan, S., Meier, G., & Vittorias, I. (2014). *US Patent No. 8,957,158*.
388. Vogt, H., Mihan, S., Mannebach, G., Fraaije, V., Mihan, S., & [illegible], H. (2015). PCT Int. Appl. 2015/0[illegible].
389. Schmitz, H., & Mihan, S. (2015). *US Patent No. 8,[illegible]*.
390. Vega, W. M., & Munoz Escalona, A. (2003). *US Patent No. [illegible]*.
391. Lopez, M., Morais, M. C., Tortosa, A. O., Sancho, E. [illegible], Fabra, P., et al. (2008). *European Patent No. 2,[illegible]*.

Chapter 2
Support Designed for Polymerization Processes

Jonas Alves Fernandes and Anne-Lise Girard

2.1 General Considerations

In the current polyolefin manufacture industry the polymerization of ethylene and propylene are mostly carried out by coordination catalyst, among these, Ziegler–Natta (ZN) and Phillips catalysts still correspond for more than 60% of polyethylene (PE) production and ZN catalyst alone is responsible for almost the entire isotactic homo-polypropylene (*i*PP) and its copolymers production. Beyond these catalyst systems, metallocenes are used to produce about 20–25% of polyethylene and still gain space in the manufacture of *i*PP. Since most of the commercial polymerization occurs in a heterogeneous medium, which simplifies the polymerization process technologies, these catalyst systems are formed by metal active center supported on an inorganic particle. Thus, the support preparation and the synthesis of the catalyst are often crucial steps to the polymerization process and on tailoring polymer properties.

With this in mind, this chapter aims to discuss some important technical aspects of the commercial support related to each catalyst technology toward the production of different polyolefin grades. It approaches several industrial factors that are defined by the fine tuning between support and catalyst preparation, such as the catalyst fragmentation, the control of the final particle morphology and the morphological replication phenomena, the heat and mass transfer limitations during the polymerization reaction, and the uses of prepolymerization stage. The understanding of these parameters and the limits of the polymerization processes are the first step for designing a suitable support for a heterogeneous polymerization catalyst. Aspects such as chemical

J. A. Fernandes (✉)
Catalysis Research Group, Braskem America Inc., Pittsburgh, PA, USA
e-mail: jonas.fernandes@braskem.com

A.-L. Girard
Department of Science, Robert Morris University, Pittsburgh, PA, USA
e-mail: girard@rmu.edu

A. R. Albunia et al. (eds.), *Multimodal Polymers with Supported Catalysts*,
https://doi.org/10.1007/978-3-030-03476-4_2

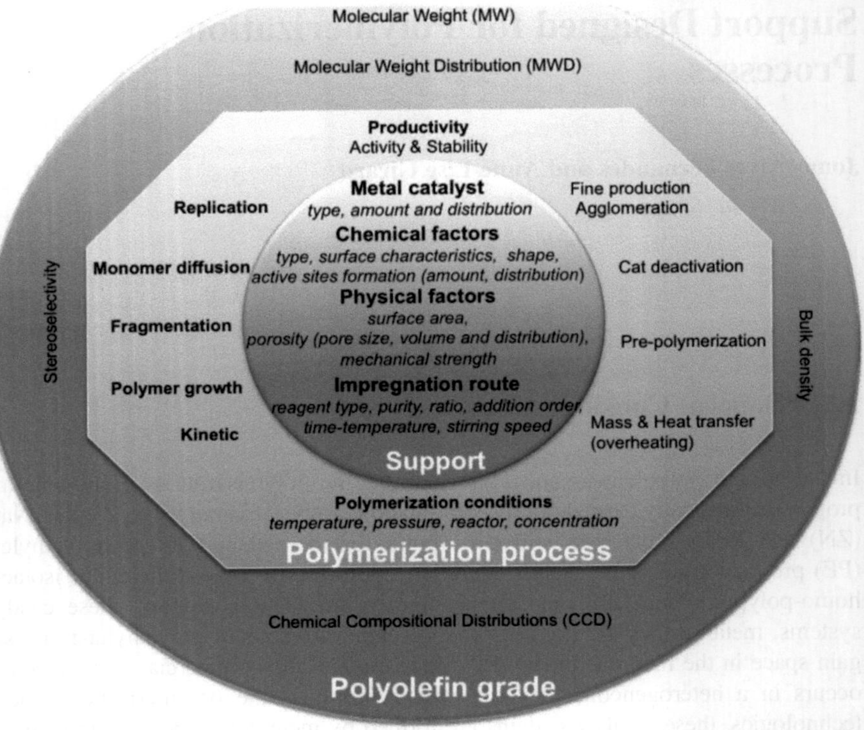

Fig. 2.1 Summary of the general parameters in polyolefin production

composition of the support, its surface characteristics, morphology (surface area, particle size, particle size distribution, and porosity), mechanical strength, and other characteristics of relevance for the two most common supports in the polyolefin industry—$MgCl_2$ and SiO_2—are discussed as well. Figure 2.1 displays a summary of the most important parameters to be considered in the design of a suitable catalyst support within their influence in critical aspects of the polyolefin production.

2.2 $MgCl_2$-Based Support

The use of $MgCl_2$ as inorganic carrier for heterogeneous ZN catalyst enabled the global industrialization of *i*-PP. The development of the ZN catalyst for *i*-PP, the so-called third-generation, consisted in the introduction of $MgCl_2$ as a support for $TiCl_3$ and the use of internal and external electron donors for stereocontrolled propylene polymerization. The support was prepared by ball-milling $MgCl_2$ with internal electron donor (ethyl benzoate), which led to the formation of a structurally disordered δ-$MgCl_2$ small crystallite with stacking faults of triple layers [1, 2]. The

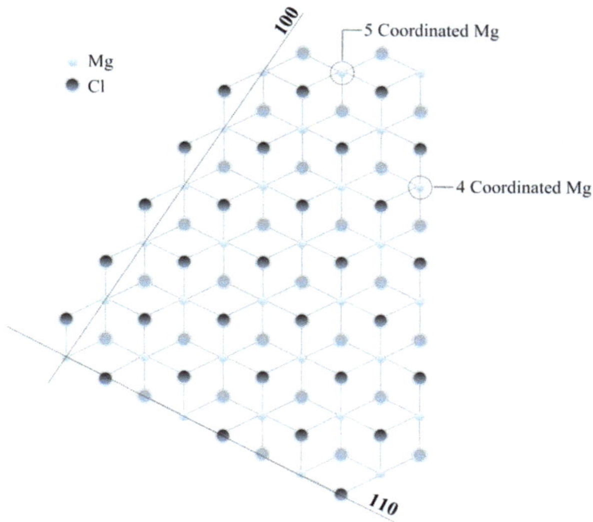

Fig. 2.2 Representation of a monolayer $MgCl_2$ crystal

structure disordered is reported to arise from the translation and rotation of the structural Cl–Mg–Cl layers with respect to one another that destroy the crystal order in the stacking direction [3]. Well defined lateral cleavage surface with a different coordination number of magnesium was characterized experimentally and the results suggest that preferential lateral cut correspond to (110) and (100) faces, as represented in Fig. 2.2 [4]. The Lewis acid character of the lateral cuts is accounted to contribute to the active site formation due to the interaction with the electron donors and $TiCl_4$, as discussed in Chap. 1.

In contradiction with the third-generation ZN catalyst, which the support is prepared by mechanical process, the fourth-generation ZN catalyst the support is prepared via chemical method. This technique enabled the morphology control of the catalyst particle and triggered the development of several different processes to control catalyst support morphology and some of them will be exemplified later in the methods of preparation part. The control of the catalyst morphology strongly impacted the ZN catalyst performance and boosted the manufacture of polyolefin. This enabled the preparation of a support for high performance ZN catalyst containing almost perfect spherical morphology. The increase in specific area of the support was important to obtain catalysts with high activity, after the preparation with $TiCl_4$ and phthalate as internal electron donor.

2.2.1 $MgCl_2$ Methods of Preparation

One of the critical steps of the preparation of $MgCl_2$ support is to control morphology and porosity, which is an important parameter in the preparation of the catalyst and therefore production of polymer with predictable shape, bulk density, and particle size distribution due to the replica phenomenon. Additional aspects should be considered during the support design, such as mechanical strength of the catalyst particle and its ability to enable uniform fragmentation stressed by the growing polymer chain during polymerization. Therefore, the preparation of support is the crucial step in the catalyst synthesis and the reagents purity, reagents ratio, order of addition, position of addition, time–temperature profile, and stirring speed are the key parameters in the preparation of a suitable catalyst support [5]. Thus, since the development of the supported ZN catalyst several methods to produce solid particles with different morphology and shape were explored by the petrochemical industry. An overview of the most common techniques developed to prepare $MgCl_2$-based supported catalyst is described below.

Emulsion

The $MgCl_2$ support prepared by an emulsion of molten $MgCl_2$ and ethanol in oil is one of the most used techniques to prepare spherical support. The resulting emulsion is transferred through a tube to a non-solvent medium at low temperature and undergoes a rapid cooling or quenching. This method enables production of spherical $MgCl_2$ particles with controlled size, but present poor surface area and porosity [6, 7]. An additional dealcoholation step, usually by fluidization of the support under hot nitrogen flow, increases porosity and surface area of the support and catalyst [8, 9].

Differently to the emulsion of molten $MgCl_2$ and alcohol in an oil medium, the system can also use other hydrocarbon liquids. This technique is followed by a cooling step under high speed stirring condition until the system reaches room temperature, allowing the formation of mostly spherical solids. The solid particles can be recovered by filtration and dried under inert atmosphere [10] or by a standard catalyst preparation work up in presence of $TiCl_4$ without the quenching step [11].

ZN catalyst can also be prepared without the use of an external support like the systems described in the procedures above. In this synthesis route based on an emulsion system, the in situ support is prepared by an initial solution of Mg-complex composed of magnesium alkoxide and internal donor. Thus, the Mg-complex is treated with $TiCl_4$ in the presence of an inert solvent followed by the emulsification of the system for stabilizing the catalyst droplets and precipitation of spherical catalyst particles at lower temperature. This method enables the preparation of very low porosity ZN catalyst particles with an extremely small surface area and narrow particle size distribution [12, 13].

Spray Drying

The spray-drying technique is typically carried out by preparing a solution or slurry of the magnesium compound, generally $MgCl_2$, fillers (optional), and alcohol, or THF, or the mixture of alcohol and THF. The resulting system is then heated and sprayed through a nozzle together with an inert gas to form droplets into a chamber with a counterflow of heated inert gas to facilitate rapid removal of solvent or slurry diluent to form solid particles. The characteristics of the solution or slurry, including type of filler, temperature of the inert gas (flow/counterflow), feed rate, and type of nozzle are important parameters to design the support particle size and morphology. Generally, the spray-drying technique can produce spheroidal particles with controlled porosity and surface area, which can be defined by the amount and nature of the filler used in the starting slurry. Common fillers are silicas, which are generally inert to the other components of the catalyst and to polymerization conditions [14, 15].

Reactive Precipitation

Reactive precipitation can be generally described as a rapid formation of a solid phase from a liquid solution phase. It is common to refer reactive precipitation as crystallization because in both processes a solid phase is formed due to the existence of supersaturation in a liquid phase. However, the methods used to generate the supersaturation may have certain differences. Most commonly, supersaturation is created by a physical process in crystallization systems, such as cooling, evaporation, or solvent gradient. For precipitation processes however, chemical reactions lead to a simultaneous and fast occurrence of nucleation, crystal growth, and agglomeration. The speed at which these processes occur could make the separation and independent investigation very difficult at both mechanistically and experimentally level [16, 17]. Thus, an example of a commercial magnesium-based supported catalyst obtained by this process starts by the complete dissolution of $MgCl_2$ in a system comprising an organic epoxide, an organic phosphorous, and an optional inert diluent. The $MgCl_2$ solution reacts with excess of $TiCl_4$ at 80 °C followed by the growth of $MgCl_2$ microcrystals resulting in the formation of crystalline rods with a cross-section of approximately 5×10^5 Å^2. The final particle is obtained by the fusion of the growing $MgCl_2$ primary particles to form tight agglomerates of catalyst particles, as shown in Fig. 2.3 [18].

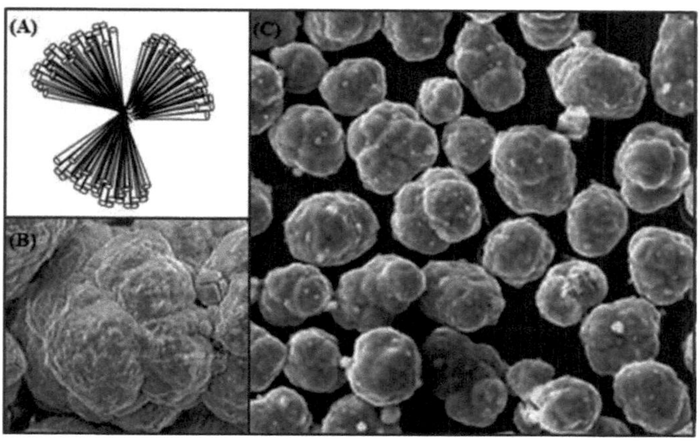

Fig. 2.3 (**a**) The model of crystal growth resulting in formation of crystalline rods, (**b**) and (**c**) SEM images of granular catalyst particles obtained by recrystallization (Figure provided by W. R. Grace & Co.)

2.3 SiO_2-Based Support

Silica-based catalysts have been widely studied in both academia and industry. It is one of the crucial components in a considerable number of heterogeneous catalyst systems such as metallocenes, Phillips, and ZN for olefin polymerization [19]. A proper design of the support involves the understanding of its physical properties, which is influenced by the type of the chemical reaction and experimental condition during the preparation of the SiO_2 particles. Thus, these important features controlled during the manufacture of the silica have a major impact on the preparation of heterogeneous catalyst because it defines the chemical composition of the particle on its surface, particle size and morphology, and particle strength of the support.

There are several advantages in the use of silica particle as support for catalyst polymerization. Its stability at high temperatures, high specific surface area, and good mechanical strength summed to the low cost of the silica makes it an important support for the use in a commodity industry such as the production of polyolefin. Furthermore, its tunable pore volume and pore size with appropriate average grain diameter enable the preparation of high performance catalyst with high activity and controllable polymer morphology [20].

2.3.1 Synthesis

Granular Silica Particles

Acidification-Polymerization

Commercial porous silica gels are prepared by neutralization of an alkali metal silicate with a mineral acid yielding silichydroxide $Si(OH)_2$. The acidification promotes the condensation of the silichydroxide to form polysilic acid units $[Si(OH)_2]_nO$, as exemplified in the Scheme 2.1. The resulting product of an extended condensation of the polysilic acid leads to the formation of a hydrosol containing micelle of few nanometers in diameter originating the primary particle constituted by bulk siloxane and silanol groups on the surface. These micelles are strictly depended of the pH of the solution and they coexist due to the electrical repulsion and therefore poor interaction with each other under specific conditions [21, 22].

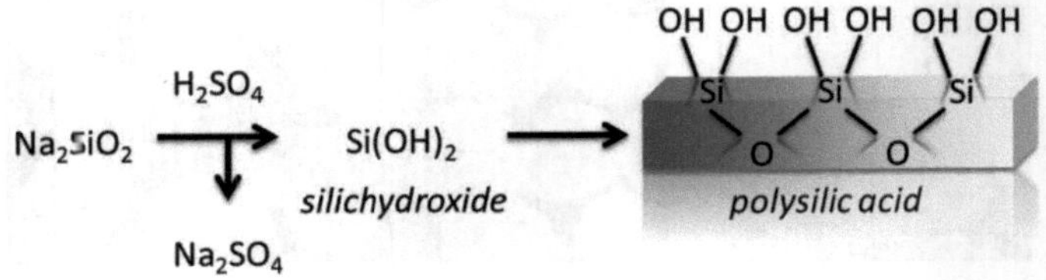

Scheme 2.1 Preparation of polysilic acid from sodium silicate (Na_2SiO_2)

Gelation

As soon as the pH is properly adjusted a 3D network is formed once the "primary particles," described previously, collide and stick to each other via hydrogen-bridges among the hydroxyl groups on the surface of the particles. The pH, stirring, and temperature conditions all influence the kinetics of the hydrogel formation and the strength of the gel. It results in a network of primary particles holding water in the space between them by capillary. An additional hydrothermal treatment promotes smaller particles to dissolve and re-precipitated on the larger particles and among themselves resulting in the strengthening of the whole construction [23, 24].

Aging

Several methods are commonly used to increase pore volume and particle strength of silica particles during the gelation step. The aging step aims to connect primary particles of silica via additional condensation of silanol groups and in the extreme case by a partial dissolution and redeposition of silica at the points of contact (Oswald ripening). For example, by treating the hydrogel at neutral pH and at temperature between 40 and 80 °C for 1 h enable to modify the final pore volume of the dried catalyst without changing the surface area. If the aging step is done under alkaline conditions (pH 9–10) for longer time and at higher temperatures, it can result in a major loss of surface area because of the coalescence of the primary silica particles. As a result, a stronger network that better resists the compression of the drying step is formed [25].

Washing and Solvent Exchange

Another important step is washing, which aims promotes the removal of the salts remained in the hydrogel. As reported by Severn (2008), these impurities can affect the electrical and thermal properties of the silica particles and in some cases reduce the thermal stability of the silica resulting in an undesired sintering of the primary particles when treated at high temperature [23]. Alternatively, the water in the hydrogel can be exchanged by an organic solvent of different surface tension, such as alcohols. This procedure is carried out until the water is completely or partially removed.

Drying

In the drying step hydrogel samples yield various degree of compression during which the pore volume decreases. For example, when a silica particle is obtained from a weak hydrogel and evaporation of water, it results in a low porosity particle due to a strong shrinkage of the silica network. Thus, the use of a low surface tension medium in the hydrogel minimizes the compression of the inner part of the silica in the drying step. Alternatively, supercritical drying using supercritical CO_2-based system prevents the shrinkage of the silica particles by eliminating the surface tensions. Therefore, the medium of the hydrogel and the drying technique is very important to define the porosity of the final silica particle [21].

Spheroidal Silica Particles

A spheroidal silica support is obtained by spray-drying fine fragments of silica suspended on "wet silica" or colloidal silica and then sprayed through a nozzle into a countercurrent of dry gas to form silica particles within the range of 10–100 μm [26, 27]. The spray dried silica gels have a distinctive secondary bulk structure which derives from the cementation of larger granule fragments. Thus, control of the particle size and particle size distribution, the void space and its distribution in the inner part of the particle are dependent on the employed silica fragments in the suspension and on the spray-drier conditions.

2.3.2 Thermal Modification of the Silica Particle

The chemical properties of the silica particles are generally governed by the chemistry of its surface, especially by the presence of the silanol groups. It has been reported that silanol covers a surface of pure silica gel at a maximum concentration of eight OH groups per nm^2 and they are mostly found as germinal or isolated pairs [27]. Thus, the thermal treatment is normally applied to modify the chemical composition of the silica particles surface, while at the same time controls the physical properties of the silica (pore volume and mechanical strength). For example, water molecules can be adsorbed through hydrogen bonds to silanol groups or through physical adsorption. A physically adsorbed water molecule desorbs from 25 °C up to 105 °C, and hydrogen-bonded water between 105 °C and 180 °C. In the course of heating at higher temperature a partial dehydroxylation of the silanol groups occurs followed by the formation of siloxane, as exemplified in the Scheme 2.2. The final OH density depends on calcination condition and hold time at a specific temperature. Commonly, silica particles calcined at temperatures between 600–800 °C the OH content ranges between one and five OH/nm^2 along with its chemical composition, as represented in Scheme 2.3 [27, 28].

The thermal treatment of the silica particles are normally conducted in fixed or fluidized-bed ovens, multiple hearth furnaces, or rotary calcination ovens. The atmosphere of the calcination is typically dry air, oxygen or an inert gas, or a combination of them (air calcination, inert gas-cooled). The process is conducted in such a way to avoid sintering of the primary particles and it consists of three

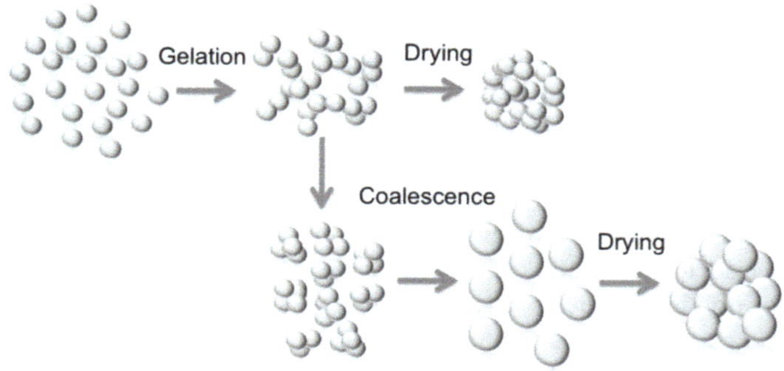

Scheme 2.2 Illustration of granular silica support preparation

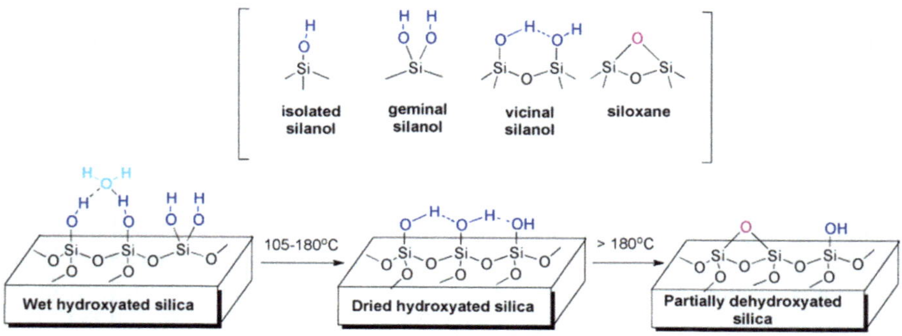

Scheme 2.3 Thermal treatment of silica particles

phases: heating, calcination, and cooling. In each phase, three parameters have to be controlled to obtain a consistent support material with desired properties: the rate of temperature increase/decrease, the hold time at calcination temperature, and/or stirring [21]. Differently, the modification of the silica particles surface can also be performed via reaction with certain surface modifiers, such as chloro- or alkoxy-silanes or disilazanes. As a result, the final composition of silica surface will also be determined by the range of hydroxylated (4 mmol OH/g for silica with an approximately 300 m^2/g surface area), partially-dehydroxylated (2.3 mmol OH/g) or dehydroxylated silicas (0.7 mmol OH/g) groups, which will be employed in the preparation of catalysts for polymerization.

2.4 Influence of Support Design on the Fragmentation Process

The replication of catalyst physical characteristics is directly linked to the growth and expansion of polymeric grains in the catalytic matrix during polymerization, which influences significantly the polymerization process. Thus, controlled fragmentation of the initial support is achieved by a continuous and gradual expansion of the particle, especially in the early stages of polymerization [20, 29, 30]. This process ensures a uniform expansion of the particle and avoids overheating and melting of the polymer, which could result in the particles adhering to form sheets on the reactor walls and agitator, or large chunks of polymer which disturb fluidization or product discharge. Rapid fragmentation due to the presence of highly active sites or uneven distribution of active sites are common causes of the particle overheating, which can lead to hot spots, especially in gas phase polymerization, causing the softening of polymer particles [31, 32]. An unsatisfactory or inadequate fragmentation can also cause the formation of fines [29, 33].

Typically supported catalysts are prepared in a wide range of particle size with diameter in the order of 10–100 μm, as discussed previously. Such catalysts are generally prepared from $MgCl_2$ and SiO_2 and composed by a large number of primary micro-particles (diameter 0.01–0.10 μm) [34]. The monomer diffuses through the support particle in order to reach the catalyst active sites and initiate the polymerization [35]. The inorganic particle suffers a mechanical stress, as the polymer grows initially in the larger pores (the most favorable pore size for monomer diffusion), followed by a very quick fragmentation [36]. These small fragments (primary particles) are dispersed uniformly leading to a formation of the polymer particle and to an expansion of the particle size up to 1 or 2 mm [4, 37]. These phenomena occur by a replication of the catalyst structure because the catalyst fragments are held together due to the adhesive forces of the formed polymer molecules. Therefore, the catalyst particle has to be fragile enough to develop the fragments as the growing polymer chain fill up the catalyst pores, but relatively stable to avoid disintegration of the particle and consequently polymer fines formation.

Two fragmentation behaviors and a combination of both are typically used to represent the fragmentation mechanism, as shown in Fig. 2.4 and described below:

1. The layer-by-layer fragmentation, also called "shell-by-shell," is commonly observed with catalysts supported on silica having small mean pore size with a uniform pore size distribution. This fragmentation begins in the outermost layer and progresses toward the center.
2. In a progressive coarse fragmentation, also called "Sectioning" or "Multi Grain," the initial olefin polymerization takes place at the surface of the catalyst particle followed by the fragmentation occurs first on the large pores and then on smaller ones until the support has completely broken up.
3. The combinations of layer-by-layer and progressive coarse fragmentation are used to explain a different evolution of the particle morphology, common when silica-based catalyst is formed by a few large pores. Thus, these large pores enable the diffusion of the monomer to the interior of the particle facilitating the coarse fragmentation. Then, this process is followed by the layer-by-layer fragmentation of the substructures inside the particle.

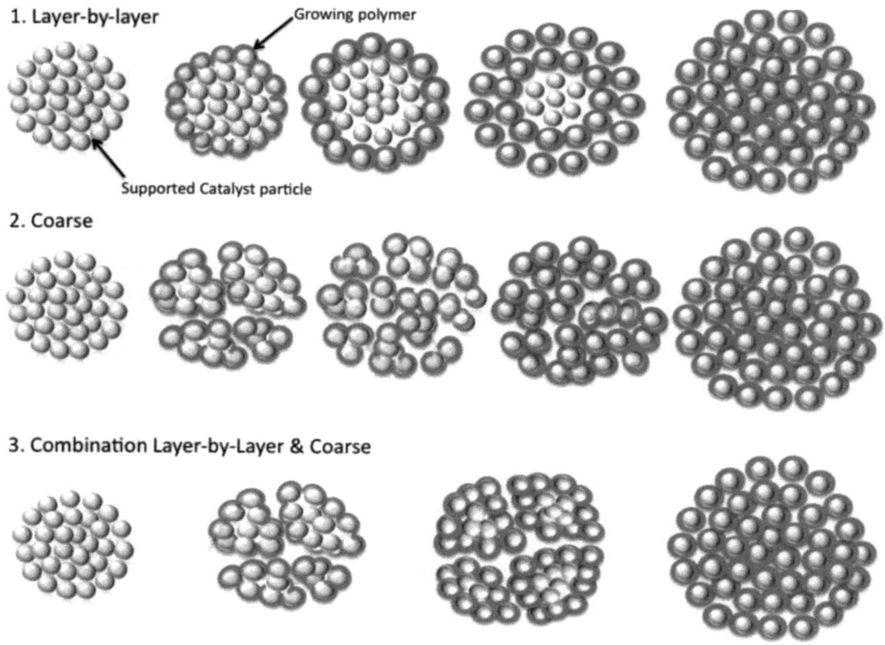

Fig. 2.4 Simplified models for the fragmentation behavior (1) layer-by-layer (2) coarse (3) combination

After all these considerations, it is worth noting that the fragmentation mechanism is very dependent on the nature and structure of the support material, catalyst preparation, and polymerization conditions. Certainly, the combination of these parameters will define the catalyst fragmentation in the early stages of the polymerization [38]. However, certain aspects are inherent to the chemical nature of the inorganic support, as for example the higher friability of $MgCl_2$ particles compared to SiO_2 particles. As a consequence, isotropy $MgCl_2$ fragmentation occurs throughout the whole particle right from the start or even after a few seconds under mild conditions, while the SiO_2 particles are likely to fragment from the outer layer and slowly proceed toward the inside in several seconds or even longer depending on the polymerization conditions [39].

2.4.1 Polymer Particle Growth

Another important aspect during the particle growth mechanism, regardless of the type of supported catalyst used, is the quantification of heat and mass transfer during the course of the polymerization. Efficient heat removal from the particles is very important to prevent undesired thermal events, such as catalyst deactivation, polymer melting, and particle agglomeration. On the other hand, the co-catalysts and monomers must diffuse through the catalyst particle to form and access the active sites, respectively, in order to enable the polymerization.

One of the first models applied to describe particle growth within heat and mass transfer is the multigrain model (MGM). It takes an assumption that all the catalyst particles are spherical (macrograins) and uniformly constitute by spherical microparticles (micrograins), which are kept together by the growing polymer. The MGM represents a theoretic base to predict diffusion coefficient of the catalyst particle under a specific polymerization process. However, as reported by McKenna and Mattioli (2001), the MGM assumes complete fragmentation at the early stage of polymerization [40]. As the reaction proceeds each grain acts as a nucleus around which further polymer growth takes place in core-shell morphology [41–43]. Using this assumption, the MGM is deficient to simulate the polymerization process for SiO_2-based catalyst. Fink et al. (2000) reported the polymer particle growth using SiO_2-supported metallocenes [27]. The initial low polymerization rate was appointed as a consequence of a gradual fragmentation of the support in the early stages of polymerization. In this case, the particle growth model adopted involved a shell-by-shell fragmentation. This process was more evident at long induction times when polymerization was carried out at relatively low temperature (30 °C) [39].

2.4.2 *Physical Properties of the Support Particles*

Physical properties of the support particles play an important role in limiting the productivity of catalysts active sites, particularly with supports having low friability. During the polymerization process, non-fragmented particles can lead to a diffusion limitation after the polymer formation. Thus, the pores of the catalyst particles might plug with polymers if the stress is too low or if the support is too strong to undergo particle fragmentation.

Therefore, one of the important aspects for an efficient particle fragmentation and polymer growth, at least for most heterogeneous catalyst systems, is the presence of pores through the entire polymerization course [44]. The pore sizes are classified into three different families:

- Micropores with a diameter less than 2 nm,
- Mesopores with diameters of 4–200 nm,
- Macropores with diameters greater than 200 nm.

As discussed above, the number and distribution of these types of pores in the supports are very important for mass and heat transfer, as well as, the resulting fragmentation behavior of the catalyst particle. Generally, pore size and pore size distribution are controlled during and after the manufacturing of the inorganic support, depending on the particle composition [45]. In the very low porosity catalyst, which, as discussed in the methods of preparation, undergoes an in situ support preparation during the catalyst synthesis, the formed particles were reported to have a surface area under the detection limit of *Brunauer–Emmet–Teller* (BET), which nevertheless undergoes a uniform fragmentation and particle growth. Thus, it would be acceptable to assume the existence of a different chemical composition in this heterogeneous catalyst compared to those prepared on $MgCl_2$ and SiO_2, which enable a better permeation of the monomers into the particle with such small pore channels. However, the mass transport mechanism and fragmentation for the low-porosity catalyst is not yet understood, this short analysis might suggest that we have to consider more than physical aspects of the catalyst particle during the fragmentation step and particle growth in this case.

2.4.3 Impact of Catalyst-Support System on the Polymerization Technology

Commercial polyolefin is manufactured in a variety of polymerization conditions carried out in different reactors and in a broad range of polymerization processes in order to produce a large number of polymer grades. In that matter, there are very intimate relationships between the catalyst type and the polymerization conditions, which strongly influence the extent of the fragmentation behavior and the replication of the polymer particle [35]. Thus, important considerations in the designing of new supported catalyst should be taken from the polymerization process limits carrying out polymerization tests under sufficiently stringent polymerization conditions. Under these circumstances, the balance between catalyst kinetics and physical aspects of the particle is very important to analyze the overall catalyst performance in a given process. Polymerization carried out at mild conditions, low temperatures and low concentrations of monomers (low pressures), may lead to poor polymer yields simply due to incomplete support fragmentation and a monomer diffusion limitation. Therefore, the low friability nature of the catalyst particle can overshadow the catalyst performance [46]. On the other hand, polymerization catalyst with high initial activity may have a high risk that the thermal energy generated by the chain propagation reaction will not dissipate enough to maintain the polymer particle with a uniform temperature. It may lead to the undesired formation of fines in the reactor due to the burst of the catalyst particle [47]. To overcome this issue, it is very common to use a prepolymerization stage to minimize the increase of intraparticle temperature and improve morphology control [48].

Another important aspect for particle overheating is the influence of the polymerization process [49, 50]. According to Kittilsen et al. (2001), the existing difference between the process of fragmentation in gas phase and in slurry polymerization can be attributed to the presence of the liquid phase (slurry) [51]. It provides a better control in temperature in the initial stage of polymerization, while overheating issues are more critical in gas phase due to the poor thermal characteristics of the gas phase composition [42]. For this reason, the gas phase polymerization takes place over small residence times [52] and reduced mean residence times result in lower catalyst mean yields. In addition to the process limitation, the design of large catalyst particles for gas phase polymerization should also be avoided since large catalyst particles are more prone to overheating as a result of low surface to volume ratios and high activity. The use of a catalyst with narrow particle size distribution and controlled catalyst morphology is also preferred to obtain a better fluidization and uniform heat transfer in gas phase reactors.

2.4.4 Kinetics

The productivity of heterogeneous polymerization is generally lower than that of the homogenous version. However, one advantage of immobilized catalysts is that intermolecular catalyst deactivation processes, which may be highly detrimental in solution, are usually frozen on surfaces. Therefore, if a good productivity can be achieved it tends to be maintained for a longer reaction time [53]. Excellent plant operability demands controlled kinetics as much as good polymer morphology from the viewpoints of process economy and viability [43, 54].

In order to meet these requirements, several aspects of the catalyst support should be considered. As reported by Tait et al. [55], the preparation of $MgCl_2$ by physical method, such as ball-milled, for ZN catalyst show high initial activity followed by a rapid activity decay throughout the polymerization course. On the other hand, ZN catalyst using $MgCl_2$ supports prepared by chemical methods present mild activation, long-standing activity, and a slow decay (phthalate base ZN-catalyst). The catalyst obtained from the chemical methods supports enabled a uniform distribution of the active sites, from the inside to the outermost part of the particle, while the ball-milled support active sites are mostly on the surface of the catalyst, which can be attributed to the difference in the kinetic profile illustrated in Fig. 2.5. Thus, the spatial distribution of the Ti species on the particle plays an important role in the catalyst kinetics because it assists the fragmentation of the catalyst particle.

Another important aspect for the catalyst kinetic profile is the nature of the catalyst support, which will also depend on the particle morphology, type of catalyst, and polymerization conditions, as discussed previously for the fragmentation mechanisms. The combination of kinetics, fragmentation steps and individual phases of polymer growth have been very well exemplified in a typical “layer-by-layer” fragmentation of a heterogeneous metallocene catalyst supported on SiO_2 for ethylene polymerization, as represented in Fig. 2.6 [20, 56–58].

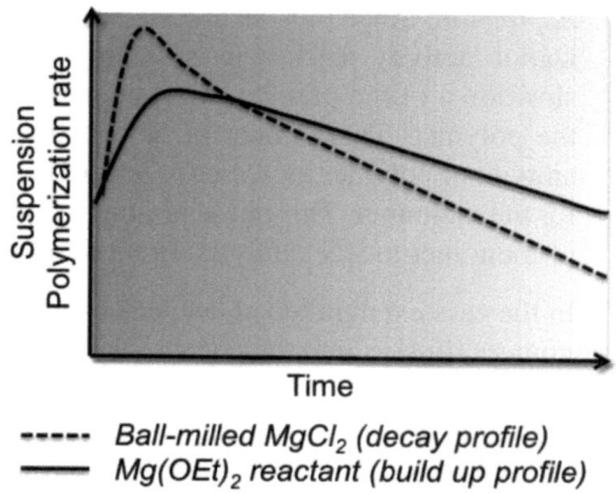

Fig. 2.5 Kinetic profiles of propylene polymerization with ZN catalyst produced from different methods of support preparation

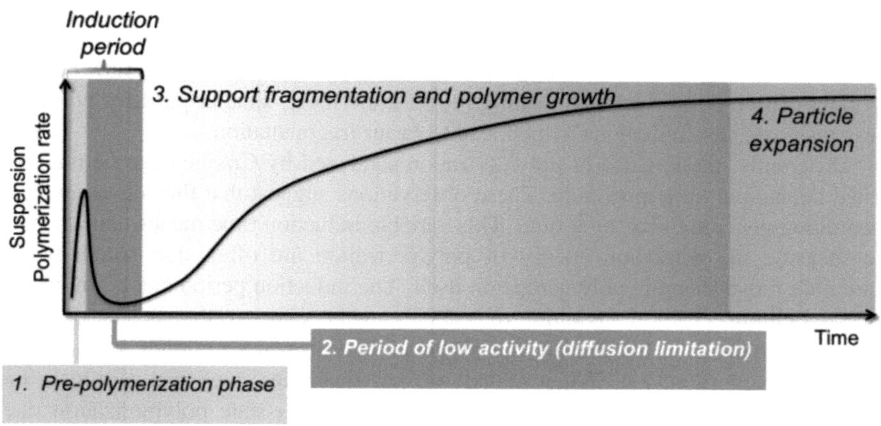

Fig. 2.6 Model of kinetic profile for SiO_2-supported metallocene

1. *Prepolymerization phase*: A few minutes after exposure to the monomer in the initial prepolymerization phase, a thin shell of highly crystalline polymer is formed just below the outer surface of the silica support, which acts as a diffusion barrier for the monomer. The polymerization rate reaches an initial maximum.
2. *Diffusion*: Consequently to the rate limiting effect of monomer diffusion through the crystalline polymer layer, the high initial polymerization activity decreases sharply after a few minutes and a period of reduced activity is reached. During this period, the polymer is growing toward the center of the carrier, thus breaking down increasing portions of the latter. Its length depends on particle diameter but also temperature and monomer concentration.
3. *Fragmentation*: As more and more active centers in the inner part of the particle become exposed to the monomer, the polymerization rate rises again and a broad maximum of activity is reached. Fragmentation of the SiO_2 support occurs from the surface to the interior because of the hydraulic forces of the growing polymer. This process continues toward the center of the particle. This phase stops when the fragmentation process is over.

4. *Expansion*: Since most active sites are now accessible to the monomer, polymerization activity reaches its maximum with continuous particle expansion. The slowdown of the particle expansion is due to the longer diffusion paths through the polymer. The number of active centers does not increase but in return the amount of polymer increases in accordance with the lengths of the diffusion paths for the monomer. This phase is similar to that observed immediately after addition of monomer to ZN catalysts on more easily fragmented $MgCl_2$ supports.

In the case exemplified above, catalysts with high porosity present higher rates of monomer diffusion into the pores of the catalyst particle and a more even distribution of monomer within the particle. This leads to faster polymerization with less induction time compared with the compacted particle in the case of catalyst with lower porosity. It has been shown that the diameter of the silica-supported metallocene/MAO has also an effect on the kinetics behavior [20]. The smaller particles showed a shorter induction period at the beginning of polymerization. Consequently, smaller particle sizes have higher activity due to high pore volume and high specific surface area, less diffusion resistance, which maximizes the number of accessible active sites per unit volume with the same catalyst loading [47]. Furthermore, smaller particles will have less mass transfer limitations, which lead to faster fragmentation.

Differently, the kinetics of polymerization catalyzed by Cr/silica can be manipulated depending on temperature. These observations suggest that the active-site concentration generally rises over time. This variable induction time means that attempts to measure the activation energy of polymerization are often dominated by the initiation rather than the polymerization itself. The induction period can be shortened or even eliminated by the addition of reducing agents either to the catalyst or to the reactor. Particularly effective are the alkyls or hydrides of aluminum, boron, zinc, lithium, magnesium, etc. When added in ppm quantities, they can eliminate the induction time of Cr(VI)/silica and also raise the steady-state polymerization rate. Some metal alkyls can remove poisons and redox by-products, helping reduce the Cr (VI) [perhaps to Cr(IV)] and help alkylate the chromium, similar to the chemistry of Ziegler catalysts [59]. Another way of shortening the induction time, and thus increasing the overall reaction rate, is to reduce the transition metal before it is added to the reactor. Treatment of activated Cr(VI)/silica in CO at 350 °C, leads to Cr(II)/silica. In contrast to Cr(VI)/silica, the Cr(II)/silica catalysts can initiate polymerization at 25 °C [60].

2.5 Influence of Catalyst Support on Tailoring Polyolefin Grades

2.5.1 Ziegler–Natta Catalyst

The most typical catalytic system for production of polyolefin is the heterogeneous ZN catalyst. Its low cost production and versatility have driven the production of linear PE and *i*PP for the past decades. Generally, heterogeneous ZN catalyst is prepared by supporting $TiCl_4$ on $MgCl_2$ or SiO_2 impregnated with $MgCl_2$. As discussed in this chapter, the main role of the support for the heterogeneous ZN catalyst system is mostly related to the catalyst performance in the polymerization process parameters rather than tailoring polyolefin grades. Interaction involving supported active sites, monomers, and its response to the polymerization condition are the basis for the production of different polymers properties. However, there are two important aspects to be considered during the support design for ZN catalyst, which is very relevant to the development of polyolefin grades. First, is for the production of reactor-made high impact polypropylene (hiPP), and second, is the difference in the support design for the production of LLDPE and HDPE.

High Impact Polypropylene

One of the limitations in the use of homogeneous polypropylene is its low impact resistance at low temperature due to its low glass transition temperature (T_g) [61]. Several approaches have been used to overcome this limitation, such as (a) blending various elastomers like ethylene–propylene rubbers (EPR) or ethylene–propylene–diene elastomers (EPDM) with the *i*PP resin, (b) by adding nucleating agent to reduce the average dimensions of spherulites [62], and (c) via sequential copolymerization of propylene with ethylene or other olefin in reactor. This last method results in a well-dispersed heterophasic morphology of rubber domains embedded in a semicrystalline PP matrix. Such heterophasic polymer is produced in a multi-reactor process with several reactors in series. The *i*PP matrix is produced in the first reactor, bulk slurry or gas phase, and the rubber, usually EPR, is produced in the second reactor inside the PP matrix in a gas phase reactor. Thus, as illustrated in Fig. 2.7, one of the most common model suggests that the rubber occupies the pores and voids among the polymer grains of the *i*PP matrix particle to determine the amount of rubber to be incorporated and how this rubber will be dispersed in final material [63]. On the other hand, once these voids and pores are filled with rubber it starts to migrate to the surface of the particle, which can cause severe sticking problems and fouling of the reactor.

Therefore, the formation of the *i*PP matrix particle in the first reactor is crucial for the production of hiPP. Due to the replica phenomena, the particle morphology of the *i*PP matrix is determined by the morphology of the ZN catalyst particle, which is dependent on the design of the support. Traditionally, catalyst supports prepared by adduct of $MgCl_2$ and ethanol followed by dealcoholation step produce highly pores particles suitable for hiPP production with high rubber content, up to 40–50 wt% [64]. On the other hand, ZN catalyst prepared by in situ support with very low measured porosity and surface area was reported to incorporate about 25 wt% of EPR rubber before it starts to migrate to the surface. It is not obvious how such features of the polymer matrix enable the diffusion of monomers into the particle in the second reactor. The authors suggested that the diffusion mechanism of the ethylene and propylene monomers occur through a non-crystalline part of the PP matrix filling up pores and voids of the particles enabling the production of hiPP

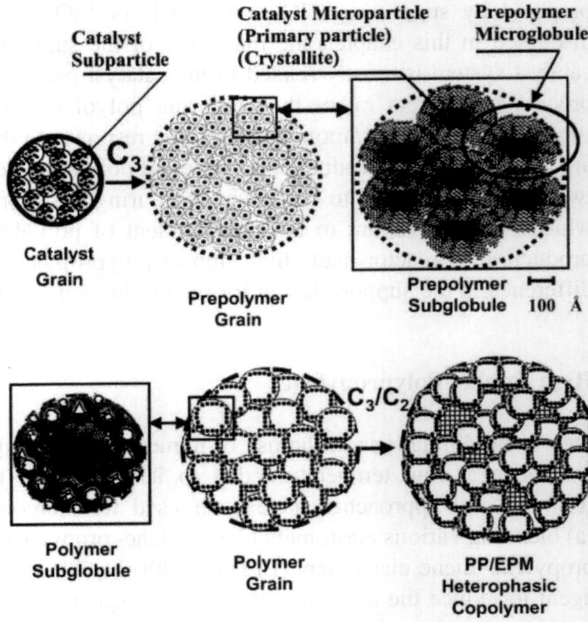

Fig. 2.7 Mechanism of polypropylene growth over spherical $MgCl_2/TiCl_4$ catalyst: prepolymerization step, main polymerization step, main polymerization, and sequential propene–ethene copolymerization (reprinted with permission from Wiley) [63]

[65]. In a different experiment the incorporation of silica nanoparticles during the production of in situ support increased slightly the measurable surface area and porosity of the catalyst, but increased in about 80% of the upper limit of the rubber content in the *i*PP matrix. Indeed, the design of the catalyst support plays an important role on the polymerization process (monomer diffusion and rubber uptake) and on the morphology (rubber dispersion and particle size) of the hiPP [66, 67]. However, aspects such as microstructure and chemical composition of the rubber are defined by the nature of the active site and polymerization conditions.

Differences in Support Design for HDPE and LLDPE

After the discovery of a new way to polymerize ethylene under low pressure and moderate temperature, compared to the traditional high pressure reactors, it results in a gradual expansion in the polyolefin industry [68]. The development of polymerization processes and catalyst performance to manufacture a wide range of polymer properties with different densities and MWD was crucial to the commercial success of PEs [69]. In this scenario, several differences in the polymerization process needed to be optimized in order to produce the PEs discussed in this section (HDPE and LLDPE). In the case of gas phase process, the reactor operating temperature should be lower than the melting point of the polymer produced. For HDPE production, the operating temperature is 90–110 °C while LLDPE, which is co-polymerized with greater α-olefins such as 1-hexene and 1-octene, is preferred to operate at temperature about 90 °C or lower [70, 71]. Another interesting aspect of the process is related to the formation of fines, which are more commonly observed in the production of HDPE than LLDPE. Soares and McKenna (2012) reported that increasing the temperature slightly during the particle growth is able to reduce the fines formation [72]. This result suggests that the temperature increase might soften the HDPE particle making it less fragile. Another important aspect in the production of HDPE and LLDPE is the type of the catalyst support used in the polymerization process. HDPE is preferable polymerized using $MgCl_2$-based ZN catalyst while LLDPE is normally polymerized by ZN catalyst using SiO_2 impregnated with $MgCl_2$. A plausible approach to the use of these two different catalyst support systems is related to the fragmentation process during the formation of the polymer particle. Catalysts supported on SiO_2 particles are known to have generally higher mechanical strength than $MgCl_2$ particles as discussed previously in this chapter. It is also known that is common in gas phase polymerization a very rapid reaction in the initial contact between ethylene and catalyst leading to a poor heat diffusion, which can cause reactor fouling or low bulk density of the particle. Thus, in order to obtain a controllable expansion of the polymer particle for a low crystalline polymer, such as LLDPE, is preferable to use SiO_2-based ZN catalyst, or alternatively prepolymerized $MgCl_2$-based ZN catalyst [73].

2.5.2 Chromium Catalyst

Chromium oxide catalyst supported on an amorphous material such as silica is mainly used to produce HDPE in a slurry process, although some LLDPE is also manufactured. This system, known as Phillips catalyst, was patented in 1958 by Hogan and Banks [74] and it still remains industrially important corresponding for about 40% of the global production of HDPE [60]. In fact, Phillips catalyst is able to make a wide range of specialized HDPE grades tailoring process characteristics of the polymer and its final application. The preparation of the Phillips catalyst requires a thermal treatment, which is commonly carried out by calcination of SiO_2 in dry air with CrO_3. This reaction with silica tends to fix the chromium and stabilize it at high temperatures by esterification to form chromate and perhaps also dichromate species at temperatures of 150–350 °C, whereby each chromium atom is directly bonded to the support, as illustrated in Scheme 2.4 [75]. Generally, the polymers properties, such as MW, the MWD, branching type and amount, branching distribution within MWD, and in molecular architecture are influenced by the catalyst. In this extend, the design of silica support should be accounted by its chemical interaction with chromium and the resulted physical characteristics of the catalyst particles. As reported by McDaniel and Collins (2008), it is the degree of structural reinforcement of the silica matrix, rather than one physical measurement of the particle that influences the polymer properties. Thus, the overall physical structure of the chromium-silica catalyst has an important influence in the polymerization process and in the final polymer properties [76].

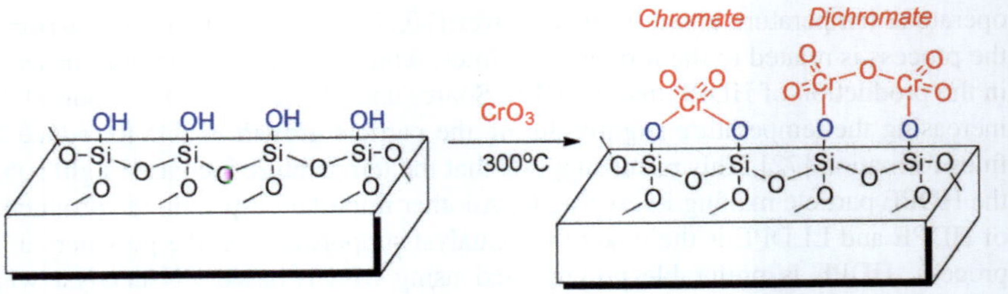

Scheme 2.4 Chromium impregnation on silica surface

Influence of Catalyst Porosity and Surface Area

The variation of pore volume in the chromium-silica catalyst is determined by the amount of shrinkage that occurs during the drying step in the preparation of the silica particles. Shrinkage is especially severe if the pore liquid during drying is water, which has a much higher surface tension than organic liquids, which was also discussed earlier in the synthesis of granular silica. For catalyst particles with identical chemical composition, the pore volume influences both catalyst activity and polymer properties. For instance, the activity has been reported to increase with higher pore volume and average pore diameter of the catalyst particle containing similar surface area until it eventually reaches a plateau [77]. Indeed, porosity has an important contribution in the polymerization activity, such that an active chromium-silica catalyst requires particle fragmentation shortly after polymerization begins, as usual for heterogeneous catalyst systems. Differently from ZN and metallocene catalyst, the Phillips catalyst can be carried out with or without a preliminary activation step using organometallics or reducing agents [78]. However, in the absence of a typical reducing agent such as CO the overall activity will be diminished due to the longer induction time. This behavior is independent of the catalyst porosity, although higher porous volume catalyst still presents better activity [77–79].

Another aspect of the pores volume to be considered is the effect in the chain transfer rate on chromium-silica catalyst. Generally, lower pore volume silica support gives higher MW polymer and therefore low melt flow index (MFI) [77]. There is no obvious explanation for this behavior, however an aspect that appears to be relevant is that the dehydroxylation in the calcination step is less efficient in silica particles containing low pore volume and pore average diameter. Since high MW polymers are known to be produced with silica supports treated at lower calcination temperatures, it suggests that the surface chemistry of small pores silica might be similar to the particles that were calcined at low temperature [78, 80]. On the other hand, catalyst activity tends to increase as function of the calcination temperature from 500 °C up to 900 °C due to the isolation of monomeric chromium species in the silica support and then it declines at higher temperatures followed by the dropping in porosity and surface area due to the process of sintering of the primary particle [59, 81].

The examples above described the effect of the catalyst pore volume and average pore diameter with similar surface area to the polymer properties. A different way to design silica support is through the treatment of the silica network by alkaline aging, which promote the coalescence and fusion of primary particles and decreasing the surface area. Thus, average pore diameter and pore volume are not proportional. This technique, also known as Ostwald ripening reinforces the silica matrix. McDaniel (2008) has reported the effect of the polymer MW and MWD using commercial grades of silica varying surface area and pore size with small difference in pore volume [82]. The MW of the polymers decreased with increasing the pore diameter range of the catalyst up to 200 Å reaching its minimum at pore diameter range of about 200–250 Å. However, polymers produced with catalyst particles containing larger pore diameter (higher than 250 Å) and lower surface area showed an increase in MW. Other examples of silica reinforced by sintering the structure at a high temperature followed by a rehydroxylation step showed similar dependence of MW and pore diameter [82]. On the other hand, MWD of polymers tend to become narrower when produced by a catalyst with larger pore diameter (up to 200 Å) and lower surface area. Therefore, increasing the pore volume and pore size range up to approximately 200 Å of the catalyst particles has been reported to decrease polymer MW reaching a plateau at constant surface area. However, the sintering primary particles of the silica support triggers a production of polymer with higher MW using catalyst particles with lower surface area and pore volume, but higher pore diameter.

Another important aspect of the polymer architecture that can be influenced by the design of the silica support during the preparation of Phillips catalyst is the level of PE long chain branch (LCB). Chromium-silica catalyst can produce very low amounts of LCB. However along with MW and MWD, LCB levels are an important factor to the melt elasticity of the PE, which defines its commercial processing characteristics. In a very simplistic manner, chromium-silica catalysts tend to reduce the formation LCB levels in PE due to the strengthening of the silica particle as a result of coalescence of the primary particles during the silica particle preparation, which consequently reduce the overall surface area of the silica particle [77].

2.5.3 *Metallocenes*

As discussed previously in this chapter, the support preparation and its morphology and mechanical strengths influence the polymerization process and the polymer morphology due to replication phenomenon. Clearly the understanding of the silica particle surface chemistry and controlling particle size and types of pores formed in the support particle is fundamental to the impregnation method used and therefore in the catalyst performance and resultant polymer properties. Severn (2008) has elegantly summarized interactions of MAO with pre-catalyst and silica surface including the innumerous combination of supporting a single pre-catalyst complex [23]. Thus, the result of a complex interaction during the immobilization of the pre-catalyst tailors the polymer grade, but their description is beyond the scope of this chapter.

2.6 Conclusion

This chapter aims at discussing some relevant aspects in the preparation of the support for the three major classes of catalyst for the polyolefin industry. As one can realize the design of a proper support is dependent of several factors, such as the type of catalyst, balance of productivity and operability in a desired process, kinetic profile, and polymer morphology. It demonstrates the complexity of developing a commercial catalyst even using a common inorganic carrier, such as SiO_2 and $MgCl_2$. The success achieved with these systems has transformed the polyolefin industry due to a very reliable heterophasic polymerization process at low cost. To this extent, the maturity of the process technology dictates the challenges to design an alternative support material for olefin polymerization nevertheless it could offer unique features on the development of new polymeric materials.

References

1. Galli, P., Barbè, P., Guidetti, G., Zannetti, R., Martorana, A., Marigo, A., Bergozza, M., & Fichera, A. (1983). The activation of $MgCl_2$-supported Ziegler-Natta catalysts: A structural investigation. *European Polymer Journal, 19*, 19–24.
2. Spitz, R., Duramel, L., & Guyot, A. (1988). Supported Ziegler-Natta catalyst for propene polymerization: Grinding and co-grinding effects on catalyst improvement. *Makromolekulare Chemie, 189*, 549–558.
3. Gerbasi, R., Marigo, A., Martorana, A., Zannetti, R., Guidetti, G. P., & Baruzzi, G. (1984). The activation of $MgCl_2$-supported Ziegler-Natta catalyst-II correlation between activity and structural disorder. *European Polymer Journal, 20*, 967–970.
4. Albizzati, E., Giannini, U., Collina, G., Noristi, L., & Resconi, L. (1996). Catalysts and polymerizations. In E. P. Moore Jr. (Ed.), *Polypropylene handbook* (pp. 11–111). New York: Hanser Publishers.
5. Ye, Z.-Y., Wang, L., Feng, L.-F., Gu, X.-P., Chen, H.-H., Zhang, P.-Y., Pan, J., Jiang, S., & Feng, L.-X. (2002). Novel spherical Ziegler–Natta catalyst for polymerization and copolymerization. I. Spherical $MgCl_2$ support. *Journal of Polymer Science Part A: Polymer Chemistry, 40*, 3112–3119.
6. Ferraris, M., Rosati, F., Parodi, S., Gianneti, E., Motroni, G., & Albizzati, E. (1983). *Catalyst components and catalyst for the polymerization of alpha-olefins*. US Patent 4-399-054.
7. Dil, E. J., Pourmahdian, S., Vatankhah, M., & Taromi, F. A. (2010). Effect of dealcoholation of support in $MgCl_2$-supported Ziegler-Natta catalyst on catalyst activity and propylene powder morphology. *Polymer Bulletin, 64*, 445–457.
8. Govoni, G., Ciarrocchi, A., & Sacchetti, M. (1993). *Crystalline olefin and copolymers in the form of spherical particles at high porosity*. US Patent 5-231-119.
9. Forte, M. C., & Coutinho, F. M. B. (1996). Highly active magnesium chloride supported Ziegler-Natta catalyst with controlled morphology. *European Polymer Journal, 32*, 223–231.
10. Martin, J. L. (1996). *Process to make small, discrete, spherical adducts*. US Patent 5-500-396.
11. Ishimaru, N., Kioka, M., & Toyota, A. (1991). *Process for polymerizing olefins and polymerization catalyst therefor*. US Patent 4-990-479.
12. Leinonem, T., & Denifl, P. (2006). *Preparation of olefin polymerization catalyst component*. European Patent 1273595.

13. Abboud, M., Denifl, P., & Reichert, K.-H. (2005). Study of the morphology and kinetics of novel Ziegler-Natta catalyst for propylene. *Journal of Applied Polymer Science, 98*, 2191–2200.
14. Hamer, A. D., & Karol, F. J. (1983). *Spheroidal polymerization catalyst, process of preparing, and use for ethylene polymerization.* US Patent 4-376-062.
15. Jorgensen, R. J., Spriggs, T. E., Turner, M. D., & Lacks, D. J. (2011). *Robust spray-dried Ziegler-Natta procatalyst and polymerization process employing the same.* PCT/US2005/028238.
16. Karpinski, P. H., & Wey, J. S. (2002). Precipitation processes. In A. S. Myerson (Ed.), *Handbook of industrial polymerization* (pp. 141–160). Woburn: Butterworth-Heinemann.
17. Pikturna, J. T. (2004). *Particle size prediction in reactive precipitation processes* (PhD Thesis). Iowa State University, Ames.
18. Chang, M., Liu, X., Nelson, P. J., Munzing, G. R., Gegan, T. A., & Kissin, Y. V. (2006). Ziegler-Natta catalyst for propylene polymerization: Morphology and crystal structure of a fourth-generation catalyst. *Journal of Catalysis, 239*, 347–353.
19. Choi, Y., & Soares, J. B. P. (2012). Supported single–site catalysts for slurry and gas–phase olefin polymerisation. *Canadian Journal of Chemical Engineering, 90*, 646–671.
20. Fink, G., Steinmetz, B., Żechlin, J., Przybyla, C., & Tesche, B. (2000). Propene polymerization with silica-supported Metallocene/MAO catalysts. *Chemical Reviews, 100*, 1377.
21. Iler, R. K. (Ed.). (1955). *Colloid chemistry of silica and silicates.* Ithaca, NY: Cornell University Press.
22. Iler, R. K. (1979). *The chemistry of silica.* New York: Wiley.
23. Severn, J. R. (2008). Methylaluminoxane (MAO), silica and complex: The "Holy Trinity" of supported single-site catalyst. In J. R. Severn & J. C. Chadwick (Eds.), *Tailor-made polymers* (1st ed., pp. 95–138). Weinheim: Wiley.
24. McDaniel, M. P. (2010). Supported chromium catalyst and its commercial use for ethylene polymerization. *Advances in Catalysis, 53*, 123–606.
25. Ellsworth, G. A. (1970). *Process for increasing pore volume of intermediate density silica gel.* US Patent 3-526-603.
26. Ito, Y., Higashiogawa, T., Matsuura, M., Orii, K., & Yamaguchi, Y. (1991, July 2). US Patent 5-028-360.
27. Fink, G., Steinmetz, B., Zechlin, J., Przyblya, C., & Tesche, B. (2000). Propene polymerization with silica-supported Metallocene/MAO catalyst. *Chemical Reviews, 100*, 1377–1390.
28. Trébosc, J., Wiench, J. W., Huh, S., Lin, V. S.-Y., & Pruski, M. (2005). Solid-state NMR study of MCM-41-type mesoporous silica nanoparticles. *Journal of the American Chemical Society, 127*, 3057.
29. McKenna, T. E., & Soares, J. B. P. (2001). Single particle modelling for olefin polymerization on supported catalysts: A review and proposals for future developments. *Chemical Engineering Science, 56*, 3931–3949.
30. Duchateau, R. (2002). Incompletely condensed Silsesquioxanes: Versatile tools in developing silica-supported olefin polymerization catalysts. *Chemical Reviews, 102*, 3525–3542.
31. Merquior, D. M., Lima, E. L., & Pinto, J. C. (2003). Modeling of particle fragmentation in heterogeneous olefin polymerization reactions. *Polymer Reaction Engineering, 11*, 133–154.
32. Merquior, D. M., Lima, E. L., & Pinto, J. C. (2005). Modeling of particle fragmentation in heterogeneous olefin polymerization reactions. *Macromolecular Materials and Engineering, 290*, 511–524.
33. Böhm, L., Bilda, D., Breuers, W., Enderle, H. F., & Lecht, R. (1995). The microreactor model-guideline for PE-HD process and product development. In G. Fink, R. Mülhaupt, & H. H. Brintzinger (Eds.), *Ziegler catalysts* (pp. 387–400). Berlin: Springer.
34. Boor Jr., J. (1979). *Ziegler-Natta catalysts and polymerizations.* New York: Academic Press.
35. Zheng, X., Smit, M., Chadwick, J. C., & Loos, J. (2005). Fragmentation behavior of silica-supported metallocene/MAO catalystin the early stages of olefin polymerization. *Macromolecules, 38*, 4673–4678.

36. Ferrero, M. A., Koffi, E., Sommer, R., & Conner, W. C. (1992). Characterization of the changes in the initial morphology for $MgCl_2$-supported Ziegler-Natta polymerization catalysts. *Journal of Polymer Science, 30*, 2131–2141.
37. Rönkkö, H. L., Korpela, T., Knuuttila, H., Pakkanen, T. T., Denifl, P., Leinonen, T., Kemell, M., & Leskelä, M. (2009). Particle growth and fragmentation of solid self-supported Ziegler–Natta-type catalysts in propylene polymerization. *Journal of Molecular Catalysis Chemical, 309*, 40–49.
38. Martin, C., & Mckenna, T. F. (2002). Particle morphology and transport phenomena in olefin polymerisation. *Chemical Engineering Journal, 87*, 89–99.
39. Klapper, M., Joe, D., Nietzel, S., Krumpfer, J. W., & Müllen, K. (2014). Olefin polymerization with supported catalysts as an exercise in nanotechnology. *Chemistry of Materials, 26*, 802–819.
40. McKenna, T. F., & Mattioli, V. (2001). Progress in describing particle growth for polyolefins: A look at particle morphology. *Macromolecular Symposia, 173*, 149–162.
41. Noristi, L., Marchetti, E., Baruzzi, G., & Sgarzi, P. (1994). Investigation on the particle growth mechanism in propylene polymerization with $MgCl_2$-supported Ziegler–Natta catalysts. *Journal of Polymer Science, Part A: Polymer Chemistry, 32*, 3047–3059.
42. Pater, J. T. M., Weickert, G., Loos, J., & van Swaaij, W. P. M. (2001). High precision prepolymerization of propylene at extremely low reaction rates—kinetics and morphology. *Chemical Engineering Science, 56*, 4107–4120.
43. Cecchin, G., Marchetti, E., & Baruzzi, G. (2001). On the mechanism of polypropene growth over $MgCl_2/TiCl_4$ catalyst systems. *Macromolecular Chemistry and Physics, 202*, 1987–1994.
44. Ashri, A. Y. (2012, April). *The effects of silica support on kinetic behavior and polymer properties of heterogonous metallocene catalyst*. Thesis dissertation, Department of Chemical Engineering, Queens University, Kingston, ON.
45. Webb, S. W., Weist, E. L., Chiovetta, M. G., Laurence, R. I., & Conner, W. C. (1991). Morphological influences in the gas phase polymerization of ethylene by silica supported chromium oxide catalysts. *The Canadian Journal of Chemical Engineering, 69*, 665–681.
46. Galli, P. J. (1999). The reactor granule technology: The ultimate expansion of polypropylene properties? *Journal of Macromolecular Science Part A: Pure and Applied Chemistry, 36*, 1561–1586.
47. Floyd, S., Heiskanen, T., Taylor, T. W., Mann, M. G., & Ray, W. H. (1987). Polymerization of olefins through heterogeneous catalysis. VI. Effect of particle heat and mass transfer on polymerization behavior and polymer properties. *Journal of Applied Polymer Science, 33*, 1021–1065.
48. Mülhaupt, R. (2003). Catalytic polymerization and post polymerization catalysis fifty years after the discovery of Ziegler's catalysts. *Macromolecular Chemistry and Physics, 204*, 289–327.
49. Hutchinson, R. A., & Ray, W. H. (1987). Polymerization of olefins through heterogeneous catalysis. VII. Particle ignition and extinction phenomena. *Journal of Applied Polymer Science, 34*, 657–676.
50. Hutchinson, R. A., & Ray, W. H. (1990). Polymerization of olefins through heterogeneous catalysis. VIII. Monomer sorption effects. *Journal of Applied Polymer Science, 41*, 51–81.
51. Kittilsen, P., Svendsen, H., & Mckenna, T. F. (2001). Modeling of transfer phenomena on heterogeneous Ziegler catalysts. IV. Convection effects in gas phase processes. *Chemical Engineering Science, 56*, 3997–4005.
52. Zacca, J. J., & Debling, J. A. (2001). Particle population overheating phenomena in olefin polymerization reactors. *Chemical Engineering Science, 56*, 4029–4042.
53. Gahleitner, M., & Severn, J. R. (2008). Designing polymer properties. In J. R. Severn & J. C. Chadwick (Eds.), *Tailor-made polymers* (pp. 1–39). Weinheim: Wiley.
54. Galli, P., Barbè, P. C., & Noristi, L. (1984). High yield catalysts in olefin polymerization. General outlook on theoretical aspects and industrial uses. *Angewandte Makromolekulare Chemie, 120*, 73–90.

55. Tait, P. J. T., Zohuri, G. H., Kells, A. M., & McKenzie, I. D. (1995). Kinetic studies on propene polymerization using magnesium dichloride supported Ziegler-Natta catalyst. In G. Fink, R. Mülhaupt, & H. H. Brintzinger (Eds.), *Ziegler catalyst* (pp. 343–362). Berlin: Springer.
56. Zechlin, J., Steinmetz, B., Tesche, B., & Fink, G. H. (2000). Development of a refined poly (propylene) growth model for silica supported metallocene catalyst systems. *Macromolecular Chemistry and Physics, 201*, 515–524.
57. Zechlin, J., Hauschild, K., & Fink, G. (2000). Silica supported metallocene/MAO-systems: Comparison of the polypropylene growth during bulk phase polymerization with slurry phase experiments. *Macromolecular Chemistry and Physics, 201*, 597–603.
58. Alexiadis, A., Andes, C., Ferrari, D., Korber, F., Hauschild, K., Bochmann, M., & Fink, G. (2004). Mathematical modeling of homopolymerization on supported metallocene catalysts. *Macromolecular Materials and Engineering, 289*, 457–466.
59. McDaniel, M. P. (1985). Supported chromium catalyst for ethylene polymerization. *Advances in Catalysis, 33*, 47–98.
60. Groppo, E., Estephane, J., Lamberti, C., Spoto, G., & Zecchina, A. (2007). Ethylene, propylene and ethylene oxide in situ polymerization on the Cr(II)/SiO2 system: A temperature- and pressure-dependent. *Catalysis Today, 126*, 228–234.
61. Cecchin, G., Morini, G., & Pelliconi, A. (2001). Polypropylene product innovation by reactor granule technology. *Macromolecular Symposia, 173*, 195–209.
62. Zhang, X., Zhang, D., & Liu, T. (2012). Influence of nucleating agent on properties of isotactic polypropylene. *Energy Procedia, 17*, 1829–1835.
63. Cecchin, G., Morini, G., & Pelliconi, A. (2001). Propylene product innovation by reactor granule technology. *Macromolecular Symposia, 173*, 195–209.
64. Debling, J. A., Zacca, J. J., & Ray, W. H. (1997). Reactor residence-time distribution effects on the multistage polymerization of olefins-III. Multi-layered products: Impact polypropylene. *Chemical Engineering Science, 52*, 1969–2001.
65. Vestberg, T., Denifl, P., & Wilén, C.-E. (2013). Increased rubber content in high impact polypropylene via a Sirius Ziegler-Natta catalyst containing nanoparticles. *Journal of Polymer Science, Part A: Polymer Chemistry, 51*, 2040–2048.
66. Tan, H., Li, L., Chen, Z., Song, Y., & Zheng, Q. (2005). Phase morphology and impact toughness of impact polypropylene copolymer. *Polymer, 46*, 3522–3527.
67. Jiang, T., Chen, H., Ning, Y., Kuang, D., & Qu, G. (2006). Study on morphology of high impact propylene prepared by in situ blending. *Journal of Applied Polymer Science, 101*, 1386–1390.
68. Böhm, L. L. (2003). The ethylene polymerization with Ziegler catalyst: Fifty years after the discovery. *Angewandte Chemie, International Edition, 42*, 5010–5030.
69. Gahleitner, M., Resconi, L., & Doshev, P. (2013). Heterogeneous Ziegler-Natta, metallocene, and post-metallocene catalysis: Success and challenges in industrial application. *MRS Bulletin, 28*, 229–233.
70. Malpass, D. B. (2010). An overview of industrial polyethylene processes. In D. B. Malpass (Ed.), *Introduction of industrial polyethylene* (pp. 85–96). Salem: Scrivener.
71. Xie, T., McAuley, K. B., Hsu, J. C. C., & Bacon, D. W. (1994). Gas phase ethylene polymerization: Production processes, polymer properties, and reactor modeling. *Industrial and Engineering Chemistry Research, 33*, 449–479.
72. Soares JBP, McKenna TFL (2012) Particle growth and single particle modeling. In: Soares JBP, McKenna TFL (eds) Polyolefin reaction engineering modeling. Wiley, Weinheim, 271–309.
73. Chama, C., & Daire, E. (1992, June 23). *Supported polyolefin catalyst for the (co-) polymerization of ethylene in gas phase.* US Patent 5-124-296.
74. Hogan, J. P., & Banks, R. L. (1958, March 4). *Polymers and production thereof.* US Patent 2-825-721.
75. Fouad, N. E. (2000). Formation of Cr(II) species in the H_2/CrO_3 system. Parameter control. *Journal of Thermal Analysis and Calorimetry, 60*, 541–547.

76. McDaniel, M. P., & Collins, K. S. (2008). The influence of porosity on the Phillips Cr/silica catalyst 2. Polyethylene elasticity. *Journal of Polymer Science, Part A: Polymer Chemistry, 47*, 845–865.
77. McDaniel, M. P. (2011). Influence of catalyst porosity on ethylene polymerization. *ACS Catalysis, 1*, 1394–1407.
78. Groppo, E., Lamberti, C., Bordiga, S., Spoto, G., & Zecchina, A. (2005). The structure of active centers and the ethylene polymerization mechanism on the Cr/SiO_2 catalyst: A frontier for the characterization methods. *Chemical Reviews, 105*, 115–183.
79. Kim, C. S., Woo, S. I., Jeong, D. J., & Oh, J. S. (1992). Comparison of ethylene polymerization catalyzed over Cr(IV)/silica and Cr(II)/silica catalysts. *Polymer Bulletin, 29*, 205–212.
80. Atiqullah, M., Akhtar, M. N., Moman, A. A., Abu-Raqabah, A. H., Palackal, S. J., Al-Muallem, H. A., & Hamed, O. M. (2007). Influence of silica calcination temperature on the performance of supported catalyst SiO_2–nBuSnCl$_3$/MAO/(nBuCp)$_2$ZrCl$_2$ polymerizing ethylene without separately feeding the MAO cocatalyst. *Applied Catalysis. A, General, 320*, 134–143.
81. Thüne, P. C., Loos, J., Wouters, D., Lemstra, P. J., & Niemantsverdriet, J. W. (2001). The CrOx/SiO2/Si(100) catalyst – A surface science approach to supported olefin polymerization catalysis. *Macromolecular Symposia, 173*, 37–52.
82. McDaniel, M. P. (2008). Influence of porosity on PE molecular weight from the Phillips Cr/silica catalyst. *Journal of Catalysis, 261*, 34–49.

Chapter 3
Fragmentation, Particle Growth and Single Particle Modelling

Timothy F. L. McKenna and Muhammad Ahsan Bashir

3.1 Introduction

In processes that rely on the use of heterogeneous catalysis as the major means of production, it should be quite obvious that understanding how the catalyst particles evolve will play an important role in many aspects related to quality and reactor performance. At the risk of oversimplifying things, the principal roles of the heterogeneous catalyst particles used in olefin polymerisation can be seen as being (1) to carry the active sites upon which the polymer is formed; and (2) to provide a structure for creating "solid" particles that can be easily transported, recovered and processed. It is therefore important for us to understand how the process used to make the polymer impacts the particle and the active sites (and vice versa!). From the schema in Fig. 3.1, where these concepts are applied to a heterogeneously catalysed olefin polymerisation process, it can be seen that one needs to consider many different length scales, from the reactor which has volumes on the order of several tens of cubic metres, to the catalyst and polymer particles with characteristic diameters on the order of 10^{-6} to 10^{-3} m, and finally the active sites with characteristic sizes on the order of Ångströms. The figure also suggests that in many ways one can consider the catalyst and polymer particles as being at the heart of a polymerisation process. This is of course not to over-simplify the technological challenges of correctly operating the reactors, nor to assume that we have totally mastered the behaviour of the active sites either. However, as we shall see below, the very fact that we are using heterogeneous catalysts implies that mass transfer limitations can eventually limit the concentrations of active species at the active sites, or that the quality of the polymer (sticky/hard, brittle/flexible) can have a major

T. F. L. McKenna (✉) · M. A. Bashir
Chemistry, Catalysis, Polymers and Processes – UMR5265, Université de Lyon, Villeurbanne Cedex, France
e-mail: Timothy.mckenna@univ-lyon1.fr

A. R. Albunia et al. (eds.), *Multimodal Polymers with Supported Catalysts*,
https://doi.org/10.1007/978-3-030-03476-4_3

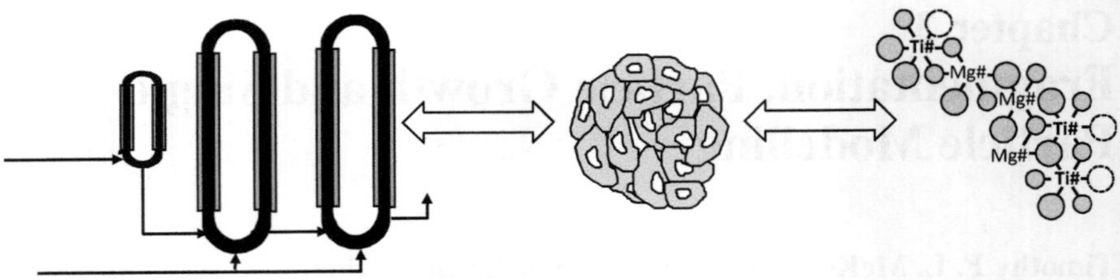

Fig. 3.1 The principal length scales in an olefin polymerisation process

impact on reactor behaviour. For these, and many other related reasons it is therefore of importance to understand what happens to the particles injected into the reactor during the polymerisation.

Of course, the overall quality of the polymer made using a supported catalyst will be the result of complex interactions between "chemical" and "physical" effects. By chemical effects we mean the nature of the active sites, the choice of cocatalyst, method of supporting the sites; these will all be critical factors in determining not only the rate of reaction, but just as importantly, properties related to the structure of the polymer molecules, such as the molecular weight distribution, copolymer composition distribution, long and short chain branching, crystallinity, etc. Physical effects can be thought of essentially in terms of phenomena such as mass and energy transport into, through and out of the particles, and particle–particle and particle–reactor interactions. These phenomena are important because they will have an impact on the local concentrations and temperatures inside the particles, at the active sites where the polymerisation really takes place, and will therefore also have an impact on the real rate of polymerisation and the polymer properties.

If we consider Fig. 3.2, it should be clear that when we use a heterogeneous catalyst we always have gradients of temperature (T) and concentration (C) inside them. So once again, factors that impact heat and mass transfer—i.e. local temperatures and concentrations—will have an impact on what occurs at the active sites we have so carefully designed and supported. In other words, if the gradients are important, the rates of polymerisation and polymer properties can be different at different spots in the catalyst, so even if we have an accurate kinetic model and parameters for our system, it is essential that we be able to predict the significance of the gradients as the particle evolves in the reactor train.

In one sense, calculating concentration and temperature profiles inside porous particles is a standard chemical engineering problem. One can do mass and energy balances around a particle to obtain general equations of the form:

$$\frac{\partial[A]}{\partial t} = -(\nabla.N_A) + R_i \tag{3.1}$$

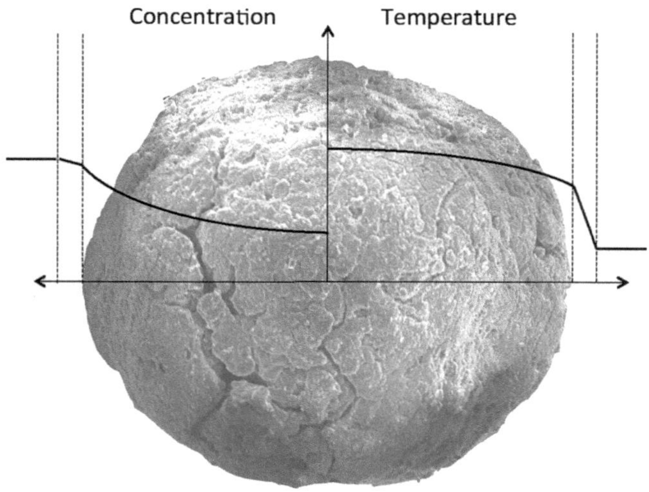

Fig. 3.2 Gradients of concentration (left) and temperature (right) in a growing polymer particle. Spatial variations in temperature and concentrations mean that polymerisation rates and polymer properties can be different at different points in the particle

$$\rho C_p \frac{\partial T}{\partial t} = -\nabla .q + \left(-\Delta .H_p\right) R_i \tag{3.2}$$

where Eq. (3.1) is the mass balance of species A around a polymerising particle that gives us the concentration of A inside the particle as a function of position and time, N_A is the molar flux of species A, and R_i the rate of production of the same. Equation (3.2) is the analogous enthalpy balance giving the temperature profile inside the particle, also as a function of time and position. Here, q is the heat flux through a medium with density and heat capacity ρ and C_p, respectively.

It is clear that in order to solve these coupled equations it is necessary to first and foremost identify the shape, boundaries and internal structure of the particles, as well as how they evolve as a function of time and reaction conditions. Clearly, assuming that the catalyst and polymer particles are spherical is not unreasonable. However, as we shall see below, the boundary of the particles is not fixed, and we need information about the internal structure of the particles, and how it evolves during the polymerisation. If one can build a representation of the physical nature of the particles, and can specify the terms and parameters associated with the fluxes, then it is possible to calculate the concentration and temperature profiles inside the particle. Thus, before we look at the different solution strategies that have been adapted to coupled mass and energy balances it is useful to think about what happens to the catalyst particles once they are injected into the reactor.

3.2 Particle Fragmentation and Growth

Industrial supports for Ziegler–Natta and metallocene catalysts include $MgCl_2$ and Silica (c.f. Chap. 2). These inorganic supports are highly porous, with specific surface areas on the order of several 100 square metres per gram of support, and typical particle diameters in the range of 10–100 μm. A closer look at the particles of both a $TiCl_4/MgCl_2$ Ziegler–Natta catalyst and a silica support with scanning electron microscopy (SEM) reveals that each catalyst particle, or macrograin, is composed of a cluster of smaller particles commonly known as microparticles or micrograins, as shown in the Fig. 3.3. While the overall shape and size of the micrograins are different for the two materials, the same general two tier structural organisation is similar for both.

It is assumed that the active sites of the catalyst are distributed more or less homogeneously throughout the supports and that they are found on the surface of the

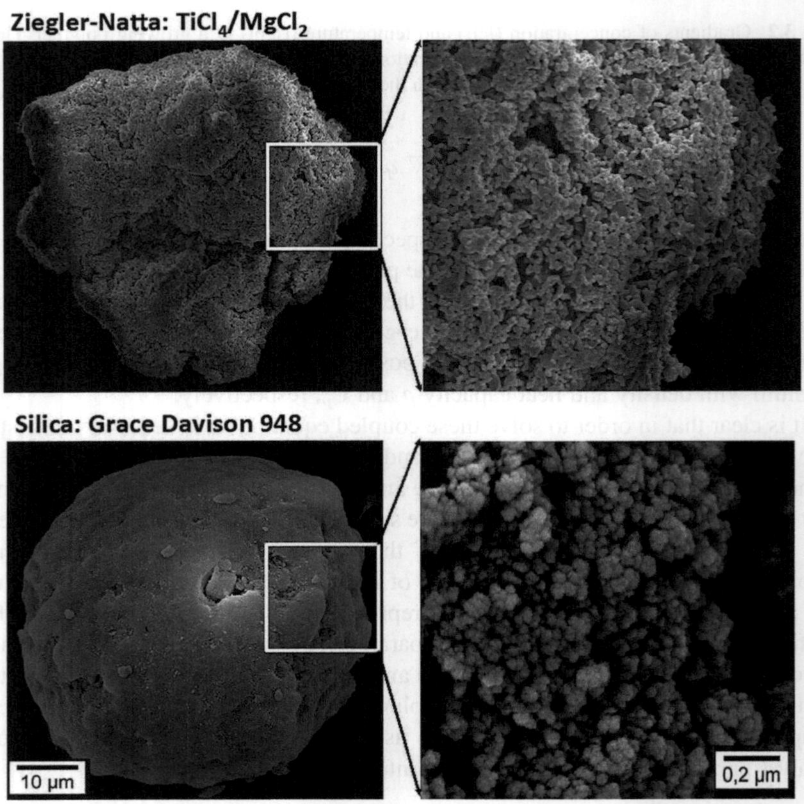

Fig. 3.3 SEM images of a commercial Ziegler–Natta $TiCl_4/MgCl_2$ catalyst particle (top) and a Grace Davison 948 silica particle (bottom) showing the two level structure assumed to be common to most standard olefin polymerisation catalysts

micrograins. This is likely true (with some limitations) of silica supports, whereas, the situation is a little more complex in the case of Ziegler–Natta catalysts on $MgCl_2$ since the active sites are actually integrated in the support itself, so additional sites might become "liberated" when the support breaks up, or fragments (see below). In either case, it is fair to say that the higher the surface area, the greater the potential number of active sites there are per gram of support, and thus the higher the potential activity of the catalyst will be.

This concept of two levels of organisation for the catalyst particle serves as a convenient starting point for the development of a strategy for the resolution of Eqs. (3.1) and (3.2). However, if we are to develop heat and mass transfer models to predict the impact of particle structure and reactor conditions on the final polymer properties we still need to understand how the morphology of the particles, and more particularly the morphology of the inside of the growing polymer particles change during the polymerisation. The next sections describe the morphological changes in the porous supported catalyst particles under polymerisation conditions.

3.2.1 The Fragmentation Step

The transformation of the catalyst particle into a growing polymer particle occurs almost immediately upon its injection into the reactor environment. The major steps and trends are shown in the schema in Fig. 3.4. In this figure, the catalyst, composed of an aggregate of micrograins is injected into the reactor, monomer (and other compounds) begins to diffuse towards the particle, into the pores where it reacts at the active sites on the surface of the micrograins (grey; step 1) to make polymer (black; step 1). As polymer accumulates this creates hydraulic stresses inside the particle, which leads to the rupture of the physical bridges that originally held the micrograins together (step 2). This step where the original support material breaks up

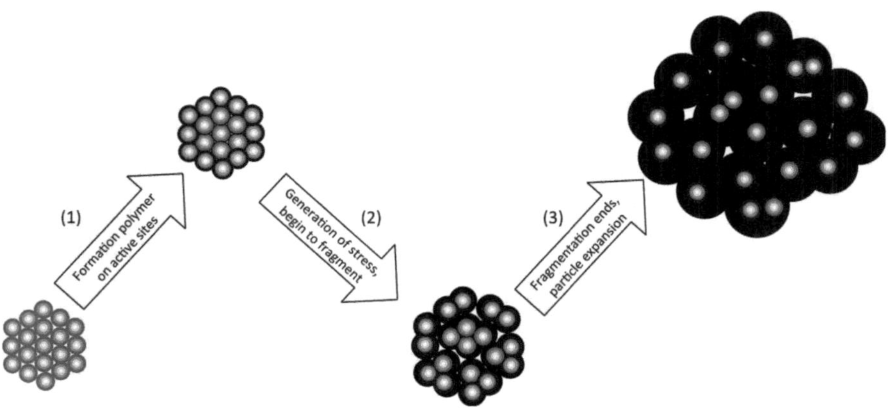

Fig. 3.4 Outline of how the particle fragments and grows upon injection into the reactor

is referred to as fragmentation, and is an extremely important phase in the growth of the polymer particles. By the end of fragmentation step, the original two phase particle (i.e. the solid particle and pore space) is converted into a triphasic particle containing catalyst-loaded solid fragments of the original starting particle, a continuous polymer phase into which the catalyst-loaded solid fragments are embedded and the pores through which the reactants are transported. If the fragmentation proceeds as expected, one catalyst particle will lead to one polymer particle, which will continue to grow by expansion as new polymer (step 3), formed at the active sites "pushes" neighbouring parts of the particle away. Thanks to very high (i.e. upwards of 50–100 kg $gcat^{-1}$ h^{-1}) activities of the modern polyolefin catalysts, there is no need to separate support fragments from the produced polymer grade.

The fragmentation step itself typically takes on the order of 10^{-1} to 10^{2} s depending on the nature of support and catalytic system, with $MgCl_2$ fragmenting more quickly and into smaller fragments than silica. Note also that it is entirely possible that steps (1) and (2) in Fig. 3.4 can occur concurrently—in fact it is highly likely that for a short period of time they are concurrent. However, despite the fact that this fragmentation process is typically much shorter than the average residence time of an industrial reactor, it remains very important in determining how the reaction will proceed (perhaps even *the* most important step in determining particle morphology). Ideally the solid support breaks up thoroughly and quickly, and if there is a balance between the rate of rupture of the support and the rate of accumulation of polymer, the solid fragments of the original support are held together by the polymer phase, and one catalyst particle yields one polymer particle. In this case all the catalyst particles undergo a more or less similar transformation, and continue to grow in size due to polymer formation unless the reaction is stopped. In this case the general shape of the catalyst particle is replicated in the final shape of the polymer particles as shown in Fig. 3.5a, b.

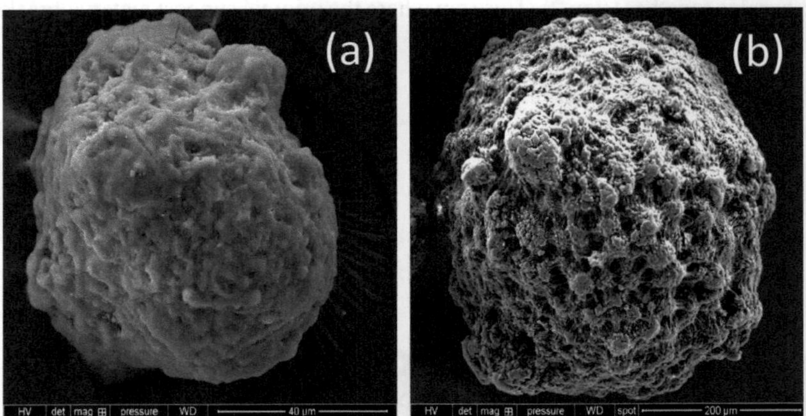

Fig. 3.5 SEM micrographs of (**a**) silica supported catalyst particle, (**b**) HDPE particle produced in heptane slurry polymerisation at 8 bar ethylene pressure and 80 °C

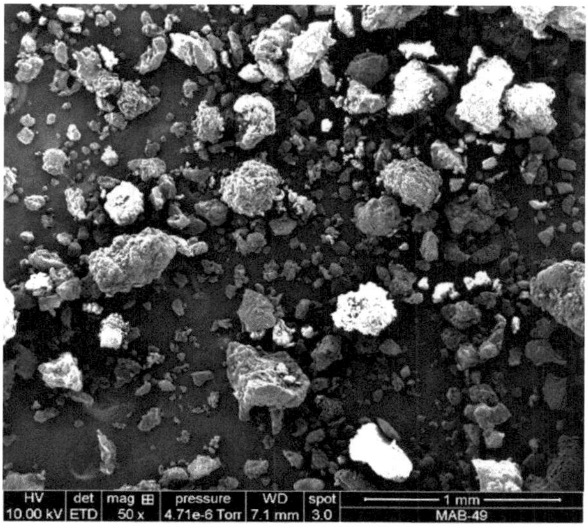

Fig. 3.6 Poor polymer morphology and fines produced in a laboratory-scale gas-phase reactor

In certain cases, for instance if the support is too fragile with respect to the rate of generation of the polymer phase during the initial instants (either because the support is poorly designed, the reaction is too rapid, or both), then somewhere between steps (1) and (2) in Fig. 3.4 the particle can lose its integrity, and fragment into many different pieces. This results in the production of "fines" (typically defined as particles smaller than 100 microns in the product stream). Fines are particularly undesirable for many reasons, including poor particle morphology, reactor fouling, static electricity, and plugging of the recycle compressor and inlet distribution grid. An example of the poor control of morphology and the generation of fines is shown in Fig. 3.6.

On the other hand, if the catalyst support is strong enough to bear all of the hydraulic forces generated by the polymer build-up it will not fragment. In this case the pores will eventually become plugged with the polymer, and the polymerisation will simply stop.

There are different ways to avoid uncontrolled fragmentation of the catalyst particles in the initial stages of olefin polymerisation. Pre-polymerising the catalyst particles in a separate reactor at conditions milder than those applied in the full reactor is one of the most widely applied industrial methods for controlling the fragmentation. This step leads to the accumulation of several tens of grams of polymer per gram of support, but at rates that are slower than in the main polymerisation step. Among other things, performing a pre-polymerisation allows us to slow down the rate of stress generation during the fragmentation step, and thus to give the particle the chance to accumulate enough polymer to act as a "glue" before the stress leads to the rupture of the original support. In addition, and as we will see in more detail below, it is more difficult to remove the heat of polymerisation from small fresh particles than it is from larger ones. Thus, pre-polymerisation also gives the

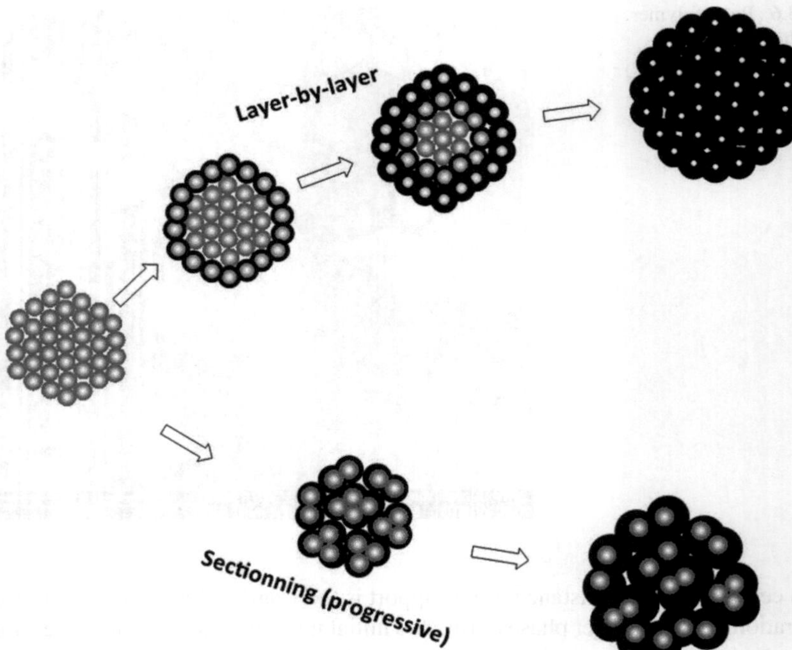

Fig. 3.7 Two routes of catalyst particle fragmentation

particles time to develop enough surface area at low rates of polymerisation (and thus of heat generation) before they are injected into the main reactor. Finally, the slower particle expansion in the early stages of polymerisation allows one to obtain better shape replication and higher final powder bulk densities.

The pre-polymerised powder is then injected into the full reactor (typically a gas-phase reactor) where rest of the polymerisation takes place on these particles.

Returning to fragmentation, it is extremely difficult to say for certain exactly how it occurs [1]. There are many reasons for this, including experimental challenges to study fragmentation in situ linked to the extreme sensitivity of polymerisation catalysts to impurities, the rapidity of the step itself, and the need to work at high pressures. In addition, there will be a very strong dependence of the fragmentation step on the initial morphology of the catalyst particles and on the specific reaction conditions.

If one considers the different studies present in the literature, people often assume that fragmentation occurs by one of two routes, shown in the illustration in Fig. 3.7. Many early studies considered that since the catalyst particle has a two-tier structure, a roughly concentric organisation of the micrograins would lead to a "layer-by-layer" (onion skin) fragmentation step [2]. Here, the porosity of the particle does not play an explicit role in the fragmentation. It is assumed that the micrograins nearest to the surface of the particle would be exposed to monomer first, and that polymer would form most rapidly here. This would cause the outer layer to break first, thereby exposing the next to last layer of micrograins to the monomer, and so on and so forth, leading to a reasonably homogeneous structure where one micrograin in the catalyst particle gives one micrograin in the polymer particle where the active sites are covered by a layer of polymer. In the second route, sometimes called progressive fragmentation or sectioning, polymer forms rapidly inside macropores in the catalyst particle structure, thereby causing the support to break up into chunks that can contain many micrograins. Once the particle is fully broken up along the macropores, polymerisation takes place in the mesopores, and then once the mesopores have fragmented, the micopores break up etc.

While at first glance this might appear to be a "one-or-the-other" situation, these two alternatives are not contradictory. If one considers the "layer-by-layer" fragmentation scheme, it is not difficult to imagine that if the catalyst particles were spatially homogeneous, with small, well-defined pores, then this method of fragmentation seems realistic enough, and it is easy to envisage that it would lead to rather homogeneous, well-defined polymer particles with low porosity. On the other hand, if the spatial organisation of the original catalyst particle were somewhat inhomogeneous, with a distribution of pore sizes, it is reasonable to suppose that polymer could form in the macropores as rapidly as it would on the surface of the particles, and that these pores would fill up and fragment first. This would then be followed by fragmentation along smaller and smaller pores until the process is as complete as it will be.

The reasons that both of these routes have some credibility are linked to the fact that not all supports are created equal: different supports have different morphologies. This can be easily seen from the images in Fig. 3.8, where the cross section of the Grace Davison silica particles appears to be much less morphologically homogeneous than the PQMS silica. The reason for this is most likely that the two types of silica are made with different processes. While very little information about the impact of the silica fabrication process is available in the open literature, this appears to be a very important point! In addition, it can be seen in Fig. 3.9 that even within a given batch of silica the macroscopic morphology might be very different from one particle to another. Different supports will have different mechanical properties and therefore, be able to resist the build-up of forces within the particle differently. It is generally thought that $MgCl_2$ supports are more friable than silica. And finally, the polymerisation conditions experienced during fragmentation (and afterwards!) can have a strong impact on the polymer properties, so not only does the rate of stress build-up impact how the particles break-up, but the deformability of the polymer (i.e. the physical properties of the polymer) will also play a role. Different temperatures and species present in the reactor (e.g. induced condensing agents that can plasticize the polymer) will change the local, instantaneous polymer properties.

In fact it is likely that both the layer-by-layer and sectioning mechanisms might occur in the same particle. Computed X-ray Tomography (CXRT) images of polymer particles produced using a silica supported metallocene after 75 s of gas-phase polymerisation are shown in Fig. 3.10. This technique allows one to generate a density map of a selection of particles, with denser regions (i.e. silica-rich regions)

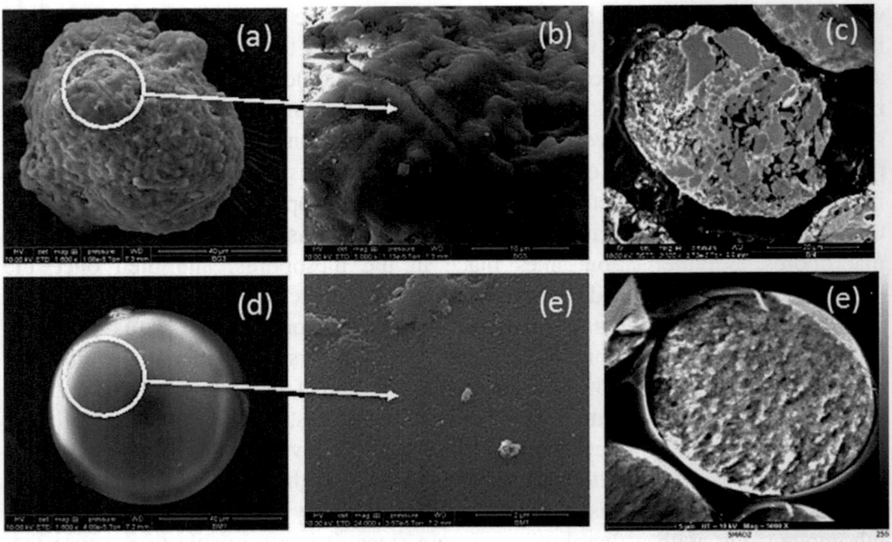

Fig. 3.8 Two commercial silica and their different internal structure. (**a**) SEM image of a Grace-Davison 948 silica particle; (**b**) SEM of the surface of the same and (**c**) microtomed cross section of same. (**d–f**) same series of images for a PQMS 3040 silica

Fig. 3.9 Computed X-ray tomography (CXRT) image of Grace Davison silica particles showing different levels of homogeneity in different particles

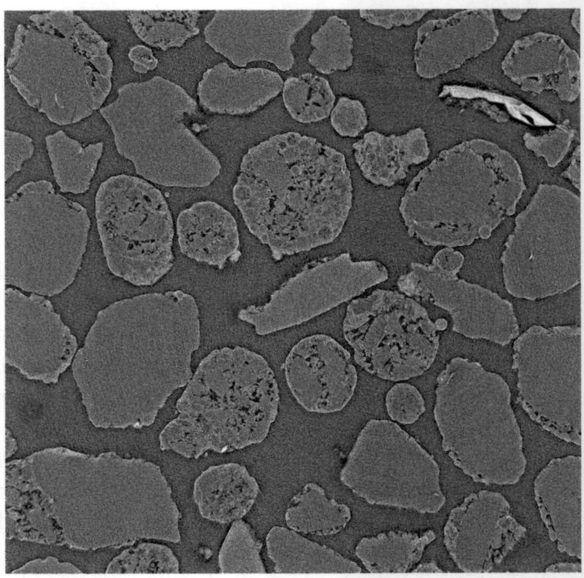

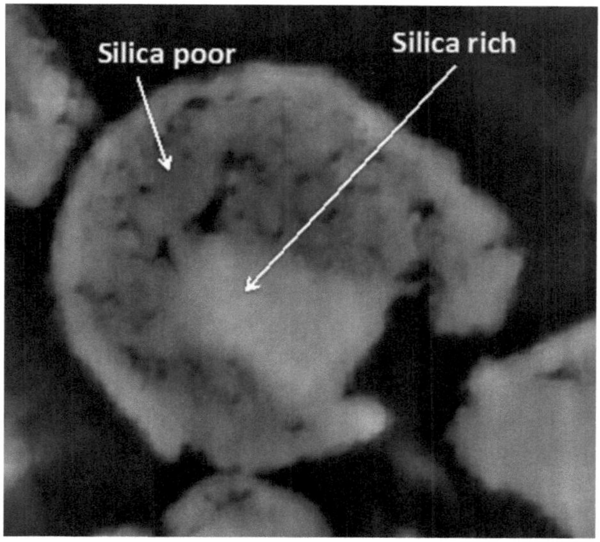

Fig. 3.10 Computed X-ray tomography (CXRT) image of a polymer particle produced on Grace Davison silica supports after 75 s of gas-phase polymerisation at 70 °C and 6 bars of ethylene. Brighter regions have higher density

appearing brighter than the less dense parts. It can be seen here that the particle appears to have completely fragmented in the porous, but not yet in the compact sections. This supports the idea that if there are macropores, then fragmentation will occur rapidly along the pores, whereas it will occur more slowly from the outside towards the centre in more compact, homogeneous regions.

Even if we admit that the macroscopic morphology of the support can influence the fragmentation step, the role of porosity and the pore size distribution remains unclear. Many studies have shown that the pore size distribution can influence the observed activity of different catalysts, and in particular those supported on silica particles. A study from our research group used the evolution of the crystallinity of HDPE at very short times as an argument to support the idea that the polymer forms first inside the smallest, most confining pores [3]. However, the importance of nanopores and mesopores on the fragmentation step is unclear.

On the one hand, if one considers conventional Ziegler–Natta or silica-supported catalysts, high specific surface areas appear to be a sine qua non for successful polymerisation, and the smaller the pores one has, the higher the specific surface area of the support, and thus the more the active sites one can pack into a given mass of support. From a productivity standpoint this makes sense, but how important is this in terms of morphology development? It has also been pointed out that the pore size can have an impact on the activity of a given catalyst. For instance McDaniel has shown that pore volume can have an impact on the activity of silica supported chromium oxide catalysts, even for constant specific surface area [4]. Nevertheless, the role that the pores and specific surface play remains unclear.

If one considers that self-supported catalysts such as Borstar Sirius catalysts have specific surface areas lower than 5 $m^2\ g^{-1}$ and thus almost no porosity, the importance of having porosity is an open question. These catalysts, which incorporate the active sites uniformly throughout the continuous phase of the particles, can be used to produce polymers on a commercial scale, which means that they have to fragment in order to allow the monomer to access the sites [5–7]. However, they are typically made in an emulsion-based process, so will have material properties very different from those of the more conventional silica- or $MgCl_2$-supported catalysts. In the case of silica-supported catalysts, the active sites are deposited on an existing support, meaning that we need high surface areas and correspondingly small pores in order to pack a sufficiently high concentration of active sites into the particles. It should also be pointed out here that especially in the case of silica supports the method used to prepare the support and to deposit the active sites will also be important. For instance, an uneven distribution of catalyst precursor caused either by the synthesis or by constraints imposed by the pore size will have an impact on the way in which the morphology evolves.

In the case of $MgCl_2$ the sites are an integral part of the support itself, but porosity is absolutely necessary if the monomer is to arrive at any active sites since without porosity not enough monomer would get to the active sites to make polymer. In addition, one could speculate that neither silica nor $MgCl_2$ would be friable enough without porosity. On the other hand, if the material of the self-supported catalysts is made permeable to monomer, then one might be able to produce polymer very close to the internal surface in an initial time. This in turn would promote the formation of local porosity, thereby allowing access to deeper parts of the particle, and so on and so forth until the entire catalyst particle is transformed into an expanding polymer particle. If this is true, then the particle break-up could be described as being a truly layer-by-layer process as shown in the top part of Fig. 3.4.

Before moving to the next section, it can be concluded that:

- Conventional supports for olefin polymerisation catalysts are porous and their morphology can be divided into two levels: the overall particle (macrograin) and the smaller primary substructures (micrograins).
- All homogeneous (in terms of catalyst distribution inside the support) catalysts undergo fragmentation. Fragmentation begins locally in the particles at places most accessible to the monomer, and where the polyolefin accumulates. This is where the resultant hydraulic forces are strong enough to rupture the support.
- If the supported catalyst has cracks or big pores on the surface and the internal pores are not well connected, fragmentation will probably start from these easily accessible macropores.
- Supported catalysts with a homogeneous or well-connected external and internal pore structure without easily accessible macropores will fragment by layer-by-layer mechanism and a fragmentation front will be formed which moves towards the centre of the catalyst particles.
- Not all catalyst particles are created equal! This can even be true within a given batch of catalyst.

3.2.2 Particle Growth

As mentioned above, fragmentation is a relatively rapid phenomenon with respect to the total residence time of a commercial process for the production of polyolefins (typically on the order of several hours). And while fragmentation and particle expansion can occur simultaneously, once fragmentation is complete the particles will continue to expand due to accumulation of polymer at the active sites. If we look at Fig. 3.4, even though this is only a schematic representation, it is not difficult to imagine that the shape the particle has soon after fragmentation can depend on the morphology generated during this step. Thus, both the starting point (i.e. what the particle looks like after fragmentation) and how the particle expands during the polymerisation step, will determine the morphology of the polymer particles in the final powder. As can clearly be seen in Fig. 3.11 SEM images of different polyolefin samples shows that particles clearly evolve very differently in the reactor, even if they all start life as a catalyst particle with two reasonably well defined levels of structural organisation.

As with the fragmentation step, the subsequent expansion of the particle is caused by the build-up of hydraulic forces due to the local accumulation of polymer and the dissipation of this energy by the deformation of the particle. This deformation/dissipation of energy will in fact lead to both the expansion of the particle, as well as to changes in local structure (e.g. compression of the polymer substructures). The resulting particle morphology will depend on the rate at which forces build up, the

Fig. 3.11 SEM images of different particle morphologies. (**a**) HiPP = High impact polypropylene, (**b**) = Polypropylene homopolymer, (**c**) HDPE = high density polyethylene produced by using silica-supported Chromium catalyst and (**d**) HDPE produced with supported metallocene catalyst

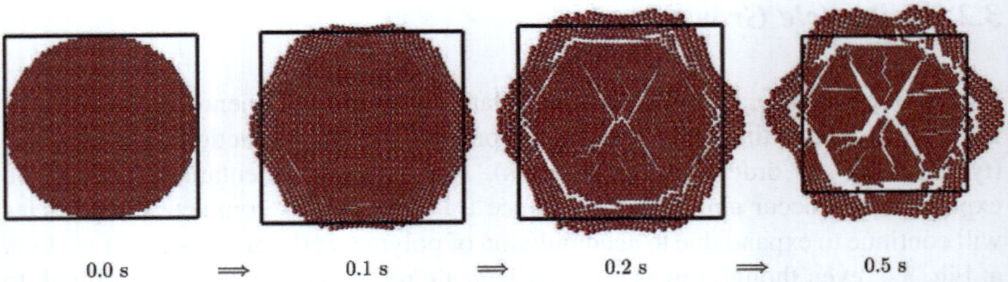

Fig. 3.12 Onion-like break-up of an expanding particle. Reprinted with permission from Industrial and Engineering Chemistry Research. Copyright (2005) American Chemical Society [9]

rate at which energy is dissipated, and on the physical properties of the polymer during the reaction. An excellent example of what we mean here can be found in a series of papers that explored the impact of changing polymer properties [8] and reaction rates [9, 10] on the morphology of expanding particles. An example of this type of study is shown in Fig. 3.12, where we can see the formation of a somewhat dense "crown" around a more loosely structured core. Structures of this type appear to be common with many Ziegler–Natta polymerisations (c.f. the cross-sectional images of the PP particles in Fig. 3.11b).

While of course having certain inherent limitations in terms of being able to predict à priori the final morphology of the particle, these studies allow us to better understand how certain parameters described below impact the evolution of morphology:

- Polymer properties: The more ductile the polymer, the more easily it deforms and the less likely we are to observe the formation of fines. Properties can of course be influenced by many factors, not least of which the temperature. In fact a patent from INEOS [11] shows that contrary to common practice, increasing the temperature in a gas-phase FBR process for HDPE can lead to a decrease in the amount of fines. Other factors that will come into play here include diluent types in slurry polymerisation, the presence of induced condensing agents in gas-phase systems, and of course any comonomers that are incorporated into the polymer.
- Spatial variations in reaction rate: Spatial variations in the local rate of polymerisation inside the particles can lead to spatial variations in the expansion rate of polymer particle. Inhomogeneities in the stress generation can cause the particle to rupture at different spots [12]. Such spatial variations can be caused by mass transfer resistance, or by improper catalyst preparation or even activation.
- Initial morphologies and instants of polymerization: Very uniform morphologies and fragmentation of the original catalyst particle produced polymer particles with structures that closely replicate those of the original particles. Structural irregularities in the initial particles led to less homogeneous particles.
- Reaction rate: The faster the rate, the less densely organised and homogeneous the final morphology is, and the more rapidly particles fragment. If the initial rate is too fast, the particles can break-up and form fines. High rates (yields) could also lead to the formation of hollow particles.

This is of course a rapid overview of a very vast subject. What we should retain from this short discussion is that the evolution of morphology is a complex issue that intervenes at several levels, and not just in terms of polymer quality. For instance, it is known that characteristic mass transfer times within particles will have a major impact on local concentrations of reactive species within the particle, and thus on the observed rates and final powder properties. This brief discussion should be kept in mind when we begin to look at how to model the temperature and concentration profiles, and thus polymerisation rate and particle growth in the next section.

3.3 Single Particle Modeling

Ultimately the goal of a serious modelling effort in the field of olefin polymerisation would be on that allows us to predict the impact of the reaction conditions on all of the particles present in the reactor at a given instant in time. However, if we reconsider Fig. 3.1, it should be apparent that any good process model requires a description of what occurs:

- At the active site;
- In a single, polymerising particle;
- At the level of the reactor;
- Interactions between the different length scales.

Kinetic (active site) models and reactor models are discussed in Chap. 4 of this book. Clearly a good single particle model is an essential part of a full process model, but on its own it is not sufficient if we want to develop a quantitative description of the production process. However, a full, multiscale approach is beyond the scope of the current discussion, so we will limit ourselves to developing the underlying approach to modelling what happens at the level of the single particle in this chapter. And in particular, we will focus on a discussion of how people attempt to calculate the evolution of concentrations and temperature inside the growing particles. Once these quantities are known, one can use the aforementioned kinetic models to calculate the molecular weight distribution, copolymer composition distributions and any other physical properties for which the models exist.

3.3.1 The Multigrain and Polymer Flow Models

A set of mass and energy balance equations (Eqs. 3.1 and 3.2 above), coupled by a reaction term, imposed on a morphological description of a polymerising particle is known as a single particle model (SPM). The solution of the SPM provides the temperature and concentration of all the reactants inside the growing polymer particle at each point (commonly known as temperature and concentration profiles inside a growing particle). If we can precisely calculate these local temperature and

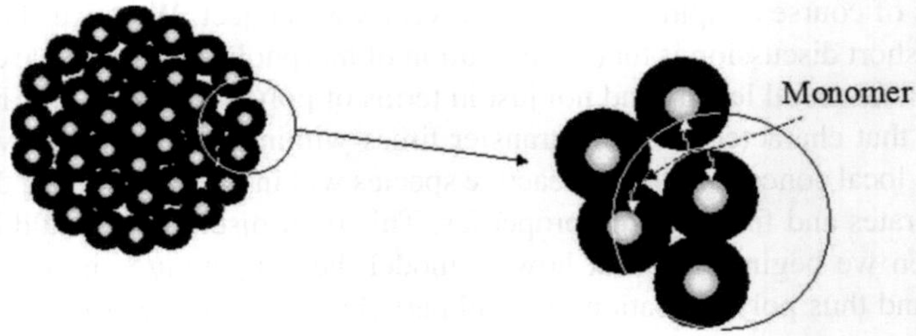

Fig. 3.13 Mass transfer as described by the MGM. Here monomer diffuses from the exterior, through the boundary layer and into the pores of the macrograin. The dashed black arrow represents mass transfer through the bulk phase and pores, and the white arrows represent the monomer sorbed in the polymer layer that diffuses to the active sites at the surface of the micrograins

concentration profiles, and if we have an accurate kinetic model, we can predict the polymerisation rates, the evolution of the molecular weight and chemical composition distributions, which enables one to predict the properties of final polymer product (of course the quality of these predictions would be enhanced by an accurate description of the reaction kinetics, as well as of thermodynamics). "All" that remains is to define how we will represent the morphology of the particle in order to be able to identify the form of the heat and mass balances, as well as the boundary conditions. As we just saw above, this is not a trivial matter.

One of the most logical places to start thinking about the how to write the equations and boundary conditions is to look at the geometry of the catalyst particles. As we saw above (c.f. Fig. 3.3), a good approximation for this would be to say that the catalyst particles are spheres composed of an agglomeration of other, much smaller spherical particles, that is, the micrograins and macrograins. This is the basis for the Multigrain model (MGM), which can be considered to be the first realistic approach to model the growth of a polyolefin particle by expansion due to polymer formation and that takes two levels of morphology into consideration [2, 13]. A representation of how mass transfer takes place in the MGM is shown in Fig. 3.13.

Here the monomer is transferred from the bulk phase of the reactor, through the boundary layer and into the pores of the macrograin. From there it diffuses through the pores of the macroparticle towards its centre, and is simultaneously absorbed by, and diffuses through, the polymer layer of the micrograin where it polymerises at the active site. Thus, we have a classic reaction-diffusion engineering problem to solve. If we assume that the macrograins are isotropically similar, mass transfer occurs only in the radial direction. Thus, the characteristic length scale for mass transfer in the macrograins is the radius of the growing particle, and that of mass transfer in the micrograins will be the thickness of the polymer layer. Similarly, heat transfer occurs by a process of heat generation at the active sites, transport through the polymer phase of the particles (it is easy to demonstrate that heat transfer in the pores of the growing particles is not very important), to the surface of the macrograin, then to the bulk phase of the reactor via the thermal boundary layer.

This representation of the particle is interesting for certain types of studies, for example modelling the fragmentation step for uniform catalyst particles [14], or to help conceptualise the impact that polymer properties might have on monomer transportation in the growing particles. The problem with the MGM is that it relies on many assumptions that are either difficult to prove or are outright incorrect [15]. The most important of these is that the original micro/macrograin structure of the catalyst particle is perfectly replicated in the polymer particle. However, if one considers the images in Fig. 3.11, it is quite obvious that none of those particles exhibits a morphology that can be well described by the MGM.

Furthermore, it turns out that even for highly productive catalysts, if this representation is correct, the thickness of the polymer layer in the micrograins is rarely enough to see the generation of the substantial concentration or temperature gradients at this length scale. Given that the morphological representation of the MGM is not particularly correct, in order to avoid over-complicating the problem, engineers turned to a more simplified version of the MGM, referred to as the Polymer Flow Model (PFM). With the PFM, the user neglects heat and mass transfer resistances at a microparticle level, and treats the macroparticle as a pseudo-homogeneous medium in which the active sites are distributed uniformly throughout a fictitious continuous medium composed of polymer and pores. The mass and energy balances for the PFM are given in the right-hand columns of Tables 3.1 and 3.2, respectively. The major difference between the macrograin models used by the MGM and PFM lays in the fact that in the latter, transport resistances at the microparticle level are neglected and one assumes that the concentrations in the polymer phase and at the active sites at any given radial position in the macrograin is in equilibrium with the concentration of this same species in the pores of the macrograin. This means that the diffusion coefficients in the two models will be different. In the case of the MGM, this parameter is the diffusivity of the species in the pores of the macroparticle ($D_{\mathrm{eff,MGM}}$); in the PFM it is an effective diffusion coefficient in the pseudo homogeneous mixed polymer/pore medium ($D_{\mathrm{eff,PFM}}$). In principle, one could make the same observation for the thermal conductivities of the macrograin energy balances for the two different models; however, as mentioned earlier, very little energy is transported in the pore space of the particle (MGM model), so for all intents and purposes there is no need to differentiate between the two. Otherwise, at the level of the macrograin, the initial and boundary conditions for the macrograin balances are the same for both models.

Generally speaking, most scientists and engineers who use a SPM tend to use the PFM simply because it requires fewer (hard to identify) parameters and the complexity of solving equations at two levels of morphology with the MGM brings very few (if any) advantages to the solution. Given the nature of the phenomena involved, it is unlikely that either model really gives a perfect representation of the changes that occur in particle morphology. Note that while even very sophisticated models of particle morphology such as those discussed above are useful tools for understanding the impact that certain parameters/phenomena *can* have on particle morphology, the authors themselves conclude that we are still not yet at a point where that type of model can be used to predict *à priori* the evolution of particle morphology as a function of the reaction conditions directly.

Table 3.1 Mass balance equations for MGM and PFM

Multigrain model (MGM)		Polymer flow model (PFM)	
Macroparticle		*Macroparticle*	
$\frac{\partial [M_i]}{\partial t} = \frac{1}{{r_L}^2}\frac{\partial}{\partial r_L}\left({r_L}^2 \frac{\partial\left(D_{\mathrm{eff,MGM}}[M_i]\right)}{\partial r_L}\right) - R_i$	(3.3a)	$\frac{\partial [M_i]}{\partial t} = \frac{1}{{r_L}^2}\frac{\partial}{\partial r_L}\left({r_L}^2 \frac{\partial\left(D_{\mathrm{eff,PFM}}[M_i]\right)}{\partial r_L}\right) - R_i$	(3.3b)
Initial condition: for all r_L at $t = 0$ $[M_i] = [M_i^{\circ}]$			(3.4)
Boundary conditions: for all $t >$ at $r = 0$ $\frac{\partial [M_i]}{\partial r_L} = 0$			(3.5)
for all $t >$ at $r = R_L$ $\left\{ \begin{array}{l} -N_i = k_s\left([M_i]_{\mathrm{bulk}} - [M_i]_{r_L=R_L}\right) \\ \text{or} \\ [M_i]_{\mathrm{bulk}} = [M_i]_{r_L=R_L} \end{array} \right\}$			(3.6a) (3.6b)
Microparticle		*No microparticle considered in PFM*	
$\frac{\partial\left[M_{i,p}\right]}{\partial t} = \frac{1}{r_S^2}\frac{\partial}{\partial r_S}\left(r_S^2 \frac{\partial\left(D_p\left[M_{i,p}\right]\right)}{\partial r}\right)_i$			(3.7)
Initial condition:for all r_s at $t = 0$ $[M_{i,p}] = [M^{\circ}_{i,p}]$			(3.8)
Boundary conditions:for all $t > 0$, $r_s = R_c$, $N_{i,p} = -k_p[M_{i,p}]_{C*}[C^*]$ for all $t > 0$, $r_s = R_S$, $\left[M_{i,p}\right] = [M_i^{\mathrm{eq}}]$			(3.9) (3.10)

Table 3.2 Energy balance equations for MGM and PFM

Multigrain model (MGM)		Polymer flow model (PFM)	
Macroparticle		*Macroparticle*	
$\overline{\rho C_p}\frac{\partial T}{\partial t}=\frac{1}{r_L^2}\frac{\partial}{\partial r_L}\left(r_L^2 k_{f,L}\frac{\partial T}{\partial r_L}\right)-(-\Delta\overline{H_p})R_i$	(3.11a)	$\overline{\rho C_p}\frac{\partial T}{\partial t}=\frac{1}{r_L^2}\frac{\partial}{\partial r_L}\left(r_L^2 k_{f,L}\frac{\partial T}{\partial r_L}\right)+(-\Delta\overline{H_p})R_i$	(3.11b)
Initial condition:for all r_L at $t=0$ $T=T_0$			(3.12)
Boundary conditions: for all $t>$ at $r=0$ $\frac{\partial T}{\partial r}=0$			(3.13)
for all $t>$ at $r=R_L$ $\left\{\begin{array}{l} -k_f\frac{\partial T}{\partial r_L}=h(T_{r_L=R_L}-T_{\text{bulk}}) \\ \text{or} \\ T_{\text{bulk}}=T_{r_L=R_L} \end{array}\right\}$			(3.14a) (3.14b)
Microparticle		*No microparticle considered in PFM*	
$\rho C_p\frac{\partial T_S}{\partial t}=\frac{1}{r_S^2}\frac{\partial}{\partial r_S}\left(r_S^2 k_{f,P}\frac{\partial T_S}{\partial r_S}\right)$			(3.15)
Initial condition: for all r_s at $t=0$ $T_S=T_{S,0}$			(3.16)
Boundary conditions: for all $t>0$, $r_s=R_c$, $kf,p\frac{\partial T_S}{\partial r_S}=k_p\left[M_{i,p}\right]_{C*}[C^*](-\Delta H_p)$ for all $t>0$, $r_s=R_S$, $T_S=T$			(3.17) (3.18)

3.3.2 *Solving the PFM*

As we go through this part of the discussion it is important to keep in mind that SPM are useful important tools. Nevertheless, it is also important to remember that there still appears to be no way for us to mathematically and quantitatively predict the evolution of the morphology of a growing polyolefin particle as a function of the reaction conditions and catalyst particle morphology. It is, therefore, suggested that the readers attempt to understand the impact of the different parameters on the model predictions, as well as the importance of choosing boundary conditions, and that they also understand that in many instances, these limitations might impose inherent errors when interpreting the results of experimental kinetic studies in heterogeneous systems.

Choice of Key Model Parameters and Boundary Conditions

Solving the PFM obviously involves solving Eqs. (3.3–3.6) (material balance) for all of the species present in the reactor, which are coupled to Eqs. (3.11–3.14) (energy balance) in order to find the concentration and temperature profiles in the particles. To do so, one will need to specify values for a range of model parameters—some of which are harder than others to set accurately.

First and foremost, it is necessary to dispose of a set of kinetic parameters for all of the reactive species. In the current presentation, we have attempted to keep the presentation of the model as simple as possible, so will consider only one species (e.g. ethylene), expecting that the extension to multiple component systems should be obvious. In this case, we can replace $[M]_i$ with $[M]_{\text{ethylene}}$.

For a reactive species (monomer, hydrogen…), the term R_i in Eq. (3.3b) represents the volumetric rate of polymerisation at a given position and time. As an example, for an ethylene homopolymerisation, this can be calculated by:

$$\begin{aligned} &R_i = R_{\text{pol}}\left(\frac{1-\varepsilon}{\phi^3}\right) \\ &\text{Where, } R_{\text{pol}} = k_p C^* [M]_{\text{ethylene}} \\ &\text{and } \phi = \frac{r_P}{r_c} \end{aligned} \tag{3.19}$$

R_{pol} is the local rate of ethylene polymerisation at the surface of catalyst particle, k_p is the propagation rate constant at *local temperature*, C^* is local concentration of active catalytic sites, $[M]_{\text{ethylene}}$ is the concentration of ethylene at the active site that is in equilibrium with $[M]_{\text{ethylene,bulk}}$, ε is the particle porosity, ϕ is the overall growth factor, r_p represents the equivalent radius of "*polymer only*" particle at each time step without considering the existing particle porosity and r_c shows the radius of initial catalyst particle.

Assuming that one can estimate a value for the particle porosity (that will likely evolve as a function of time), even in a relatively simplified system, estimating a value for even k_p and for C^* is not always straightforward. If one is dealing with a Ziegler–Natta or a chromium-based catalyst, it is generally accepted that there is a multiplicity of active sites on a single support, with each family of sites behaving differently (see Chap. 5). This value of k_p that appears in Eq. (3.19) can be thought of as a lumped parameter; and the same goes for C^*. Also, it is widely accepted that catalysts will undergo phase of activation and deactivation (and occasionally even poisoning). This in turn means that one would need to identify an expression that takes this into account (eventually for each of the different “families” of active sites as it is not all guaranteed that they each behave in the same manner). As an example, if one assumes that catalyst activation is rapid, and that deactivation of the lumped active site concentration is of first order, then deactivation can be taken into consideration using an expression in the form:

$$C^* = \left[C_1^* \exp(-k_d t)\right] + C_2^* \tag{3.20}$$

At time $t = 0$, the total number of active sites is the sum of C_1^* plus C_2^*, where we consider C_2^* to be the number of active sites still participating in the polymerisation after a long time, and C_1^* is the fraction of sites that are deactivated with a rate constant k_d. Equation (3.20) is for first order deactivation, which is generally considered for most polymerisation reactions [16, 17]. However, it would need to be modified for a catalyst system with a different deactivation rate order.

Since olefin polymerisation is an exothermic reaction, the rate constants above will be linked to the energy balance, and it is necessary to calculate their value as a function of the local temperature using an Arrhenius type correlation (also true for any other rate constants used in the model).

Note that if one wishes to include additional species, or expressions for calculating the molecular weight distribution, a phenomenological description of catalyst activation and/or deactivation additional rate constants and species balances will be required. For a more detailed discussion on how to develop such kinetic models and estimate the required rate constants, the reader is referred to Chap. 4, or additional references [18].

Returning to the macroparticle mass balance and kinetic expression above, it is also necessary be able to calculate the concentration of monomer at the active sites *below* the layer of polymer that forms in the particles. Note that the same concentration term is also present in Eq. (3.19) for the determination of polymerisation rate, R_{pol}, and therefore, precise estimation of this term is very important. By assuming absence of any mass transfer resistance in the layer surrounding the growing polymer particle (basic premise of PFM physical model), the concentration of ethylene in the bulk phase, $[M]_{\mathrm{ethylene,bulk}}$, is considered in equilibrium with the concentration of ethylene in the polymer phase present on the active sites, $[M]_{\mathrm{ethylene}}$, of the growing macroparticle. This assumption also constitutes the first boundary condition for PFM. It remains an open question as to whether or not one can consider the particle to be in equilibrium with its environment under all pertinent circumstances, especially during moments of rapid polymerisation.

$[M]_{\text{ethylene}}$ can be estimated by using experimental solubility data of ethylene in polyethylene (the same is true for other species present in the reactor). Experimental solubility data of various olefins in different polyolefin grades (with different properties like crystallinity etc.) at conditions of industrial importance is available in the open literature [19–23]. Such experimental data of ethylene solubility in polyethylene can be either used directly or fitted with a mathematical model to estimate the value of $[M]_{\text{ethylene}}$ at the desired temperature and pressure.

Henry's law is the simplest mathematical correlation used to estimate $[M]_{\text{ethylene}}$, but it has been shown in the literature that the predictions of Henry's law strongly deviate from the experimental data at higher ethylene pressures. Henry's law also underestimates the concentration of higher α-olefins like 1-butene, 1-hexene or 1-octene in a polyolefin. Therefore, thermodynamic models like activity coefficient models (ACM) or equations of state (EoS) are to be preferred over Henry's law for the estimation of $[M]_{\text{ethylene}}$. The most widely used models include UNIQUAC functional-group activity coefficients (UNIFAC) which is an ACM, and EoS based on different versions of perturbed-chain statistical associating fluid theory (PC-SAFT) and Sanchez–Lacombe equation of state (SL EoS) [20, 24–28]. It appears that of all these models provide similar estimates under conditions of industrial relevance and therefore, selection of a specific model depends mainly upon its simplicity and ease of solution.

One of the major problems with these thermodynamic models (i.e. ACM or EoS) is that they are unable to account for the effects of polymer crystallinity on monomer sorption, and thus need to be corrected for semicrystalline polymers such as polyolefins. One of the more common ways to do this is to tune the binary interaction parameter(s) of an EoS [20, 24, 26, 27, 29, 30] or to use an elastic constraints model an ACM or EoS to account for the effects of polymer crystallites on monomer sorption [28, 31–33].

Another challenge arises when one attempts to estimate the solubility of a mixture of gases present in the reactor that are sorbing simultaneously into the polymer phase as in the case of a fluidized bed reactor for polyethylene production where the gaseous feed can typically contain a mixture of ethylene, comonomer(s), hydrogen, as well as inerts including nitrogen or propane gases, and induced condensing agents (ICA). For such multicomponent mixtures, conventional thermodynamic models like the SL EoS or PC-SAFT cannot predict precisely the solubility of each component in the polymer phase. In a ternary system such as ethylene and iso-pentane dissolving in polyethylene, there will be what is called a co-solubility effect. In this case, the presence of the heavier component enhances the solubility of the lighter gas in the polymer phase with respect to the solubility of ethylene alone at the same pressure and temperature. On the other hand, the lighter component can be an anti-solvent for the heavier one. For instance, in the case just cited, the quantity of iso-pentane absorbed in the polyethylene is actually lower than the amount that would be absorbed in the binary iso-pentane/PE system. This is also going to occur

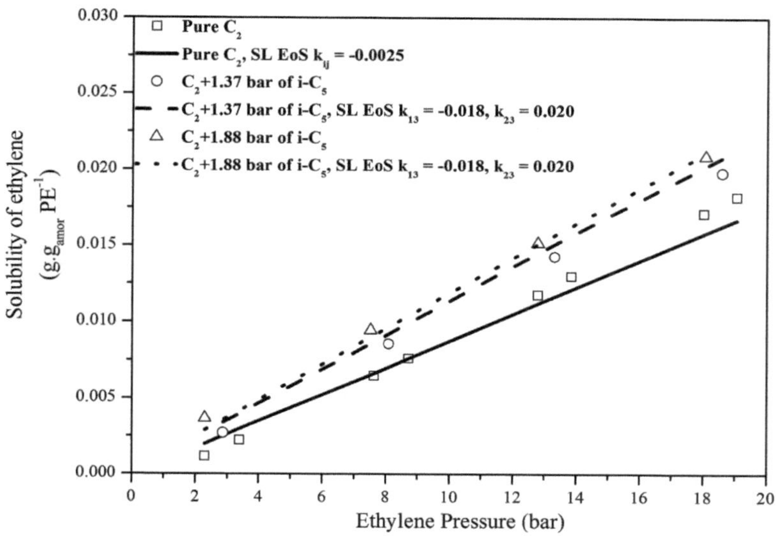

Fig. 3.14 Co-solvent effect of iso-pentane (i-C_5) on the solubility of ethylene (C_2) in polyethylene. Ternary mixture experimental solubility data obtained from Yao et al. [34] k_{13} is the SL EoS binary interaction parameter in ternary mixture for ethylene (1)/PE (3), whereas, k_{23} is the binary interaction parameter in the ternary mixture for iso-pentane (2)/PE (3)

in the case of LLDPE production when a heavier comonomer such as 1-butene or 1-hexene is used—here we will observe an increase of ethylene concentration in the polymer combined with a complex impact of the alpha-olefin comonomer on the reaction rate (the so-called comonomer effect). An example of the enhancement effect of iso-pentane on the solubility of ethylene in the amorphous phase of LLDPE is shown in Fig. 3.14.

Once the equilibrium solubility of ethylene in pure polymer at a given temperature and pressure is known, the concentration of ethylene in the pseudo-homogenous macroparticle (in mole cm^{-3} of PE) can be estimated as follows:

$$[M]_{\text{ethylene}} = \left(\frac{S_{\text{ethylene}}\rho_{\text{pol}}}{MW_{\text{ethylene}}}\right)(1-\varepsilon) \tag{3.21}$$

where S_{ethylene} is the equilibrium solubility of ethylene in pure amorphous polyolefin in g g^{-1} of amorphous polymer at the given conditions obtained by a thermodynamic model (or experimentally), ρ_{pol} is density of the swollen polyolefin in g cm^{-3} estimated also by a thermodynamic model, MW_{ethylene} is the molar mass of ethylene in g mol^{-1} and ε is the porosity of macroparticle.

Note that one of the major obstacles to fully exploiting such an approach at the time this chapter was written is that there is a severe lack of experimental data for ternary and higher order systems under realistic polymerisation conditions.

Returning once again to the macroparticle mass balance, it is also easy to identify the effective diffusivity ($D_{eff,PFM}$) a as a key model parameter. This parameter describes the process of diffusion of a given species present in the mixture through the pseudo-homogenous macroparticle assumed in PFM. Correct estimation of this parameter is clearly very important since it provides information about the relative rate of mass transfer by diffusion through the particle to the active sites. Slow diffusion in the presence of a rapid reaction through a large particle can in fact lead to situations where mass transfer limitations control the observed rate of reaction and polymer properties. Since we have defined our growing polymer particle as being a "pseudohomogeneous" phase, that is, one constituted of both polymer phase and pore space, $D_{eff,PFM}$ is a complex function of particle morphology (e.g. porosity and pore size distribution of the particle), molecular properties (e.g. molecular weight of the polymer) and physical properties like polymer crystallinity.

A dual-transport mechanism is usually associated with ethylene diffusion in a macroparticle. The first mechanism is related to the diffusion of a given species from the continuous gas phase to the polymer phase in the particle through the open pore network, which includes micropores and macropores. During the early stages of polymerisation, immediately before and just after fragmentation, the available pore volume can be as large as 0.14 $cm^3\ g^{-1}$ indicating that monomer transport can occur through open pore network of the particle [30]. Note, however, that as the polymerisation progresses this situation can change, depending on how the evolution of the overall particle morphology proceeds. The second mechanism relates the diffusion of ethylene through the amorphous fraction of the polymer in the particle, since the crystalline fraction of the polymer is impenetrable to the monomer. This mode of monomer diffusion likely becomes dominant when enough polymer has form around the growing polymer particle.

One means of taking this duality of diffusion mechanisms into account is through the use of the following combined equation [30]:

$$D_{ij} = \frac{\varepsilon}{\tau_f^2} D_{ij}^{o} + (1 - \varepsilon)(1 + 3\varepsilon) D_i^P \tag{3.22}$$

where D_{ij} is the overall binary diffusivity, D_{ij}^{o} is the gas-phase binary diffusion coefficient of the ith species in the presence of another species "j", D_i^P is the diffusivity of the ith species in the amorphous polymer phase, ε is the particle porosity and τ_f is the tortuosity factor showing the irregularities in pore structure and diffusion path through pores. Of course, one of the major limitations of the use of this form of equation is that it is necessary to know the value of the porosity term, which will undoubtedly vary as a function of time (advancement of the reaction), composition of the fluid phase and from polymerisation to polymerisation.

Nevertheless, this form of equation gives us at least an opportunity to link the morphology of the growing particle to the rate of mass transfer. The first term on the right-hand side of the Eq. (3.22) represents the diffusion through pores of the starting particle, that is, the fresh catalyst particle and the second part shows the diffusion

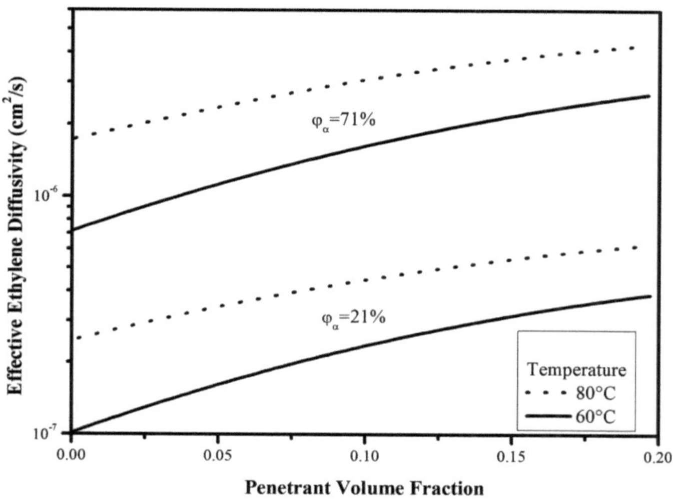

Fig. 3.15 Effect of polymer crystallinity and reactor temperature on the effective diffusivity of ethylene by using Hybrid model of Doong et al. [40] Figure taken from Yiagopoulos et al. [25], with permission

through the amorphous polymer phase. In case of binary gaseous mixtures, D_{ij}^{o} can be calculated according to the general gas diffusion theory of Chapman and Enskog [30, 35, 36]. The diffusivity, D_i^P, of the *i*th component in the polymer phase can be described using the free volume theory, proposed originally by Fujita [37] and updated many times since then. In short, this means the diffusion of a small molecule through the amorphous phase of the polymer depends on many parameters, including the temperature, the nature (size) and concentration of sorbed species, the degree of polymer crystallinity, its thermal expansion coefficient, molecular weight, and many others. As long ago as 1961, Michaels and Bixler [38, 39] proposed a semi-empirical correlation to estimate the diffusivity of different gases in polyethylene as a function of the size of the penetrant and the degree of crystallinity of the polymer. Doong et al. [40] proposed a hybrid model by combining the free-volume [41, 42] and molecular models of penetrant diffusion in polymers [43] for the estimation of the penetrant diffusivity. Yiagopoulos et al. [25] have shown the effect of polymer crystallinity and reaction temperature on the effective ethylene diffusivity in polyethylene by employing Eq. (3.22) in PFM. Figure 3.15 shows the effect of polymer crystallinity and reaction temperature on ethylene diffusivity in polyethylene (D_i^P) by considering the hybrid model of Doong et al. [40]. An increase in ethylene diffusivity can be seen with an increase in the amorphous fraction of polyethylene, the reaction temperature, and the degree of swelling of the polymer (penetrant volume fraction).

The effect of polyethylene growing particle porosity and reaction temperature on overall binary monomer diffusivity (D_{ij}) estimated by using Eq. (3.22) is shown in Fig. 3.16 where it can be observed that for highly porous polymer particles

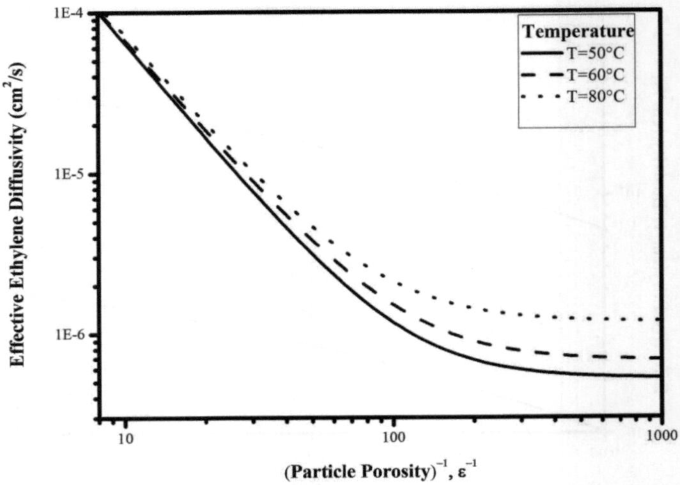

Fig. 3.16 Effect of particle porosity on the effective ethylene diffusivity (D_{ij}) at different temperatures. Figure taken from Yiagopoulos et al. [25], with permission

(i.e. $\varepsilon = 0.1$–0.2) the pore diffusion (i.e. the first term on right-hand side of Eq. 3.22) will dominate the overall effective diffusivity value. As the particle grows and the porosity decreases due to the accumulation of more polymer, the value of D_{ij} comes to be dominated by the diffusion through amorphous polyethylene (i.e. the D_i^P value).

An important point to be highlighted here is that co-solubility effects can also play a significant role in increasing the diffusivity of ethylene in the amorphous polymer phase because the co-solvent effect increases the ethylene concentration in amorphous fraction of polyethylene, and according to Fig. 3.15 the higher the ethylene concentration in the amorphous part the higher the diffusivity. However, it is hard to find any such studies in the open literature which clearly show the co-solubility effects on diffusivity of olefins in polyolefins.

There are, of course, many other approaches to modelling penetrant-polymer diffusivity that we will not discuss here. Suffice to say that the estimation of this diffusivity needs to take into account the semicrystalline nature of the polymer being produced in the very least.

It is important to understand the choice of boundary conditions for the model as well. Referring back to the macroparticle balance in Table 3.1, in the choice for an initial condition (Eq. 3.4), the monomer concentration in the particle can be set either equal to zero, or set at some prespecified finite value. Setting this quantity to zero generally results in a stiff differential equation which is difficult to solve. Generally, a pseudo-steady state concentration is assumed at $t = 0$ to avoid this problem. The zero-flux, or symmetry boundary condition at the centre of the particle is expressed with Eq. (3.5). Since the particles are almost always described as being spherical, there is not much choice here (besides, changing the shape of the particles to ellipsoids rather than spheres changes nothing in the model simulations except to add complications). Finally, the boundary condition at the surface of the particle is expressed using either Eq. (3.6a or 3.6b). Equation (3.6a) allows for the calculation of a concentration gradient in the boundary layer of the particle. In fluidised beds this can probably be neglected as it is expected that external resistance to mass transfer will be negligible, even for high rates of polymerisation, due to the significant mobility of the species in the gas phase. If this is the case (negligible external mass transfer resistance), then Eq. (3.6b) can be used [44–49]. It is possible that in slurry processes that external resistances will be significant when dealing with systems showing very high reaction rates and large catalyst particles. However, most authors opt for Eq. (3.6b), and then check to see that this is an acceptable simplification.

Moving on to the energy balances (Eqs. 3.11–3.13) it is perhaps less difficult to pin down values for the additional terms shown in Eq. (3.11). The heat capacity of the pseudo-homogeneous phase of the PFM particle can be approximated by the heat capacity of the polymer as most of the energy will be transported through the particle by conduction through the polymer phase. The precise value of the density will of course depend on how much the polymer phase is swollen by the species in the reactor, and the pore volume. However, it seems reasonable to use a value given by Eq. (3.23) as a first approximation, where ρ_{pol} is the density of the polymer.

$$\rho = \rho_{pol}(1 - \varepsilon) \tag{3.23}$$

The other parameter in Eq. (3.11) is the thermal conductivity (k_f) of the particle. This too can be approximated by that of the polymer. A measured value can be specified (eventually as a function of temperature), or a more precise value might be found using a correlation [50].

The energy balance and mass balances are coupled via the reaction term in Eqs. (3.3b) and (3.11b). The initial condition (Eq. 3.12) is straightforward as the particle entering the reactor will do so at a known temperature. The symmetry condition at the centre of the particle (Eq. 3.13) holds for the same reason as given above for the mass balance.

The most challenging part of the energy balance lies in the choice of the second boundary condition for the energy balance. In the event that one is simulating a slurry process, then the heat transfer from the particle to the fluid phase might be efficient enough that we can use Eq. (3.14b) as the boundary conditions. Nevertheless, it is probably wiser, and in fact obligatory when modelling gas-phase systems, to use boundary condition (3.14a). This boundary condition allows one to account for heat transfer resistance in the boundary layer, and to calculate the temperature gradient between the particle surface and the continuous phase. Note that it is entirely possible that the difference in temperature between the particle and the continuous fluid can be well over 10 °C for certain gas-phase polymerisations. The major challenge here, especially for gas-phase systems will be the estimation of external heat transfer coefficient "h". Many researchers [16, 17, 25, 26, 30, 45–49, 51] used Ranz–Marshall correlations for estimating the external heat transfer (as well as the film mass transfer coefficient, k_{ext}). These correlations are attractive for their simplicity and ease of use, and the Nusselt number (Nu) correlation for estimating the heat transfer coefficient is given by:

$$\mathrm{Nu} = 2 + 0.6\mathrm{Pr}^{\frac{1}{3}}\mathrm{Re}^{\frac{1}{2}} \tag{3.24}$$

where

$$\mathrm{Nu} = \frac{d_p h}{k}, \mathrm{Pr} = \frac{\mu_g C_g}{k_g}, \mathrm{Re} = \frac{d_p u_r \rho_g}{\nu_g}$$

In these dimensionless numbers, d_p is the growing polymer particle diameter, μ_g is the viscosity of gas phase, k is the thermal conductivity of the fluid, C_g is the heat capacity of the gas phase, ρ_g is the gas phase density, ν_g is the kinematic viscosity of the gas phase and u_r is the particle to fluid relative velocity.

However, this correlation (and its mass transfer analog) was developed for an isolated, falling droplet, and particles evolving in a polymerisation reactor are far from isolated. In addition, it is very difficult (to say the least) to have an accurate estimate of the local value for the gas-particle velocity (a determining factor in the estimation of this parameter) since the hydrodynamics of the reactors are complex, and they are full of dense beds of particles of different sizes. According to Wisseroth [51] a value of 1.5 cm s^{-1} is a reasonable estimate for u_r in the case of stirred bed polymerisation reactors, whereas, for fluidized bed reactors the values is in the range 2–4 cm s^{-1} [51]. A detailed comparative discussion about the results obtained by using the abovementioned correlations can be found in the work of Floyd et al. [51]

Nevertheless, if one wishes to accurately evaluate the heat transfer between a growing single particle and the continuous phase of the reactor, it would be necessary to use a much more complete Nu correlation than that given by the Ranz–Marshall model. Numerous correlations have been proposed for the calculation of reactor-average values of the external boundary layer mass and heat transfer coefficients, that take many factors into consideration [50, 52–58]. Kanellopoulos et al. [30], proposed that the correlation proposed by Nicolella et al. [59], provided more realistic results than those calculated by using the traditional Ranz–Marshall correlation in gas-phase process. This was attributed to the fact that the presence of other solids in the gas phase is taken into account via an energy dissipation term in the correlation of Nicolella et al. [59], and that this provides a better estimate for the heat transfer coefficient than does the Ranz–Marshall correlation.

Brief Overview of the Solution Methods

The mass and energy balance equations which constitute both the multigrain and polymer flow models fall in the category of partial differential equations (PDE) since they involve more than one independent variable. Among various sub-classes of

Table 3.3 Brief literature review of the numerical methods applied in solving MGM and PFM

References	Model used	Method
Ray et al. [2, 55, 56]	MGM	Finite difference
Chiovetta et al. [44, 45, 48, 57]	MGM	Finite difference
Sarkar et al. [58, 61]	MGM	Finite difference
Bhagawat et al. [62]	MGM	Finite difference
Kittilsen et al. [46]	MGM	Finite volume
Schmeal et al. [63]	PFM	Finite difference
Merrill et al. [64]	PFM	Finite difference
Galvan et al. [54]	PFM	Orthogonal collocation
Kiparissides et al. [25, 30]	PFM	Global collocation
Veera et al. [16, 17]	PFM, MGM	Orthogonal collocation
Hoel et al. [65]	PFM	Method of lines

PDEs defined on the basis of order, linearity and boundary conditions, diffusion and energy transfer equations of MGM and PFM are considered as second order, non-linear parabolic equations with Neumann type initial and boundary conditions [52] which needs to be solved simultaneously if one wants to estimate the temperature and concentration profiles at each point in a spherical growing polyolefin particle. Various numerical methods have been developed to solve such PDEs including finite difference methods, finite element methods, method of lines, collocation methods and finite volume methods [52, 53, 60]. In the field of polyolefin reaction engineering, the methods of finite differences and collocation seem to be the most widely used for the solution of both the MGM and PFM, as shown in Table 3.3. The major advantage of using a collocation method, besides its simplicity, stability and good accuracy, is that it requires lesser discretization in the spatial domain as compared to the finite difference methods. This is very important, especially at high values of Thiele modulus where the finite different methods require a very fine mesh compared to much lower number of points required by the collocation method [54]. A detailed discussion of these numerical methods is beyond the scope of this chapter but the interested reader is referred to for example Beers [60], Villadsen et al. [53] or Mostoufi et al. [52], for further understanding the applied numerical methods for solving PDEs.

It should also be pointed out that simplified versions of Eqs. (3.3–3.6) and (3.11–3.14) can be solved in order to understand what are the most important model parameters, and what general conclusions we can draw. An elegant way of doing this is to invoke what is called the Quasi Steady State Hypothesis (QSSH) [15, 51, 55]. This hypothesis simply implies that the rates of change of concentration in the particle [derivative on the left hand site of Eq. 3.3b] and of temperature [derivative on the left hand site of Eq. 3.11b] are very small compared to the terms on the right-hand side of the same equations. If we apply these hypotheses then we can draw the following (very general) conclusions about heat and mass transfer in growing polymer particles, and these, regardless of the particle morphology, are as follows:

1. We expect more mass transfer resistance through the boundary layer in slurry phase polymerisations than in gas phase.
2. We expect more heat transfer resistance through the boundary layer in gas phase polymerisations than in slurry phase (i.e. the particles have a much greater tendency to be hotter than the surrounding continuous phase in the gas phase polymerisations than slurry. In slurry the temperature of the particles will be close to that of the continuous medium).
3. Both heat and mass transfer resistances will be more important during the initial instants of polymerisation than at later stages, all other things being equal. For heat transfer this is because the surface area available for heat transfer grows as the particle grows, but the amount of heat generated per particle usually stays constant or decreases (one can of course find certain exceptions). Similarly, the surface area per unit volume of particle being lowest at the beginning of the reaction, it is harder to supply the active sites with monomer through the particle surface at this point, so we risk running into mass transfer problems.
4. Both heat and mass transfer resistances will be greater for larger fresh catalyst particles than for smaller ones. This explains in part why gas phase catalyst particles are typically smaller than slurry phase ones to avoid overheating, even though larger catalysts particles would offer many practical advantages.
5. We expect more mass transfer resistance through the macroparticle in slurry phase polymerisations than in gas phase (i.e. because the pore diffusivity of reactive species is lower in the liquid state than in the gas, there is a greater potential for concentration gradients inside the particles in slurry polymerisation than gas—however this does not mean we can neglect mass transfer resistance in the gas phase, especially for higher activities).
6. We expect negligible heat transfer resistance inside the particle for slurry phase polymerisations.
7. Generally speaking, internal particle temperature gradients in gas phase systems are much smaller than temperature gradients across the boundary layer.

The reader should be aware that these are very general conclusions and tendencies. They can be used as guidelines to understand the difficulties we expect to see, and eventually to guide us in terms of investing time and effort in identifying key model parameters.

3.4 Conclusions

Hopefully we have been able to show that particle morphology is as important as it is difficult to define. The reasons for the challenges we face when trying to predict particle morphology, and the impact of process conditions there upon basically lays in the fact that there are many, many parameters that influence this property. Very rapid reactions lead to rapid particle expansion. Depending on how this energy is dissipated, we can see particles that disintegrate, particles with poorly controlled morphology, low or high bulk densities (porosities), etc. The way the energy is dissipated can depend on the initial form of the catalyst particles and the physical properties of the polymer itself. All of these complexities explain why, at the current time, there is still no reliable way to predict the evolution of particle morphology using a mathematical model.

A rapid glance at two of the more widely used methods for modelling particle growth (and kinetics, molecular weights and copolymer composition) underlines the importance that particle morphology can have in terms of determining the rates of matter and energy transport (and thus local concentrations and temperatures) inside the growing polymer particles. For instance, mass transfer in compact, low porosity particles will be much slower than in highly porous particles. This means that the low porosity particles would have lower concentrations of monomer (and probably lower local temperatures) than their more porous counterparts. This in turn implies that even if the two particles were chemically identical at the onset of their respective polymerisations, we could end up making very different polymers in each case.

In addition to the challenges posed by morphology and its evolution during the course of the reaction, one should not forget that there are also many poorly known quantities that can have a real impact on the way in which particles grow. We can think in particular of the complex thermodynamics of these systems, where the presence of heavy components such as pentane or hexane will have an impact on the reaction rate (not to mention the morphology) through swelling and co-solubility effects.

References

1. McKenna, T. F. L., Tioni, E., Ranieri, M. M., Alizadeh, A., Boisson, C., & Monteil, V. (2013). Catalytic olefin polymerisation at short times: Studies using specially adapted reactors. *Canadian Journal of Chemical Engineering, 91*, 669–686.
2. Nagel, E. J., Kirillov, V. A., & Ray, W. H. (1980). Prediction of molecular weight distributions for high-density polyolefins. *Industrial and Engineering Chemistry Product Research and Development, 19*, 372–379.
3. Tioni, E., Monteil, V., & McKenna, T. (2013). Morphological interpretation of the evolution of the thermal properties of polyethylene during the fragmentation of silica supported metallocene catalysts. *Macromolecules, 46*, 335–343.
4. McDaniel, M. P. (2010). In C. G. A. H. Bruce (Ed.), *Advances in catalysis*. Cambridge, MA: Academic Press.
5. Abboud, M., Denifl, P., & Reichert, K. H. (2005). Advantages of an emulsion-produced Ziegler-Natta catalyst over a conventional Ziegler-Natta catalyst. *Macromolecular Materials and Engineering, 290*, 1220–1226.
6. Abboud, M., Denifl, P., & Reichert, K. H. (2005). Study of the morphology and kinetics of novel Ziegler-Natta catalysts for propylene polymerization. *Journal of Applied Polymer Science, 98*, 2191–2200.
7. Bartke, M., Oksman, M., Mustonen, M., & Denifl, P. (2005). A new heterogenization technique for single-site polymerization catalysts. *Macromolecular Materials and Engineering, 290*, 250–255.

8. Kittilsen, P., Svendsen, H. F., & McKenna, T. F. (2003). Viscoelastic model for particle fragmentation in olefin polymerization. *AICHE Journal, 49*, 1495–1507.
9. Grof, Z., Kosek, J., & Marek, M. (2005). Principles of the morphogenesis of polyolefin particles. *Industrial and Engineering Chemistry Research, 44*, 2389–2404.
10. Grof, Z., osek, J., & arek, M. (2005). Modeling of morphogenesis of growing polyolefin particles. *AICHE Journal, 51*, 2048–2067.
11. Llinas, J. R., & Selo, J. L. (2010). *Method for reducing sheeting and agglomerates during olefin polymerisation*. US20030144432A1.
12. Di Martino, A., Weickert, G., Sidoroff, F., & McKenna, T. F. L. (2007). Modelling induced tension in a growing catalyst/polyolefin particle: A multi-scale approach for simplified morphology modelling. *Macromolecular Reaction Engineering, 1*, 338–352.
13. Yermakov, Y. I., Mikhalchenko, V. G., Beskov, V. S., Grabovskii, Y. P., & Emirova, I. V. (1970). The role of transfer processes in gaseous phase polymerization of ethylene. *Plasticheskie Massy, 9*, 7–10.
14. Laurence, R. L., & Chiovetta, M. G. (1983). In K. H. Reichert & W. Geisler (Eds.), *Polymer reaction engineering: Influence of reaction engineering on polymer properties*. Munich: Hanser.
15. McKenna, T. F., & Soares, J. B. P. (2001). Single particle modelling for olefin polymerization on supported catalysts: A review and proposals for future developments. *Chemical Engineering Science, 56*, 3931–3949.
16. Parasu Veera, U., Weickert, G., & Agarwal, U. S. (2002). Modeling monomer transport by convection during olefin polymerization. *AICHE Journal, 48*, 1062–1070.
17. Parasu Veera, U. (2003). Mass transport models for a single particle in gas-phase propylene polymerisation. *Chemical Engineering Science, 58*, 1765–1775.
18. Soares, J. B. P., & McKenna, T. F. L. (2012). *Polyolefin reaction engineering*. Weinheim: Wiley.
19. Jin, H. J., Kim, S., & Yoon, J. S. (2002). Solubility of 1-hexene in LLDPE synthesized by (2-MeInd)2ZrCl2/MAO and by Mg(OEt)2/DIBP/TiCl4-TEA. *Journal of Applied Polymer Science, 84*, 1566–1571.
20. Kiparissides, C., Dimos, V., Boultouka, T., Anastasiadis, A., & Chasiotis, A. (2003). Experimental and theoretical investigation of solubility and diffusion of ethylene in semicrystalline PE at elevated pressures and temperatures. *Journal of Applied Polymer Science, 87*, 953–966.
21. Maloney, D. P., & Prausnitz, J. M. (1976). Solubility of ethylene in liquid, low-density polyethylene at industrial-separation pressures. *Industrial and Engineering Chemistry Process Design and Development, 15*, 216–220.
22. Moore, S. J., & Wanke, S. E. (2001). Solubility of ethylene, 1-butene and 1-hexene in polyethylenes. *Chemical Engineering Science, 56*, 4121–4129.
23. Yoon, J. S., Yoo, H. S., & Kang, K. S. (1996). Solubility of a-olefins in linear low density polyethylenes. *European Polymer Journal, 32*, 1333–1336.
24. Bashir, M. A., Al-haj Ali, M., Kanellopoulos, V., & Seppala, J. (2013). Modelling of multicomponent olefins solubility in polyolefins using Sanchez-Lacombe equation of state. *Fluid Phase Equilibria, 358*, 83–90.
25. Yiagopoulos, A., Yiannoulakis, H., Dimos, V., & Kiparissides, C. (2001). Heat and mass transfer phenomena during the early growth of a catalyst particle in gas-phase olefin polymerization: The effect of prepolymerization temperature and time. *Chemical Engineering Science, 56*, 3979–3995.
26. Khare, N. P., Lucas, B., Seavey, K. C., Liu, Y. A., Sirohi, A., Ramanathan, S., Lingard, S., Song, Y., & Chen, C. C. (2004). Steady-state and dynamic modeling of gas-phase polypropylene processes using stirred-bed reactors. *Industrial and Engineering Chemistry Research, 43*, 884–900.
27. Novak, A., Bobak, M., Kosek, J., Banaszak, B. J., Lo, D., Widya, T., Harmon Ray, W., & de Pablo, J. J. (2006). Ethylene and 1-hexene sorption in LLDPE under typical gas-phase reactor conditions: Experiments. *Journal of Applied Polymer Science, 100*, 1124–1136.

28. Serna, L. V., Becker, J. L., Galdímez, J. R., Danner, R. P., & Duda, J. L. (2008). Elastic effects on solubility in semicrystalline polymers. *Journal of Applied Polymer Science, 107*, 138–146.
29. Bashir, M. A., Al-haj Ali, M., Kanellopoulos, V., Seppala, J., Kokko, E., & Vijay, S. (2013). The effect of pure component characteristic parameters on Sanchez-Lacombe equation-of-state predictive capabilities. *Macromolecular Reaction Engineering, 7*, 193–204.
30. Kanellopoulos, V., Dompazis, G., Gustafsson, B., & Kiparissides, C. (2004). Comprehensive analysis of single-particle growth in heterogeneous olefin polymerization: The random-pore polymeric flow model. *Industrial and Engineering Chemistry Research, 43*, 5166–5180.
31. Banaszak, B. J., Lo, D., Widya, T., Ray, W. H., de Pablo, J. J., Novak, A., & Kosek, J. (2004). Ethylene and 1-hexene sorption in LLDPE under typical gas phase reactor conditions: A priori simulation and modeling for prediction of experimental observations. *Macromolecules, 37*, 9139–9150.
32. Bashir, M. A., Ali, M. A., Kanellopoulos, V., & Seppala, J. (2015). Combined EoS and elastic constraints models to predict thermodynamic properties for systems involving semi-crystalline polyolefins. *Fluid Phase Equilibria, 388*, 107–117.
33. Michaels, A. S., & Hausslein, R. W. (1965). Elastic constraints in solvent swollen polyethylene. *Journal of Polymer Science, Part B: Polymer Letters, 3*, 61–62.
34. Yao, W., Hu, X., & Yang, Y. (2007). Modeling the solubility of ternary mixtures of ethylene, iso-pentane, n-hexane in semicrystalline polyethylene. *Journal of Applied Polymer Science, 104*, 3654–3662.
35. Desilets, M., Proulx, P., & Soucy, G. (1997). Modeling of multicomponent diffusion in high temperature flows. *International Journal of Heat and Mass Transfer, 40*, 4273–4278.
36. Cussler, E. L. (1997). *Diffusion mass transfer in fluid systems*. Cambridge, UK: Cambridge University Press.
37. Fujita, H. (1961). *Fortschritte Der Hochpolymeren-Forschung*. Berlin Heidelberg: Springer.
38. Michaels, A. S., & Bixler, H. J. (1961). Flow of gases through polyethylene. *Journal of Polymer Science, 50*, 413–439.
39. Michaels, A. S., & Bixler, H. J. (1961). Solubility of gases in polyethylene. *Journal of Polymer Science, 50*, 393–412.
40. Doong, S. J., & Ho, W. S. W. (1992). Diffusion of hydrocarbons in polyethylene. *Industrial and Engineering Chemistry Research, 31*, 1050–1060.
41. Vrentas, J. S., Duda, J. L., & Ling, H. C. (1985). Free-volume theories for self-diffusion in polymer GÇô solvent systems. I. Conceptual differences in theories. *Journal of Polymer Science Polymer Physics Edition, 23*, 275–288.
42. Vrentas, J. S., Duda, J. L., Ling, H. C., & Hou, A. C. (1985). Free-volume theories for self-diffusion in polymer-solvent systems. II. Predictive capabilities. *Journal of Polymer Science Polymer Physics Edition, 23*, 289–304.
43. Pace, R. J., & Datyner, A. (1979). Statistical mechanical model for diffusion of simple penetrants in polymers. I. Theory. *Journal of Polymer Science Polymer Physics Edition, 17*, 437–451.
44. Chiovetta, M. G., & Estenoz, D. A. (2004). Behavior of active sites in a changing, supported metallocene catalyst particle: Modeling monomer transport and kinetics. *Macromolecular Materials and Engineering, 289*, 1012–1026.
45. Estenoz, D. A., & Chiovetta, M. G. (2001). Olefin polymerization using supported metallocene catalysts: Process representation scheme and mathematical model. *Journal of Applied Polymer Science, 81*, 285–311.
46. Kittilsen, P., McKenna, T. F., Svendsen, H., Jakobsen, H. A., & Fredriksen, S. B. (2001). The interaction between mass transfer effects and morphology in heterogeneous olefin polymerization. *Chemical Engineering Science, 56*, 4015–4028.
47. Ferrero, M. A., & Chiovetta, M. G. (1991). Catalyst fragmentation during propylene polymerization. III: Bulk polymerization process simulation. *Polymer Engineering and Science, 31*, 886–903.

48. Ferrero, M. A., & Chiovetta, M. G. (1987). Catalyst fragmentation during propylene polymerization: Part II. Microparticle diffusion and reaction effects. *Polymer Engineering and Science, 27*, 1448–1460.
49. McKenna, T. F., Dupuy, J., & Spitz, R. (1995). Modeling of transfer phenomena on heterogeneous Ziegler catalysts: Differences between theory and experiment in olefin polymerization (an introduction). *Journal of Applied Polymer Science, 57*, 371–384.
50. Van Krevelen, D. W. (1997). *Properties of polymers*. Amsterdam: Elsevier.
51. Floyd, S., Choi, K. Y., Taylor, T. W., & Ray, W. H. (1986). Polymerization of olefins through heterogeneous catalysis IV. Modeling of heat and mass transfer resistance in the polymer particle boundary layer. *Journal of Applied Polymer Science, 31*, 2231–2265.
52. Constantinides, A., & Mostoufi, N. (1999). *Numerical methods for chemical engineers with MATLAB applications*. Upper Saddle River, NJ: Prentice Hall.
53. Villadsen, J., & Michelsen, L. (1978). *Solution of differential equation models by polynomial approximation*. Upper Saddle River, NJ: Prentice-Hall.
54. Galvan, R., & Tirrell, M. (1986). Orthogonal collocation applied to analysis of heterogeneous Ziegler-Natta polymerization. *Computers and Chemical Engineering, 10*, 77–85.
55. Floyd, S., Choi, K. Y., Taylor, T. W., & Ray, W. H. (1986). Polymerization of olefins through heterogeneous catalysis. III. Polymer particle modelling with an analysis of intraparticle heat and mass transfer effects. *Journal of Applied Polymer Science, 32*, 2935–2960.
56. Hutchinson, R. A., Chen, C. M., & Ray, W. H. (1992). Polymerization of olefins through heterogeneous catalysis X: Modeling of particle growth and morphology. *Journal of Applied Polymer Science, 44*, 1389–1414.
57. Ferrero, M. A., & Chiovetta, M. G. (1987). Catalyst fragmentation during propylene polymerization: Part I. The effects of grain size and structure. *Polymer Engineering and Science, 27*, 1436–1447.
58. Sarkar, P., & Gupta, S. K. (1991). Modelling of propylene polymerization in an isothermal slurry reactor. *Polymer, 32*, 2842–2852.
59. Nicolella, C., van Loosdrecht, M. C. M., & Heijnen, J. J. (1998). Mass transfer and reaction in a biofilm airlift suspension reactor. *Chemical Engineering Science, 53*, 2743–2753.
60. Beers, K. J. (2007). *Numerical methods for chemical engineering: Applications in MATLAB*. Cambridge, UK: Cambridge University Press.
61. Sarkar, P., & Gupta, S. K. (1992). Simulation of propylene polymerization: An efficient algorithm. *Polymer, 33*, 1477–1485.
62. Bhagwat, M. S., Bhagwat, S. S., & Sharma, M. M. (1994). Mathematical modeling of the slurry polymerization of ethylene: Gas-liquid mass transfer limitations. *Industrial and Engineering Chemistry Research, 33*, 2322–2330.
63. Schmeal, W. R., & Street, J. R. (1971). Polymerization in expanding catalyst particles. *AICHE Journal, 17*, 1188–1197.
64. Singh, D., & Merrill, R. P. (1971). Molecular weight distribution of polyethylene produced by Ziegler-Natta catalysts. *Macromolecules, 4*, 599–604.
65. Hoel, E. L., Cozewith, C., & Byrne, G. D. (1994). Effect of diffusion on heterogeneous ethylene propylene copolymerization. *AICHE Journal, 40*, 1669–1684.

Chapter 4
Polymerization Kinetics and the Effect of Reactor Residence Time on Polymer Microstructure

João B. P. Soares and Vasileios Touloupidis

*Everything should be made as simple as possible,
but not simpler.*

A. Einstein

4.1 Polyolefin Microstructure

Polymers have revolutionized our world. Today, it is difficult to conceive how everyday features of modern life could ever have been developed without this class of materials [1]. Academia and industry continuously innovate to cover new needs on commodity and specialized materials, helping to shape the future with novel applications of already existing or new polymeric materials.

Polymers are macromolecules built up by linking together a large numbers of much smaller molecules [2]. For polyolefins, these small molecules, or monomers, consist of simple alkenes (also called olefins, with the general formula C_nH_{2n}). Despite this deceiving simplicity in construction, polyolefins are rather complex, made up of a population of macro-chains exhibiting a wide range of intramolecular and intermolecular distributions (differences within the same macromolecule or among macromolecules, respectively).

This wide variety of possible microstructures within the polymer population affects the final mechanical, rheological and optical properties to the extent that it

J. B. P. Soares
University of Alberta, Edmonton, AB, Canada
e-mail: jsoares@ualberta.ca

V. Touloupidis (✉)
Borealis Polyolefine GmbH, InnoTech Process Technology/Modelling & Simulation, Linz, Austria
e-mail: vasileios.touloupidis@borealisgroup.com

A. R. Albunia et al. (eds.), *Multimodal Polymers with Supported Catalysts*,
https://doi.org/10.1007/978-3-030-03476-4_4

is impossible to talk about the properties of a polyolefin as a single material (polyethylene or polypropylene, for instance) but rather about the properties window of a class of polyolefins. One should also have in mind that the final resin placed in the market may also include a number of compounded additives that further affect its performance. Having a way to describe the microstructure of these polyolefin populations help us understand the deeper structure–property relationships that ultimately determine how a polyolefin will perform in a given application, which is an essential requirement for more efficient polymer design.

The polyolefin microstructure (also called molecular architecture) describes in detail the structure of a polyolefin at the molecular level. In this sense, the molecular architecture of polyethylene (PE) can be characterized by its:

- Molecular weight distribution (MWD),
- Chemical composition distribution (CCD) or short chain branching distribution (SCBD),
- Comonomer sequence length distribution (CSLD),
- Long-chain branching distribution (LCBD).

In addition, the microstructure of polypropylene (PP) is also defined by its region- and stereo-chemical distributions because of the apparent asymmetry of the propylene molecule.

The measurement of these fundamental microstructural features is the basis of polymer characterization. All secondary polymer macroscopic properties such as strength, tear resistance, viscosity, melt flow index, and density are a reflection of a given microstructure. For example, in the polyolefin industry it is typical to use melt flow index (MFI) for product characterization or as a target in product specification. MFI is the mass of polymer, in grams, flowing in 10 min through a capillary under specified conditions. MFI is inversely proportional to the weight average molecular weight (M_w) of the polymer, but it also depends on a series of other microstructural properties such as MWD, LCBD and, to a lesser extent, SCBD. MFI is not a fundamental microstructural feature, but rather a snapshot of polymer flow behaviour. Apart from the fact that MFI can be misleading because polymers with different MWDs may have the same MFI, flowability is not a fundamental property of a polymer but the result of its underlying microstructure. Similarly, polymer density is customarily used in product design and quality control, but density is not a fundamental property of polyolefins, but an emerging property that depends on the CCD, comonomer type, and MWD. Polyolefins with the same density may have different MWDs and CCDs and entirely distinct performance properties.

Major trends in the way microstructure can affect polymer properties are illustrated in Fig. 4.1.

4.1.1 Molecular Weight Distribution

Molecular weight is the most fundamental feature of a polymer, strongly affecting its mechanical and rheological properties [2, 3]. Because of the random kinetic

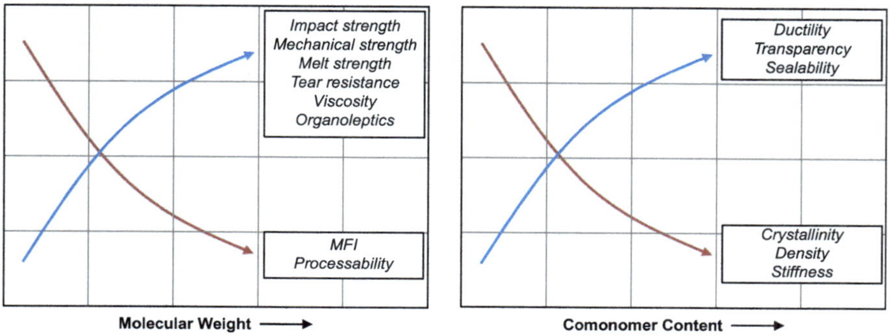

Fig. 4.1 Major effects of microstructure on end-use properties of polyolefins

mechanism involved in polymerization, we are only able to produce polyolefins that follow a molecular weight distribution with breadth that depend strongly on catalyst type and polymerization conditions.

High-temperature size exclusion chromatography (SEC), also known as high-temperature gel permeation chromatography (GPC), is typically used to measure the MWD of polyolefins. Field flow fractionation (FFF) is more suited for ultrahigh-molecular weight polyethylene (UHMWPE) resins because SEC typically exhibits an upper molecular weight limit of about 10^7 mol/g. Both SEC and FFF fractionate polymer molecules based on their size in solution [4], albeit relying on different separation mechanisms.

Typically, a MWD is characterized by its average molecular weight values and polydispersity index:

Number average molecular weight:

$$M_n = \frac{\sum N_x M_x}{\sum N_x} \tag{4.1}$$

Weight average molecular weight:

$$M_w = \frac{\sum N_x {M_x}^2}{\sum N_x M_x} \tag{4.2}$$

Polydispersity index:

$$\mathrm{PDI} = \frac{M_w}{M_n} \tag{4.3}$$

where N_x is the number of polymer chains with a molecular weight of M_x.

One should always have in mind, however, that even though these averages offer a convenient way to quantify the MWD of a polyolefin using only a few values, they

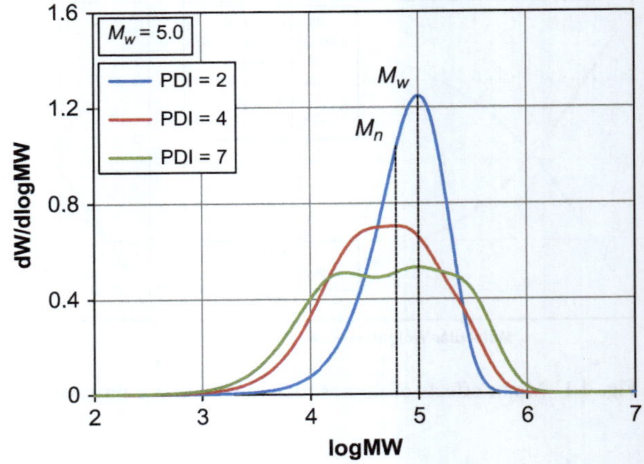

Fig. 4.2 Polymers with different MWDs may have the same M_w

cannot fully describe it. For example, polymers with different MWDs may have the same M_w. In Fig. 4.2, three polymers having different MWDs share the same M_w. The Mn value indicated in Fig. 4.2 corresponds only to the MWD of PDI = 2 and will significantly vary for the higher polydispersity value MWDs. As can be easily inferred, despite having the same M_w, these three polymers will exhibit different end-use properties.

Single-Site Versus Multi-Site Catalysts
The MWD of polyolefins made with single-site catalysts (under steady-state conditions in a well-mixed reactor) follow a Schulz–Flory most probable distribution with polydispersity index value equal to 2 [5]. In contrast, heterogeneous Ziegler–Natta (Z–N) and Phillips catalysts make polymer with broad MWDs and polydispersity indices which may typically range from 3 to 10, or even up to 20, depending on the catalyst type. These catalysts make polymers with broad MWD because they have more than one site type; intraparticle and interparticle mass- and heat-transfer resistances during polymerization may also contribute to this broadening but are not the dominant mechanism [6–8].

It is well accepted that Z–N catalysts have a limited number of different catalyst site types (typically up to six different site types), albeit their chemical nature is still a matter of debate among researchers. Each site type has a different set of polymerization kinetics parameters, producing polymer populations with different microstructural distributions. In this sense, a multi-site catalyst can be seen as a mixture of a number of different single-site catalysts, each producing polymer populations of different molecular weight averages, chemical compositions and comonomer sequence lengths. The overall microstructural distributions (MWD, CCD and CSLD) are then calculated as the weighted sum of the distributed properties produced by different site types. The number and fraction of active site types that are present in a catalyst, as well as the kinetic parameters that describe each site type, is unique for each specific catalyst system and polymerization conditions.

4.1.2 Chemical Composition Distribution

In the case of copolymers, additional microstructure information is needed to describe the fraction of comonomer in the polymer. As comonomer incorporation is influenced by the statistical nature of polymerization, even polyolefins made with single-site catalysts have a chemical composition distribution (CCD) rather than a single comonomer fraction for all molecules. The knowledge of the CCD is essential to uniquely characterize the microstructure of an olefin copolymer.

There are many experimental techniques for characterizing the CCD (or tacticity distribution for the case of PP grades), including temperature rising elution fractionation (TREF), crystallization analysis fractionation (CRYSTAF), crystallization elution fractionation (CEF) and differential scanning calorimetry (DSC). These techniques are all based on the same premise, that the crystallizability of polyolefin chains is a function of their 1-olefin comonomer content. Short chain branches (SCB) produced by comonomer addition act as defects, hindering crystallization and lowering polymer density. Calibration curves can then be generated relating the crystallization/elution/melting temperatures of polyolefins to their comonomer fractions. Molecular weight may also affect the kinetics of crystallization to some extend (typically, higher molecular weight chains have lower crystallizabilities due to slower crystallization kinetics), while co-crystallization effects need to be taken into account when interpreting experimental measurements of these crystallization based techniques used to measure CCD.

More recently, high-temperature solvent gradient interaction chromatography (HT-SGIC) and thermal gradient interaction chromatography (HT-TGIC) have been shown to be excellent alternatives to the crystallization-based techniques described above [9–12]. Both techniques rely on adsorption of polyolefin chains onto a support such as porous graphite. Since the fractionation is not based on chain crystallization, a wider range of olefin comonomer contents can be investigated using HT-TGIC or HT-SGIC. In addition, co-crystallization effects that may affect the resolution of TREF and other related techniques are not a limitation in HT-TGIC or HT-SGIC.

Orthogonal two-dimensional fractionation techniques are even more powerful methods to fully characterize the microstructure of polyolefins. Typically, the first dimension involves the fractionation of polymer chains according to their CCD. TREF, CEF, HT-TGIC, or HT-SGIC can be employed at this stage. In the second step, narrow-CCD fractions are transferred to a GPC unit to measure their MWDs [13–15]. This allows the measurement of the full bivariate molecular weight and chemical composition distribution of a polyolefin copolymer (MWD-CCD), undoubtedly the most detailed map of polyolefin microstructure available today [16]. Figure 4.3 shows the joint MWD-CCD of a polyolefin copolymer. It is also possible to fractionate polyolefins first according to their MWD and then measure the CCD of the narrow MWD fractions. This approach is more adequate for samples with broad MWD and relatively narrow CCD, such as bimodal HDPE pipe resins.

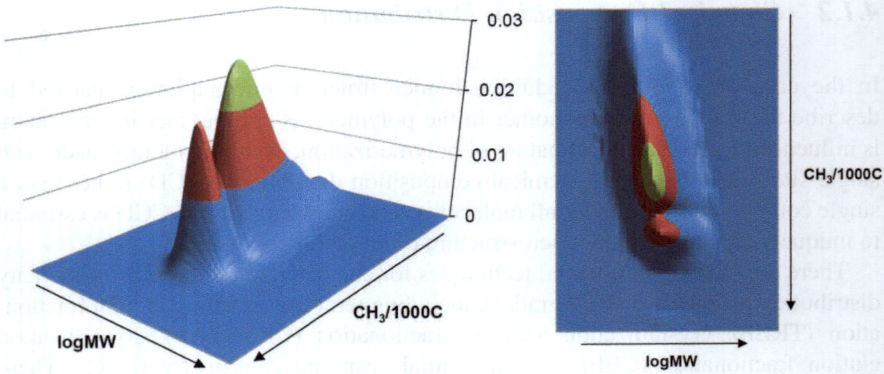

Fig. 4.3 Joint MWD-CCD of a polyolefin copolymer

The use of composition detectors coupled with MW fractionation techniques (SEC and FFF) allows one to measure copolymer compositions across the MWD. For example, an on line infrared detector (IR) coupled to SEC can measure the average copolymer composition across the MWD [4]. Based on this information, one can then attempt to calculate the CCD based on the monomer and comonomer species in use. Alternatively, it is also possible to couple a MW-sensitive detector (such as a viscometer or light scattering detector) with a composition fractionation technique. For instance, when TREF or CEF are combined with a light scattering detector, it is possible to measure how MW varies as a function of the CCD of a polyolefin. Both approaches are simple, but very powerful, ways to extend the amount of microstructural information measured by MWD- or CCD-fractionation techniques.

Single-site catalysts, contrary to their multi-site counterparts, produce (under steady-state conditions in well-mixed reactors that are not subject to significant mass and/or heat transfer resistances) polymers exhibiting uniform CCDs, that is, a CCD that follows Stockmayer distribution. Considering multi-site catalysts as a mixture of different site types, each one of them behaving as a single-site catalyst, the overall MWD is calculated as the weighted sum of MWDs produced per site type. Similarly, a non-uniform CCD can be modelled as the weighted sum of uniform CCDs made by each different site type. This behaviour is illustrated in Fig. 4.4. The distribution on the left of the left-hand side illustrates the SEC-IR of a single-site copolymer with a uniform comonomer incorporation of 20 SCB/1000 C atoms. The plot on the right-hand side shows that the comonomer distribution of the polymer made in a multi-site catalyst varies from high to low frequency as the molecular weight increases, resulting from the superposition of four different polymer populations with decreasing uniform SCB frequencies.

The density of a polyolefin made with a single-site catalyst is correlated to its average comonomer content but it is also affected by its MWD. No simple density-comonomer content relations are applicable for olefin copolymers made with multi-site catalysts because of their broad, and often bimodal, CCDs.

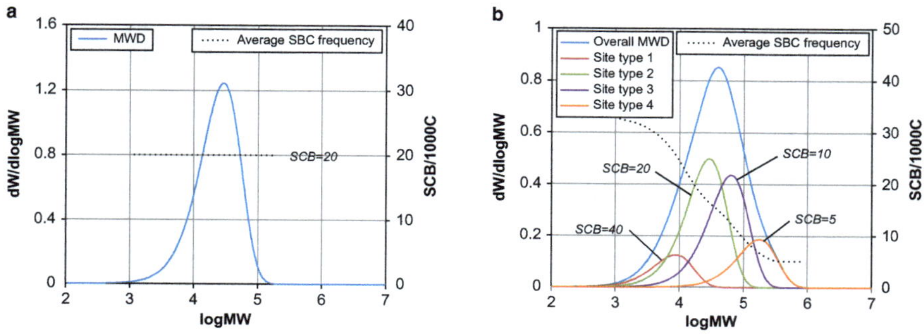

Fig. 4.4 Effect of single- and multi-site catalyst on the average comonomer content across the MWD, as measured by SEC-IR

4.1.3 Chain Sequence Length Distribution

The CCD provides information regarding the intermolecular comonomer distribution. However, at the intramolecular level, there are many options on the way comonomer molecules are incorporated to chains, leading to block, alternating or random copolymers. In Fig. 4.5, different CSLD configurations for the same branching frequency are presented. Carbon-13 nuclear magnetic resonance (^{13}C NMR) is the method of choice for determining CSLD. This technique is able to quantify dyad, triad, tetra, and longer sequence distributions along the chains, offering a deeper insight into polymer microstructure. It should be noted that the CSLD is affected primarily by intrinsic catalyst characteristics (mainly reactivity ratios) as well as by reaction conditions (monomer concentrations, polymerization temperature, etc.). It is also important to note that ^{13}C NMR gives only average values of the CSLD for the whole polymer. This should be taken into consideration when interpreting ^{13}C NMR of polymers made with multi-site catalysts [17–19].

A journey of a thousand miles begins with a single step.
Lao Tse

4.2 Reaction Kinetics

4.2.1 Multi-Scale Approach

The key to polymer design is the ability to control chain microstructure [20]. The most important objective of polymerization kinetics investigations is to explain, describe and predict polymer microstructure development throughout the polymerization.

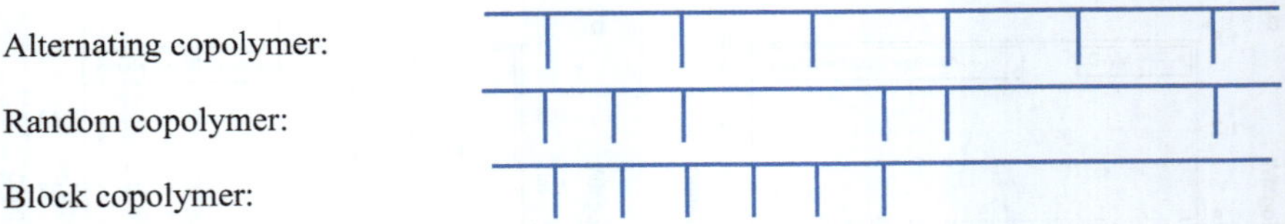

Fig. 4.5 CSLD arrangements for the same branching frequency at the same chain length

Polymerization processes include a number of complex physical and chemical phenomena taking place simultaneously in the reactor. Typically, coordination catalytic polymerizations include the following steps:

- Introduction of monomer(s) into the reaction medium.
- Diffusion and/or solubilization of monomers into the reaction medium.
- Monomer sorption onto the polymer/catalyst particles.
- Diffusion through catalyst particle pores and/or amorphous polymer phase.
- Polymerization at a catalyst active site.

A multi-scale approach can be applied [21, 22] to model different phenomena taking place during the polymerization according to their different time/length scales:

1. Micro-scale: Elementary or apparent polymerization kinetics.
2. Meso-scale: Thermodynamic equilibrium, micro-mixing, sorption phenomena, interparticle and intraparticle mass and heat transfer phenomena, and particle morphology.
3. Macro-scale: Macro-mixing, overall mass and energy balances, particle population balances, reactor residence time distribution effects, reactor dynamics and control.

Phenomena taking place in different scales are studied and modelled separately, while relevant information is transferred to models at other scales. The aim of macro- and meso-scale modelling is to follow the evolution of species concentrations from the bulk reactor phase to the active sites embedded in the polymer phase, where the polymerization takes place. In this way, polymerization kinetics studies can be treated independently from phenomena taking place at other scales (such as reactor performance and system thermodynamics), as catalyst active sites can only 'see' the local conditions and concentrations.

4.2.2 *Polymerization Kinetic Scheme*

The use of solid-supported catalysts is essential for olefin polymerization in slurry and gas reactors. The only exceptions are free-radical polymerization taking place at high temperatures and pressures, and solution polymerizations with soluble coordination catalysts [23]. Ziegler–Natta, Phillips-chromium, and supported metallocene catalysts are commercially used in slurry and gas-phase polymerizations. Table 4.1 shows a generic olefin copolymerization kinetic scheme for a multi-site catalyst [24–26]. The selected kinetic scheme refers to Z–N olefin polymerization, but it can also be applied to metallocene or chromium-based catalysts. The exponent k refers to reactions taking place in different site types ($k = 1,2...N_s$; N_s = number of site types. For single-site catalysts $N_s = 1$). The symbol $P_{n,i}^k$ denotes the concentration of live copolymer chains of length n ending in monomer unit of type i, produced at the kth catalyst site type. The reaction rate kinetic constants, k_r, of all elementary reactions, follow Arrhenius law:

$$k_r = K_0 e^{-E/RT} \tag{4.4}$$

where K_0 and E are the pre-exponential constant and activation energy, respectively, and T is absolute temperature.

Elementary reactions can be clustered into the following categories:

1. *Activation of potential sites*

   This reaction converts inactive potential catalyst sites, S_p, to reactive vacant sites, P_0^k. Activation occurs by reaction of a potential site (a chlorinated transition metal) with an alkylaluminum or aluminoxane cocatalyst. Activation by hydrogen, monomer or spontaneous activation has also been used to model some aspects of these polymerizations, but this is less common.
2. *Chain initiation*

   This is the first step towards the building of a polymer chain. A monomer molecule, M_i, reacts with an activated catalyst site, P_0^k, producing an active site occupied by a single monomer, $P_{1,i}^k$.
3. *Chain propagation*

   This is to the main polymerization reaction step. The active polymer chain of length n, $P_{n,i}^k$ attached to the catalyst site k, reacts with an additional monomer molecule, M_i, increasing its chain length by one, $P_{n+1,i}^k$. For the case of copolymerization, it is essential to distinguish between all possible reactions between the active polymer chain ending in monomer i, $P_{n,i}^k$, and monomer type M_j, $P_{n,j}^k$. Reactivity ratios are used to quantify this reaction selectivity towards different monomer types:

$$r_1 = \frac{k_{p11}^k}{k_{p12}^k} \tag{4.5}$$

$$r_2 = \frac{k_{p22}^k}{k_{p21}^k} \tag{4.6}$$

   Reactivity ratios determine the characteristics of the CCD and CLSD development for a given set of polymerization conditions.

Table 4.1 Generic catalytic olefin polymerization kinetic scheme

Site activation		Site transformation	
By hydrogen:	$S_p^k + H_2 \overset{k_{aH}^k}{\rightarrow} P_0^k$	To hydrogen:	$P_{n,i}^k + H_2 \overset{k_{tH}^{kl}}{\rightarrow} P_0^l + D_n^k$
By co-catalyst:	$S_p^k + A \overset{k_{aA}^k}{\rightarrow} P_0^k + B$		$P_0^k + H_2 \overset{k_{tH}^{kl}}{\rightarrow} P_0^l$
By monomer *i*:	$S_p^k + M_i \overset{k_{aMi}^k}{\rightarrow} P_0^k + M_i$	To co-catalyst:	$P_{n,i}^k + A \overset{k_{tA}^{kl}}{\rightarrow} P_0^l + D_n^k$
Spontaneous:	$S_p^k \overset{k_{aSP}^k}{\rightarrow} P_0^k$		$P_0^k + A \overset{k_{tA}^{kl}}{\rightarrow} P_0^l$
Chain initiation		To solvent:	$P_{n,i}^k + S \overset{k_{tS}^{kl}}{\rightarrow} P_0^l + D_n^k$
By monomer *i*:	$P_0^k + M_i \overset{k_{0i}^k}{\rightarrow} P_{1,i}^k$		$P_0^k + S \overset{k_{tS}^{kl}}{\rightarrow} P_0^l$
Chain propagation		To poison:	$P_{n,i}^k + X \overset{k_{tX}^{kl}}{\rightarrow} P_0^l + D_n^k$
By monomer 1:	$P_{n,1}^k + M_1 \overset{k_{p11}^k}{\rightarrow} P_{n+1,1}^k$		$P_0^k + X \overset{k_{tX}^{kl}}{\rightarrow} P_0^l$
	$P_{n,2}^k + M_1 \overset{k_{p21}^k}{\rightarrow} P_{n+1,1}^k$	Spontaneous:	$P_{n,i}^k \overset{k_{tSP}^{kl}}{\rightarrow} P_0^l + D_n^k$
By monomer 2:	$P_{n,1}^k + M_2 \overset{k_{p12}^k}{\rightarrow} P_{n+1,2}^k$		$P_0^k \overset{k_{tSP}^{kl}}{\rightarrow} P_0^l$
	$P_{n,2}^k + M_2 \overset{k_{p22}^k}{\rightarrow} P_{n+1,2}^k$	*Site deactivation*	
Chain transfer		By hydrogen:	$P_{n,i}^k + H_2 \overset{k_{dH}^k}{\rightarrow} C_D^k + D_n^k$
To hydrogen:	$P_{n,i}^k + H_2 \overset{k_{trH}^k}{\rightarrow} P_0^k + D_n^k$		$P_0^k + H_2 \overset{k_{dH}^k}{\rightarrow} C_D^k$
To co-catalyst:	$P_{n,i}^k + A \overset{k_{trA}^k}{\rightarrow} P_0^k + D_n^k$	By co-catalyst:	$P_{n,i}^k + A \overset{k_{dA}^k}{\rightarrow} C_D^k + D_n^k$
To solvent:	$P_{n,i}^k + S \overset{k_{trS}^k}{\rightarrow} P_0^k + D_n^k$		$P_0^k + A \overset{k_{dA}^k}{\rightarrow} C_D^k$
To monomer *i*:	$P_{n,i}^k + M_i \overset{k_{trMi}^k}{\rightarrow} P_{1,i}^k + D_n^k$	By by-product:	$P_{n,i}^k + B \overset{k_{dB}^k}{\rightarrow} C_D^k + D_n^k$
Spontaneous:	$P_{n,i}^k \overset{k_{trSP}^k}{\rightarrow} P_0^k + D_n^k$		$P_0^k + B \overset{k_{dB}^k}{\rightarrow} C_D^k$
		By monomer *j*:	$P_{n,i}^k + M_j \overset{k_{dMij}^k}{\rightarrow} C_D^k + D_n^k$
			$P_0^k + M_j \overset{k_{dMj}^k}{\rightarrow} C_D^k$
		Spontaneous:	$P_{n,i}^k \overset{k_{dSP}^k}{\rightarrow} C_D^k + D_n^k$
			$P_0^k \overset{k_{dSP}^k}{\rightarrow} C_D^k$

4. *Chain transfer*

These reactions terminate the growth of live polymer chains, $P_{n,i}^k$, forming dead polymer chains, D_n^k, and vacant active sites, P_0^k. Chain transfer can be promoted by a transfer agent (hydrogen is widely used for controlling chain length in industry), other small molecules present in the reactor, including the cocatalyst or spontaneously (β-hydride or β-methyl elimination, for instance).

5. *Active site transformation*

   This class of reaction converts one catalyst site type into another type, at the same time promoting a chain transfer reaction. Active site transformation reactions are useful modelling tools, but their existence is controversial.

6. *Site deactivation*

   This reaction step accounts for catalyst activity loss during the polymerization. Deactivation reactions form dead sites, C_D^k and dead polymer chains, D_n^k. Typically, spontaneous deactivation predominates, but deactivation by other means, such as poisoning reactions with polar impurities present in the reactor, is also possible. Some modelling approaches separate deactivation and poisoning reactions into two different reaction classes.

The polymerization kinetics mechanism shown in Table 4.1 captures the essential features but it is not the complete description of all the elementary reaction steps taking place during the polymerization of olefins. Some other reactions, may take place when olefins are polzmerized with coordination catalysts, including cocatalyst/catalyst ratio influence on catalyst activity, cocatalyst/catalyst pre-contact time effects, poison scavenger effects (including excess of cocatalyst in the reaction medium), hydrogen concentration effects on polymerization rate, comonomer rate effect enhancement, or additional reactions leading to short chain branching such as chain walking with late transition metal catalysts. Finally, donor effects on polymerization activity and polymer microstructure should always be taken into consideration in polypropylene polymerization cases [27]. In view of these exceptions, it is fair to ask the question whether the reactions listed in Table 4.1 are sufficient to allow for the description of olefin polymerization. We believe they are, since they describe most elementary steps taking place during olefin catalytic polymerization and set the framework for corresponding reaction engineering studies for predicting polymer microstructure [28].

In practice, simplified kinetic schemes (much simpler than that shown in Table 4.1) are used to reduce the number of kinetic constants that need to be estimated. The accurate estimation of kinetic parameters is a fundamental step in process modeling. Even though only apparent kinetic constants can be obtained for the vast majority of olefin polymerization catalysts because of the complexity involved in these processes, they are still very useful when designing or optimizing polymerization reactors or polymer microstructures.

After deciding which simplified polymerization kinetic scheme to use, the tuning parameters typically include:

- The number and corresponding fraction of active sites in the catalyst under investigation (for multi-site catalysts).
- The kinetic parameter values for every elementary reaction per site type.

Several mathematical techniques can be used to model olefin polymerization (such as population balances, the method of moments, Monte Carlo simulations and instantaneous analytical solutions). The statistical method of moments is typically the method of choice, by-passing the numerical difficulties associated with solving the large number of differential equations resulting from the full solution of population balances.

The statistical method of moments uses moment balances to predict average molecular properties such as number- and weight average molecular weights. The equations below are the definitions of live (λ_ν^k), dead (μ_ν^k), and bulk (ξ_ν^k) moments of the chain length distribution of a polymer population:

$$\text{Live moments}: \lambda_\nu^k = \sum_{i=1}^{N_m} \lambda_{\nu,i}^k = \sum_{i=1}^{N_m} \sum_{n=1}^{\infty} n^\nu \left[P_{n,i}^k\right];\ \nu = 0, 1, 2, \ldots \tag{4.7}$$

$$\text{Dead moments}: \mu_\nu^k = \sum_{n=2}^{\infty} n^\nu \left[D_n^k\right];\ \nu = 0, 1, 2, \ldots \tag{4.8}$$

$$\text{Bulk moments}: \xi_\nu^k = \lambda_\nu^k + \mu_\nu^k;\ \nu = 0, 1, 2, \ldots \tag{4.9}$$

where $\lambda_{\nu,i}^k$ and μ_ν^k are the νth moment of live chains ending in an $i(=1,2,\ldots N_m)$ monomer unit and the ν moment of dead chains, produced at the kth catalyst site, respectively. ξ_ν^k is the respective ν^{th} bulk moment [29–31].

Based on the above moment equations, average molecular properties can be calculated:

Cumulative copolymer composition:

$$\Phi_i = \sum_{k=1}^{N_s} \xi_{1,i}^k \Big/ \sum_{k=1}^{N_s} \xi_1^k \tag{4.10}$$

Number-average molecular weight:

$$M_n = \left(\sum_{k=1}^{N_s} \xi_1^k \Big/ \sum_{k=1}^{N_s} \xi_0^k \right) \overline{MW} = (\xi_1/\xi_0)\overline{MW} = DP_n\overline{MW} \tag{4.11}$$

Weight-average molecular weight:

$$M_w = \left(\sum_{k=1}^{N_s} \xi_2^k \Big/ \sum_{k=1}^{N_s} \xi_1^k \right) \overline{MW} = (\xi_2/\xi_1)\overline{MW} = DP_w\overline{MW} \tag{4.12}$$

where $\overline{MW}$ is the average molecular weight of the repeating structural unit in the copolymer chains, that is:

$$\overline{MW} = \sum_{i=1}^{N_m} \Phi_i MW_i \tag{4.13}$$

and MW_i is the molecular weight of monomer i, M_i.

The reconstruction of the weight chain length distribution of the polyolefin, produced at the kth catalyst site, can be achieved by using the two-parameter Schulz–Flory distribution:

$$W^k(x) = \frac{y^k \left(xy^k\right)^{z^k} e^{-xy^k}}{e^{\ln\left[\Gamma(z^k+1)\right]}} \; ; \; k = 1, 2, \ldots N_s \tag{4.14}$$

where $W^k(x)$ is the mass fraction of polymer chains with a degree of polymerization x produced at the kth catalyst site. Note that the Schulz–Flory parameters can be expressed in terms of the number- and weight-degree of polymerization or in terms of 'bulk' moments:

$$z^k = \frac{DP_n^k}{DP_w^k - DP_n^k} = \frac{\xi_1^k/\xi_0^k}{\xi_2^k/\xi_1^k - \xi_1^k/\xi_0^k} \; ; \; y^k = \frac{z^k+1}{DP_w^k} = \frac{z^k+1}{\xi_2^k/\xi_1^k} \tag{4.15}$$

Thus, for a multi-site catalyst, the overall MWD will be given by the weighted sum of the MWDs of all polymer fractions produced over the different catalyst active sites:

$$W_t(x) = \sum_{k=1}^{N_s} f_w^k W^k(x) = \sum_{k=1}^{N_s} \left(\xi_1^k/\xi_1\right) W^k(x) \tag{4.16}$$

where f_w^k is the mass fraction of polymer produced by catalyst site type k, while, ξ_1 is the total polymer mass produced over the N_s catalyst active sites $\left(\xi_1 = \sum_{k=1}^{N_s} \xi_1^k\right)$.

The MWD can also be reported in log scale, in the way SEC reports it, by multiplying the distribution by the factor $\frac{1}{\log e} = 2.3026$:

$$W_{t,GPC}(x) = 2.3026 \; W_t(x) = 2.3026 \sum_{k=1}^{N_s} f_w^k W^k(x) \tag{4.17}$$

4.2.3 Polymer Design

Single-Stage Processes

Following the multi-scale modelling approach described in the previous paragraphs, we can understand how reaction polymerization kinetics, reactor performance and system thermodynamics affect polyolefin microstructure and specify the degrees of freedom offered for polymer design:

- Catalyst design,
- Operating conditions, and,
- Process design.

However, this operating window is usually narrowed down by intellectual property rights and licenses, safety instructions and catalyst system limitations. For a given process and catalyst system, the simplest means we have in hand to affect MWD (connected to target rheological properties and melt flow index) is varying the concentration of transfer agent (typically H_2) in the polymerization reactor, while polymer density is manipulated by comonomer inflow.

Effect of Hydrogen

Hydrogen acts as a transfer agent by controlling chain length averages of the produced polymer, shifting the corresponding MWD to lower molecular weights (Fig. 4.6). For single-site catalysts, the polydispersity should be relatively insensitive to hydrogen concentration (remaining approximately 2.0), but it may change to lower or higher values when multi-site catalysts are used. When hydrogen is not also acting as activation or deactivation agent (depending on the catalyst system and monomer type in use), the polymerization rate should not be affected by hydrogen inflow rate.

Effect of Comonomer

Polymer density is directly controlled through comonomer inflow to the reactor. Thus, higher comonomer concentrations will lead to higher comonomer fractions in the polymer produced (see Fig. 4.6). Comonomer inflow may also affect polymer production rate, either lowering or increasing it, depending on monomer/comonomer rates and type of catalyst used in the polymerization. In addition, most catalysts are more sensitive to transfer to 1-olefin comonomer than to ethylene. Therefore, as the concentration of 1-olefin increases in the reactor, one expects the molecular weight averages to decrease slightly, both for polymers made with single- and multi-site catalysts.

Comonomer inflow may also trigger some thermodynamic effects independently from reaction kinetic developments. Comonomer co-solubility effects may affect the concentration of monomer available at the active site and, consequently, polymerization rate. Moreover, comonomer addition will lower polymer crystallinity by increasing the fraction of the amorphous polymer domains. As the crystallites act as a barrier for monomer diffusion (in contrast to the amorphous domains), comonomer addition may also lead to increased production rate connected to more effective monomer diffusion towards the active catalyst centers [32].

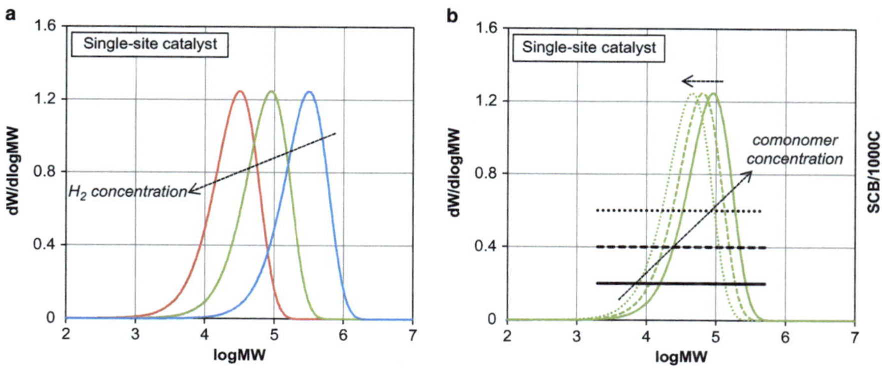

Fig. 4.6 Effect of hydrogen and comonomer concentrations or reactor inflows on the produced MWD and average SCB frequency

Multi-Stage Processes

Multi-Stage processes offer an additional degree of freedom in our efforts to reach desired polymer properties. A second or third reactor in series can be used to make polyolefins with tailored and multimodal microstructures. In what follows, we will show examples on how to tailor the polymer microstructure using two reactors in series. For the sake of simplicity, all examples refer to single-site catalyst polymerization.

A second reactor in series can dramatically increase polymer polydispersity values of the final product. Figure 4.7 shows the effect of H_2 inflow to the second reactor. Decreasing H_2 inflow increases the molecular weight of the polymer produced in the second reactor, raising the polydispersity for the overall polymer. In this way, we can overcome the restriction of having polydispersity of two when using a single-site catalyst. Moreover, some limited CCD changes may also result from this operation.

The ratio between the polymer yield in the first and second reactors (also called the split value) has a significant effect on the polymer properties. Figure 4.8 illustrates the effect of the split value on the overall MWD. By sliding the split value from 40:60 to 60:40 we can drastically affect the overall MWD shape for the polymer and, consequently, its application properties.

Using a second reactor can also drastically affect the shape of the overall CCD. By manipulating the comonomer concentration in each reactor it is possible to change the shape of the overall copolymer CCD. Figure 4.9 compares three different scenarios: (1) both reactors exhibit the same average SCB value, (2) the first reactor exhibits higher SCB value, and, (3) the second reactor exhibits higher SCB value. Three different CCD trends are generated in the final polymer by affecting the comonomer addition per reactor. Usually, the third microstructure is favored for HDPE pipe applications.

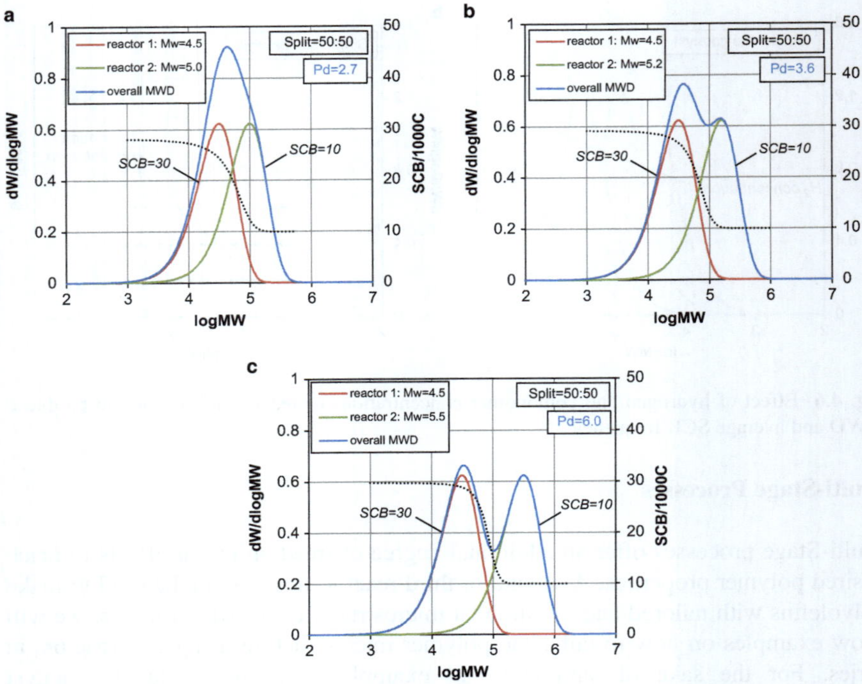

Fig. 4.7 Effect of a decrease of H_2 inflow to the second reactor on the overall MWD and average SCB frequency

4.2.4 Microstructural Deconvolution Techniques

Catalysts are the heart of olefin polymerization processes. From a polymer reaction engineering perspective, they are characterized by their number of site types and polymerization kinetic parameters per site type. These values affect polymerization yield, polymer microstructure, particle morphology, and particle size distribution. Unfortunately, these values cannot be directly measured but only estimated based on experimental measurements of the polymer produced. MWD deconvolution is a standard analysis tool used to extract polymerization kinetic/catalyst information from experimental data. It can estimate the mass fraction of polymer made on each site type, reactivity ratios and propagation/transfer ratios per site type. It is also a simple method to estimate the polymerization kinetic rate constant values for each site type in the catalyst. Despite of being a powerful methodology, the interpretation of MWD deconvolution results should be done with care, as there are many requirements needed to obtain meaningful results [4].

The MWD of the polymer produced over the total number of different catalyst site types, is given by the weighted sum of all polymer fractions produced (based on the polymer produced by each site type) over the different catalyst active site types

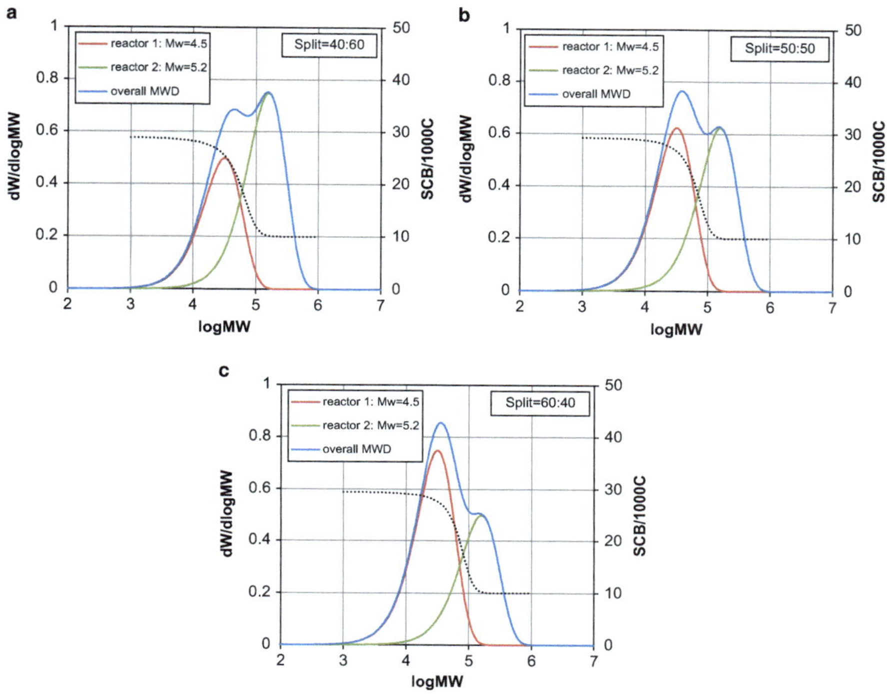

Fig. 4.8 Effect of split value on the overall MWD

(see Eqs. 4.14–4.17). The parameters involved are the number of different catalyst sites present (N_s) and its corresponding mass fraction of polymer made on each site type (f_w^k), along with the two parameters describing each Schulz–Flory distribution (i.e., y^k and z^k, or DP_n and DP_w).

Since each individual site type distribution exhibits polydispersity index equal to $2\left(\mathrm{PDI} = \frac{M_w}{M_n} = \frac{DP_w}{DP_n} = 2\right)$, Schulz–Flory distribution (Eqs. 4.14 and 4.15) is reduced to Flory's one-parameter distribution:

$$W^k(x) = x\left(\tau^k\right)^2 e^{-x\tau^k}\ ;\ k = 1, 2, \ldots N_s \tag{4.18}$$

$$\tau^k = \frac{1}{DP_n^k} \tag{4.19}$$

For a given number of different catalyst site types, N_s, an optimization procedure can be followed so that changing the two remaining parameters, that is, the mass fraction of polymer produced by catalyst site type k, f_w^k, and the distribution parameter, DP_n^k, over the different site types, the resulting overall MWD can best reconstruct the experimentally measured MWD (i.e., SEC curve). These two

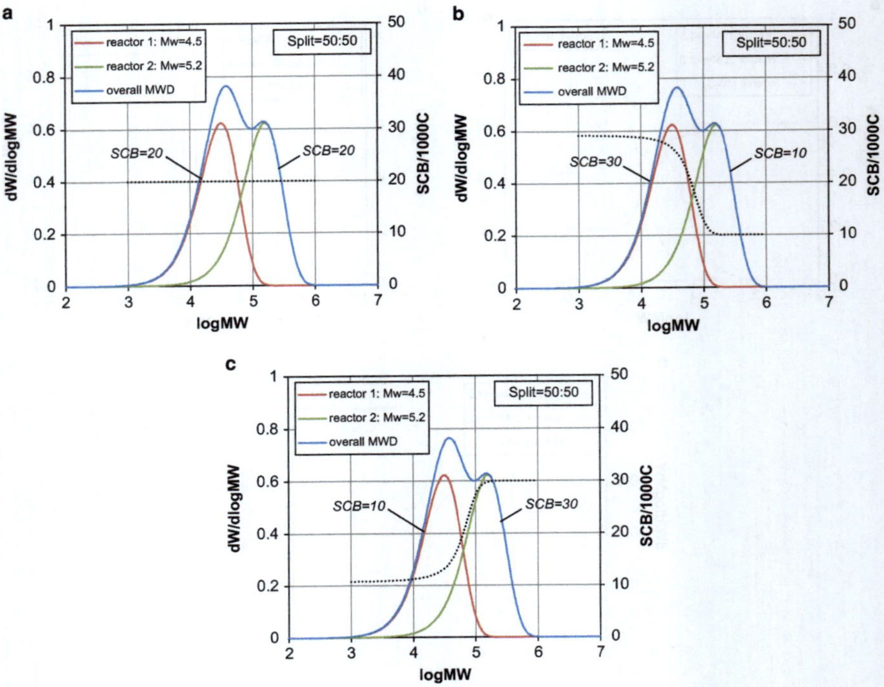

Fig. 4.9 Effect of reactor comonomer fraction on the overall CCD

parameters provide two degrees of freedom enabling the individual MDWs to be moved in both dimensions (i.e., f_w^k increases/decreases the distributions height, DP_n^k shifts the distributions over different chain lengths, see Fig. 4.10). Thus, the optimum values regarding mass fraction and degree of polymerization of the distribution coming from each catalyst site type can be easily estimated via non-linear regression.

MWD deconvolution aims at minimizing the sum of least squares between the MWD computed by Eqs. 4.14–4.17 and the measured distribution, in order to determine the adjustable parameters.

$$\text{Minimization target} = \sum_{x=1}^{N} \left(W(x)_{t,\,\text{measured}} - W(x)_{t,\,\text{estimated}} \right)^2 \qquad (4.20)$$

The constraint that has to be always valid is that the sum of mass fraction equals to 1:

$$\sum_{k=1}^{N_s} f_w^k = 1 \qquad (4.21)$$

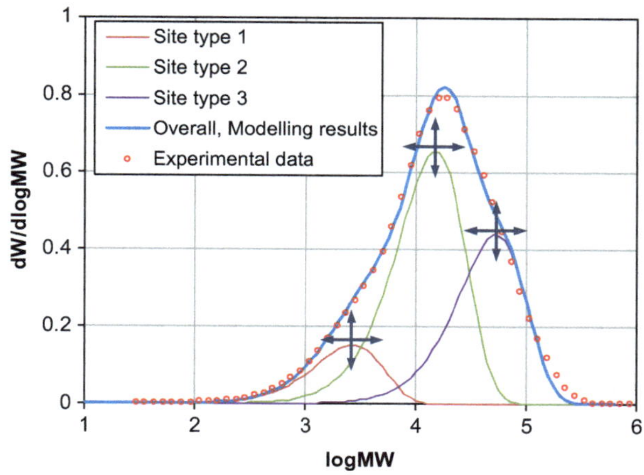

Fig. 4.10 Degrees of freedom during deconvolution

Repeating the procedure for a number of different catalyst site types, *k*, types from 1–6 (typically maximum value of different site types) the optimum parameter values for f_w^k and, DP_n^k, are estimated so that the minimization target approaches its minimum value. The optimization calculations can be performed in any programming language (Fortran, Matlab, C, etc.) or spreadsheets, using a variety of available optimization routines.

One catalyst site type is not enough to reconstruct a wide MWD, as it is limited to a polydispersity value equal to 2. By adding more catalyst site types, the value of the minimization target given in Eq. (4.20) will decrease up to a minimum limit. This is the point where we have to decide, based on a coherent criteria, whether the addition of one more site type is needed. This decision is generally easy to reach when one evaluates a series of polymers made with the same catalyst under a different set of conditions. Figure 4.11 exemplifies a MWD deconvolution procedure. The sum of least squares difference between measured and predicted MWD is used as the minimization target. By using more site types employed, lower values of the minimization target are achieved, but after a certain number of site types are included (four in the present case), the benefit of additional site types is negligible. This number of site types is referred to as the minimum number of site types required to represent a given MWD.

A detailed description of MWD deconvolution can be found in the literature [8, 33, 47]. This deconvolution technique can be further extended by including information coming from other analytical techniques such as Fourier Transform Infrared Spectroscopy (FTIR), Crystallization Analysis Fractionation (CRYSTAF) and Nuclear Magnetic Resonance spectroscopy (^{13}C-NMR) in order to provide additional information concerning comonomer incorporation [18, 23, 34, 35]. Examples found in literature, employing MWD deconvolution regarding the modelling of both slurry- and gas-phase processes, include the works of [31, 36, 37]. It is also possible to deconvolute CCDs measured by techniques such as TREF and even

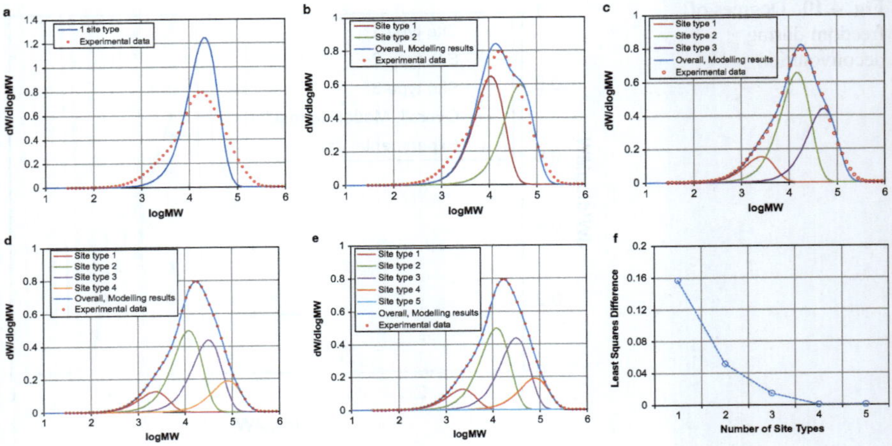

Fig. 4.11 MWD deconvolution procedure

MWD-CCD joint distributions obtained by cross-fractionation, albeit the interpretation of the results in these cases is significantly more elaborate [34].

4.2.5 Deconvolution and Estimation of Kinetic Constants

In some cases, under specific polymerization conditions and assumptions, the deconvolution methodology can lead to the direct estimation of kinetic constant ratios. To apply the proposed approach, we must assume that the MWD for the accumulated polymer is essentially the same as the instantaneous MWD. This assumption is valid when:

1. The polymerization reactor is operated at steady state conditions and the weight chain length distribution is spatially independent.
2. The ratio of transfer to propagation rates of all active sites does not change during the polymerization.
3. The relative amounts of polymer made by each site type do not change during the polymerization.
4. Mass- and heat-transfer effects are negligible, since these effects could give an instantaneous MDW that is spatially independent.

The one parameter describing Flory's distribution (τ^k) is the ratio of transfer to propagation rates, for each distribution coming from different catalyst site type k [33]. Moreover, τ^k equals to the inverse of the number average molecular weight, characterizing the distribution (Eqs. 4.19 and 4.20):

$$\tau^k = \frac{\text{transfer rate}}{\text{propagation rate}} = \frac{k_{tA}^k[A] + k_{trH}^k[H_2] + \ldots + k_{tsp}^k}{k_p^k[M]} = \frac{1}{DP_n^k} = \frac{1}{M_n^k/\overline{MW}} \quad (4.22)$$

where $k_{t,A}^k, k_{t,H}^k, k_{t,sp}^k$ are kinetic rate constants regarding transport to co-catalyst, to hydrogen and spontaneous transfer, k_p^k is the propagation rate constant. $[A]$,$[H_2]$ and $[M]$ denote the active concentration values of co-catalyst, hydrogen and monomer, M_n^k is the number-average molecular weight, DP_n^k is the number-average degree of polymerization and $\overline{MW}$ is the average molecular weight of the building polymer unit. Thus, having extracted degree of polymerization values for all different catalyst site types, via MWD deconvolution, a path is offered to directly correlate the kinetic parameter values between them [39].

> We shall not cease from exploration, and the end of all our exploring will be to arrive where we started and know the place for the first time.
> *T. S. Eliot*

4.3 Reactor Residence Time Distribution Effects

4.3.1 Bench-Scale Versus Industrial Reactors

When a new olefin polymerization catalyst is under development, it is usually tested in bench-scale reactors that are operated in semi-batch mode, as illustrated in Fig. 4.12 [40, 41]. In a typical slurry polymerization experiment (the most common bench-scale reactor set up), the diluent (usually an aliphatic solvent) is introduced in the reactor with the cocatalyst first. If a liquid comonomer, such as 1-hexene or 1-octene is needed, it is generally added to the reactor after the cocatalyst. More sophisticated reactor systems may allow for the liquid comonomer to be fed on demand through a measuring pump at the rate it is polymerized [42]. The reactor is then pressurized with the gaseous monomer (ethylene and propylene) to the target polymerization pressure. Hydrogen, acting as a chain transfer agent, is often fed at this time as well. After thermal stabilization of the system, the catalyst is added as a suspension (for supported catalysts) or as a solution (for unsupported catalysts) in a short pulse to start the polymerization. A mass flow meter measures the flow rate of the gaseous monomer to the reactor, which can be later converted to the polymerization rate. When the desired polymerization time is reached, a catalyst poison is injected to stop the polymerization and the reactor is quickly depressurized to remove unreacted monomer. Assuming proper reactor temperature control is maintained throughout the polymerization and that mass transfer resistances are negligible, apparent rate constants can be estimated from this experimental procedure.

The mode of operation described above is called semi-batch because some of the reagents are fed in batch mode at the beginning of the polymerization (diluent, catalyst, cocatalyst, and often the liquid comonomer) while others are supplied

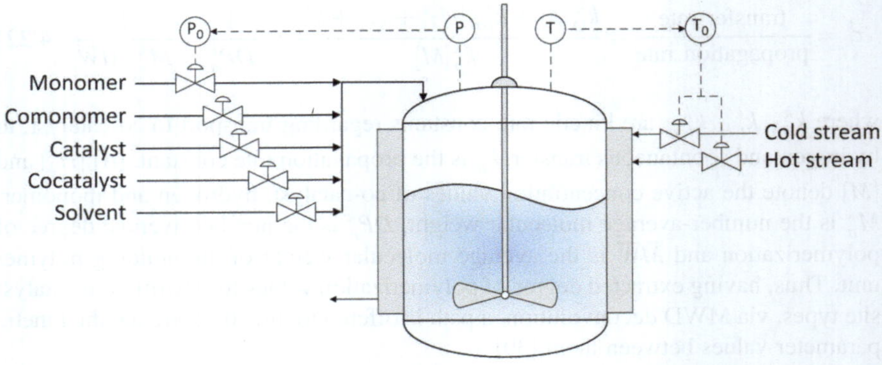

Fig. 4.12 Typical bench-scale semi-batch reactor system for olefin polymerization

continuously (gaseous monomers and, sometimes, liquid comonomers) to the reactor.

Bench-scale gas-phase reactors, albeit less common than slurry reactors, are operated similarly, but evidently without the use of a diluent. To allow for the proper fluidization and avoid agglomeration of the polymer particles, it is customary to use dried salt as a dispersing bed [43]. Finally, for liquid propylene polymerization, the monomer itself acts as the suspending media for the polymer particles. In this case, the polymerization is done in batch mode and the polymerization kinetics can be followed either by performing several polymerizations at different times and weighing the polymer formed, a very tedious and prone to error procedure, or by using a reactor calorimeter to relate the heat of polymerization to the polymerization rate [44].

More recently, the polyolefin industry has felt the need to develop new catalysts and polymer grades at a faster pace to stay competitive, leading several of them to adopt high-throughput techniques [45] that permit screening a very large number of catalysts and polymerization conditions in a short time. These reactors, albeit much smaller than the bench-scale autoclave shown in Fig. 4.12, are also operated in semi-batch mode. Successful catalyst formulations and process conditions identified with high-throughput reactors are often tested in larger semi-batch reactors before being scaled up to pilot plant or industrial scales.

What do all these bench-scale reactors have in common? Catalyst and polymer particles stay in the reactor for exactly the same time, from the beginning to the end of the polymerization. If this does not seem to be relevant, remember that polyolefins are produced commercially in reactors that have an entirely different residence time distribution (RTD) from that of a batch or semi-batch reactor. The RTD of a chemical reactor is a probability density function that describes the duration of time a fluid element (or a catalyst/polymer particle, in our case) will spend inside the reactor. Consider a population of catalyst particles that have just been injected as a pulse in a bench-scale semi-batch reactor at time $t = 0$, as shown in Fig. 4.13a.

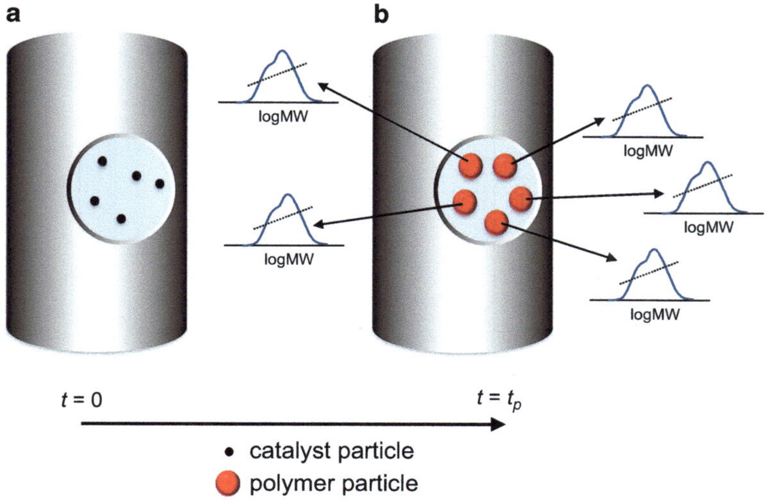

Fig. 4.13 Polymer particle time evolution in a semi-batch reactor: (**a**) $t = 0$, (**b**) $t = t_p > 0$. Catalyst particles are assumed to be uniform and free from significant heat and mass transfer limitations

Following the experimental procedure described above, all particles will be introduced into the reaction medium at the same time and will start producing polymer after activation by the cocatalyst. After the desired polymerization time, t_p, has elapsed, the polymerization is interrupted. It is obvious that all polymer particles spend the same time in the reactor in this mode of operation. Assuming that all catalyst particles have the same distribution of active sites and are not subject to severe intraparticle mass and/or heat transfer limitations, one should expect that the properties of the polymer (such as MWD and CCD) made by each catalyst particle should be the same within the expected experimental error (Fig. 4.13b).

4.3.2 Residence Time Distribution Fundamentals

The RTD, *E*(*t*), of semi-batch reactors is represented by the equation:

$$E(t) = \delta(t - t_p) \tag{4.23}$$

where t_p is the polymerization time and δ is the Dirac delta function,

$$\delta(x) = \begin{cases} \infty, x = 0 \\ 0, x \neq 0 \end{cases} \tag{4.24}$$

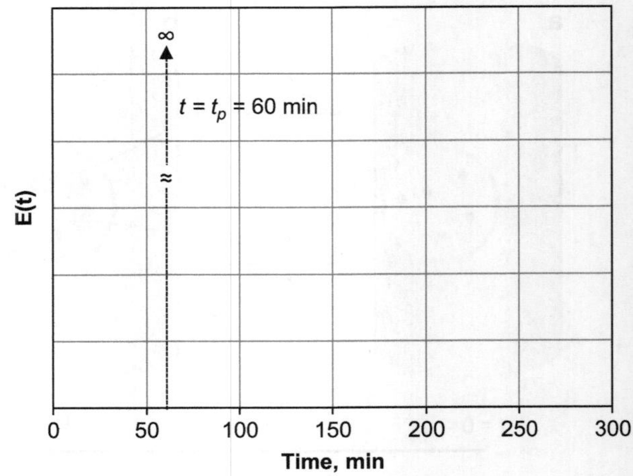

Fig. 4.14 Residence time distribution in a semi-batch reactor with $t_p = 60$ min

Equation 4.23 is the mathematical expression of the observation made above that all fluid elements stay in the reactor for exactly the same time, t_p, as illustrated in Fig. 4.14 for a reactor with $t_p = 60$ min.

There are two limiting flow configurations for the production of polyolefins (or any other chemical compound) in a continuous reactor:

1. Tubular plug flow reactor (PFR).
2. Continuous stirred-tank reactor (CSTR).

In an ideal PFR, all fluid elements (and consequently catalyst/polymer particles) move as a "plug," from the entrance to the exit of the reactor, as illustrated in Fig. 4.15. If plug flow behavior is achieved, all catalyst particles will spend the same time inside the PFR, in analogy to a semi-batch reactor:

$$t_R = \frac{V_R}{\dot{Q}} = \frac{A_R L_R}{\dot{Q}} = t_p \tag{4.25}$$

where t_R is the residence time in the PFR, $\dot{Q}$ is the reactor volumetric flow rate and V_R, A_R and L_R are the reactor volume, cross-sectional area and length, respectively. The RTD of a PFR reactor is also given by Eq. 4.23 and represented by Fig. 4.14, with $t_R = t_p$. Consequently, no scaling up issues related to RTD differences exist between a semi-batch reactor and a PFR, since they share the same RTD. Other consideration, such as axial concentration gradients in a PFR that are absent in a semi-batch reactor, may play a role in scaling up. In practice, the RTD of real tubular reactors may approximate, but it will seldom be exactly the same, as a PFR because of axial dispersion and back-mixing effects, but this is beyond the scope of our current explanation.

Unfortunately, practically no commercial grade of polyolefins made with coordination catalysts is produced in PFRs. Instead, polyolefin reactors that use Z–N,

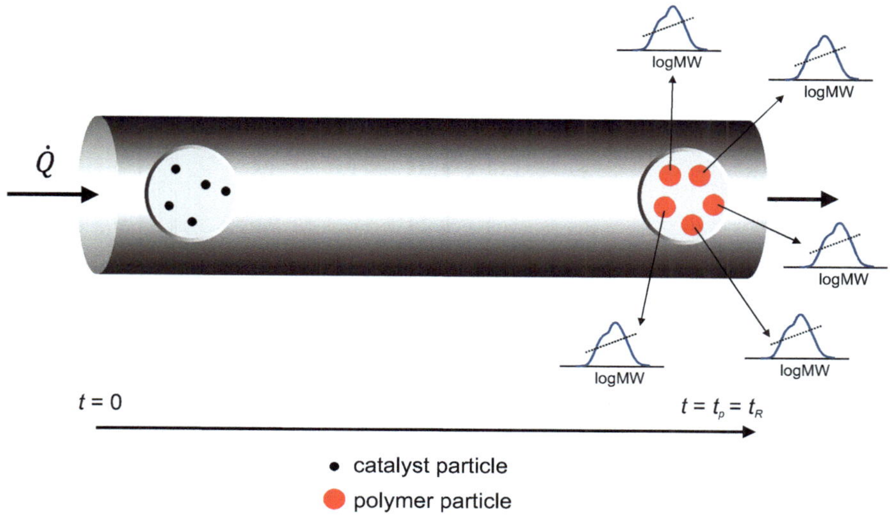

Fig. 4.15 Polymer particle time evolution in a tubular PFR with $t_R = 60$ min

Phillips, and supported metallocenes have RTDs characteristics of CSTRs. Even loop reactors, despite having a tubular configuration, have RTDs that approximate that of a CSTR because of their high recirculation ratios [26, 46]. For the case of fluidized-bed reactors, the gas phase flow may be approximated as a plug flow flowing from the distributor plate to the disengagement zone, but the solid polymer phase also follows a CSTR pattern. As we will see below, the RTD of a CSTR is completely different from that of PFRs, semi-batch and batch reactors.

Ideal CSTRs have an exponential RTD given by the equation:

$$E(t) = \frac{1}{t_R} \exp\left(-\frac{t}{t_R}\right) \tag{4.26}$$

where t is time and t_R is the average residence time in the CSTR. Equation 4.26 is represented in Fig. 4.16 for a CSTR is $t_R = 60$ min. It is apparent from Fig. 4.16 that the RTD of an ideal CSTR is broader and bears no resemblance to that of a bench-scale semi-batch reactor.

Let us now analyze the meaning of Eq. 4.26 and Fig. 4.16. Equation 4.26 is a classical chemical reaction engineering expression [47] obtained by assuming that the contents of the CSTR are perfectly mixed (the term "ideal" in ideal CSTR implies perfect mixing of all fluid elements in the reaction medium). Assume that a pulse of catalyst is fed to the reactor at a given time t_0. Since the reactor is perfectly mixed, these catalyst particles will, instantaneously, be dispersed within the reaction medium; some may exit the reactor right away, some may stay a little longer, and some may linger in the reaction much longer than the average residence time in the

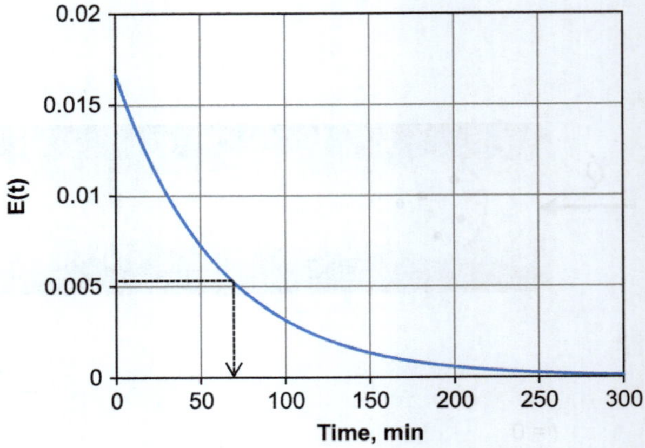

Fig. 4.16 Residence time distribution in a CSTR with $t_R = 60$ min

CSTR. The result is the broad RTD depicted in Fig. 4.16. Since a continuous flow of catalyst particles (and other reactants) is fed to the CSTR and a continuous flow of catalyst/polymer particles, unreacted reagents and products is withdrawn from the reactor, these phenomena takes place continuously throughout the polymerization.

The area under the $E(t)$ curve is 1.0, independently of the value of t_R. The area under the $E(t)$ curve in the time interval $[0, t_e]$ represents the fraction of fluid elements that stay in the reactor at least for t_e seconds, as given by the cumulative function:

$$F(t) = \int_0^{t_e} E(t)\mathrm{d}t = 1 - \mathrm{e}^{-\frac{t_e}{t_R}} \tag{4.27}$$

Figure 4.17 plots $F(t)$ for a CSTR with $t_R = 60$ min. For example, about 63% of the polymer particles stay in the CSTR no longer than the average residence time of 1 h. Similarly, approximately 86% of the polymer particles stay in the CSTR no longer than 2 h, or two average residence times. Consequently, a sizeable fraction of the polymer particle population (14%) stays in the CSTR longer than twice the average residence time of the reactor. Do you see the implications for product design from bench-scale to industrial reactors? If semi-batch experiments to make a given polymer composition were optimized for a polymerization time of 1 h in a semi-batch reactor, one should be at least doubtful that the same composition would be achieved in a CSTR with an average residence time of 1 h, considering that a substantial fraction of polymer particles will spend a much longer (or much shorter) time inside the CSTR than in the semi-batch reactor. Figure 4.18 illustrates this concept: Because of the broad RTD in a CSTR, particles that stay different times in the reactor will have different amounts of polymer and grow to different sizes. Under the conditions examined in the case study below, they may also have polymers with different microstructures.

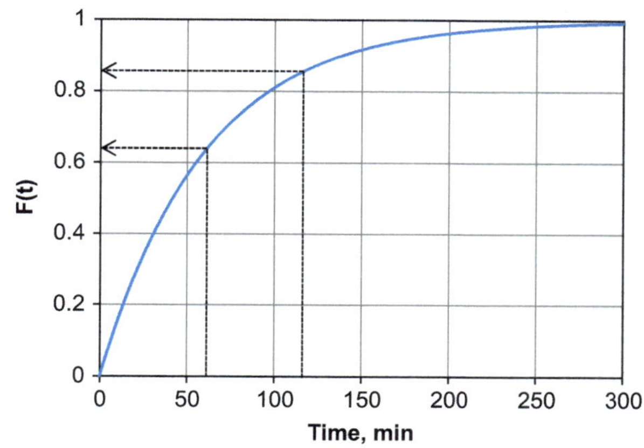

Fig. 4.17 Cumulative residence time distribution in a CSTR with $t_R = 60$ min

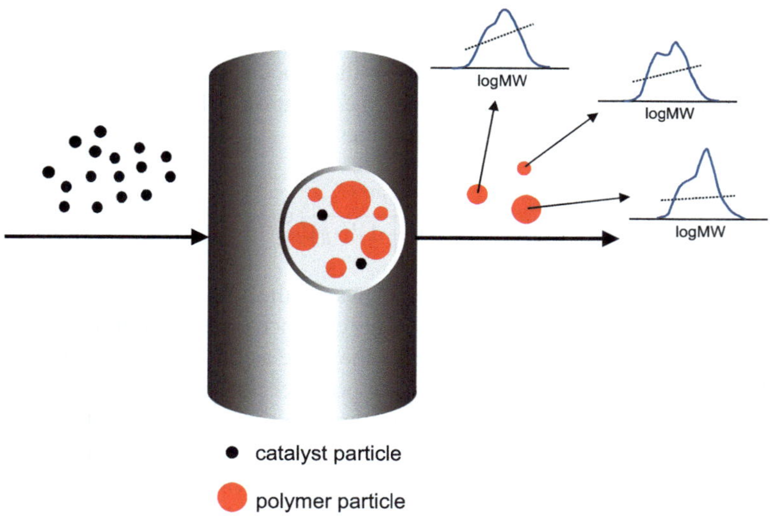

Fig. 4.18 Effect of RTD in a CSTR on polymer microstructure and particle size distribution

Actual CSTRs may have RTDs that differ from that of an ideal CSTR [47] because of imperfect mixing, presence of stagnant zones and by passing, but any CSTR will have an RTD that is different from that of a semi-batch reactor. Depending on the type of catalyst being developed, this difference may have marked consequences on the properties of the polymer being produced, as will be illustrated with a case study later in this chapter.

In the polyolefin industry, it is common to use two or more reactors in series to control polymer properties, as already discussed in this chapter. Some industrial reactors, such as gas-phase horizontal stirred reactors, also have an RTD that can be approximated by a series on 4–6 CSTRs [46, 48]. An additional consequence of

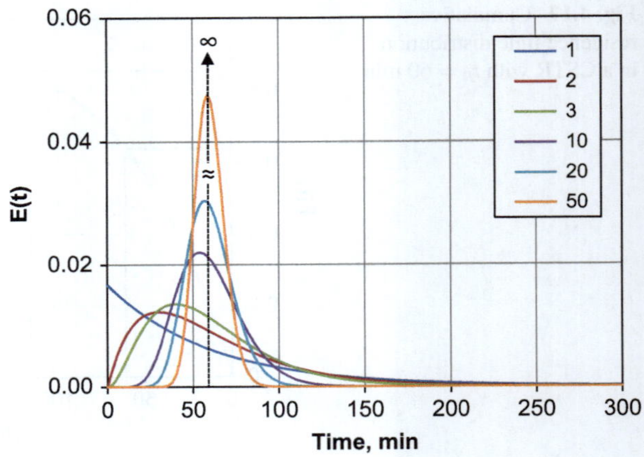

Fig. 4.19 Residence time distribution in a series of CSTRs with total residence time of $\sum t_R = 60$ min. The legends indicate the number of CSTRs in the series

using series of reactors is the narrowing of the RTD for the reactor train. Assuming that n CSTRs are operated in series, each one separately will have the RTD illustrated in Fig. 4.16, but the RTD for the reactor train will be narrower, as illustrated in Fig. 4.19 and expressed by the equation below for a series of ideal CSTRs with the same average residence times:

$$E(t) = \frac{t^{n-1}}{(n-1)!(t_R)^n} e^{-\frac{1}{t_R}} \tag{4.28}$$

Figure 4.19 shows that the RTD narrows as the number of CSTRs in series increases, keeping the same overall average residence time in the train. In the limit, as $n \to \infty$, the RTD of the series becomes the same as for a PFR [47]. It is intuitive why this should be so: as we follow a given polymer particle down the reactor train, its residence time will tend to $\sum t_R$ due to the statistical averaging of its residence times in each reactor in the series, as shown schematically in Fig. 4.20. Consequently, all other factors being the same, one should expect that the properties of polyolefins made in reactors in series should be closer to those made in semi-batch reactors than those made in a single CSTR.

4.3.3 *Case Study: Effect of RTD on MWD and Chemical Composition*

We will illustrate the effect of RTD on the properties of a hypothetical ethylene/1-octene copolymer produced with a 2-site type catalyst. The nature of the two site types is irrelevant for our purposes. They could result from supporting two different metallocene catalysts onto a carrier or they could also be the naturally occurring site types on a heterogeneously Z–N catalysts, albeit most Z–N catalysts require at least 3 or 4 site types for an adequate description of the MWD of the polymers they

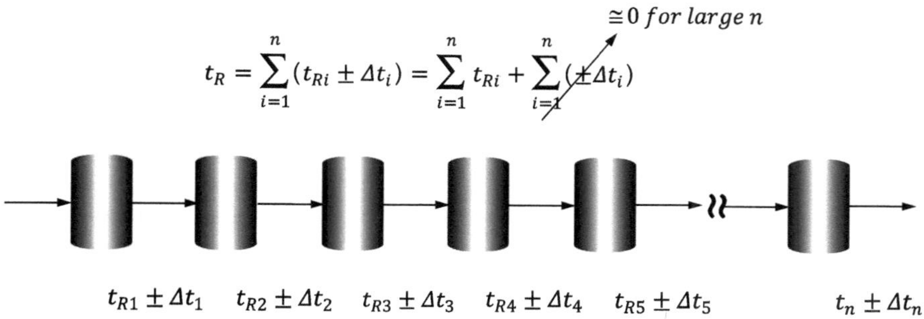

Fig. 4.20 Statistical averaging of residence times for a polymer particle flowing through a series of CSTRs

produce. The approach described above can be easily extended to more than two site types if necessary.

Figure 4.21 compares the polymerization kinetic curves for the two catalysts (C_1 and C_2) at the polymerization conditions shown in Table 4.2. C_1 has very fast activation rate but deactivates rapidly, while C_2 has a slow activation rate but does not undergo any appreciable deactivation during the average polymerization time of 60 min. The polymerization kinetic parameters used to describe these two curves are listed in Table 4.3. The meaning of these polymerization parameters will be explained below.

The MWD of a polyolefin made with the two catalysts in a CSTR is given by the expression:

$$\overline{W_t} = \int_0^\infty E(t)\frac{R_p^1(t)W^1 + R_p^2(t)W^2}{R_p^1(t) + R_p^2(t)}\mathrm{d}t$$
$$= \int_0^\infty E(t)\left[f_w^1(t)W^1 + \left(1 - f_w^1(t)\right)W^2\right]\mathrm{d}t \tag{4.29}$$

where $f_w^1(t)$ is the instantaneous mass fraction of polymer made by C_1, $R_p^1(t)$ and $R_p^2(t)$ are the instantaneous rates of polymerization for C_1 and C_2, respectively, and W^1 and W^2 are the MWDs of the polyolefins made on each site type, as given by Schulz–Flory distribution.

Equation 4.29 can be rearranged to the more computationally convenient form:

$$\overline{W_t} = W^1\int_0^\infty E(t)f_w^1(t)\mathrm{d}t + W^2\int_0^\infty E(t)\left(1 - f_w^1(t)\right)\mathrm{d}t$$
$$= W^1\overline{f_{w,\mathrm{CSTR}}^1} + W^2\left(1 - \overline{f_{w,\mathrm{CSTR}}^1}\right) \tag{4.30}$$

Table 4.2 Polymerization conditions used to generate the plots in Fig. 4.21

Variable	Units	Value	Definitions
t_r	min	60	Average reactor residence time
T	°C	60	Polymerization temperature
$[M_1]$	mol L^{-1}	1.5	Concentration of ethylene
$[M_2]$	mol L^{-1}	0.5	Concentration of 1-hexene
$[H_2]$	mol L^{-1}	1.0×10^{-3}	Hydrogen concentration
$[A]$	mol L^{-1}	1.0×10^{-2}	Cocatalyst concentration
$[C_{0,1}]$	mol L^{-1}	1.0×10^{-4}	Initial concentration of C_1
$[C_{0,2}]$	mol L^{-1}	1.0×10^{-4}	Initial concentration of C_2

Table 4.3 Polymerization kinetic parameters for C_1 and C_2

		Catalyst 1 (C_1)		Catalyst 2 (C_2)	
Rate constant	Units	K_0	E (kcal mol^{-1})	K_0	E (kcal mol^{-1})
k_a	L mol^{-1} min^{-1}	2.5×10^{36}	25	1.5×10^{17}	28
k_d	min^{-1}	1.0×10^{18}	30	1.0×10^{10}	35
k_{p1}	L mol^{-1} min^{-1}	3.5×10^{10}	8.5	6.0×10^{10}	9
k_{p2}	L mol^{-1} min^{-1}	3.5×10^{9}	8.5	1.0×10^{10}	9
k_{tH}	L mol^{-1} min^{-1}	2.0×10^{12}	12	3.0×10^{11}	12
$k_{t\beta}$	min^{-1}	2.0×10^{9}	12	3.0×10^{8}	12

where $\overline{f^1_{w,\mathrm{CSTR}}}$ is the cumulative mass fraction of polymer made by C_1 in the CSTR taking into account its RTD.

In contrast, for a semi-batch reactor where all catalyst/polymer particles stay the same polymerization time, t_R, in the reactor:

$$\begin{aligned}\overline{W_t} &= \int_0^{t_R} \frac{R_p^1(t)W^1 + R_p^2(t)W^2}{R_p^1(t) + R_p^2(t)}\mathrm{d}t = \int_0^{t_R} \left[f_w^1(t)W^1 + \left(1 - f_w^1(t)\right)W^2\right]\mathrm{d}t \\ &= W^1\overline{f^1_{w,\mathrm{SB}}} + W^2\left(1 - \overline{f^1_{w,\mathrm{SB}}}\right)\end{aligned} \tag{4.31}$$

As mention above, it is nowadays becoming routine to equip SEC instruments with an infrared detector (IR) that can measure the average copolymer composition as a function of polymer molecular weight. Since the average chemical composition of copolymers made in with single-site catalysts does not vary with chain length [4], it is possible to model the IR-measured compositions for the polymer made with the dual-site catalyst under study with the with the following expression:

$$\overline{\Phi_1}(MW) = \frac{\overline{f_w^1}W^1\Phi_1^1 + \left(1 - \overline{f_w^1}\right)W^2\Phi_1^2}{\overline{f_w^1}W^1 + \left(1 - \overline{f_w^1}\right)W^2} \tag{4.32}$$

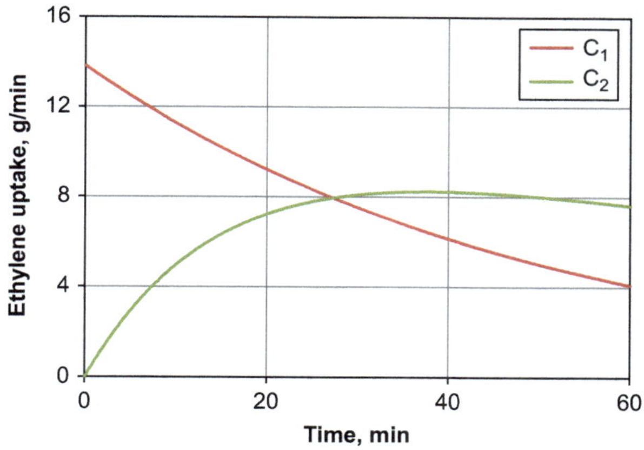

Fig. 4.21 Polymerization rates with two different single-site catalysts

where Φ_1^1 and Φ_1^2 are the average molar fraction of monomer type 1 (ethylene) in the copolymer made by C_1 and C_2, respectively, and $\overline{\Phi_1}$ is the average molar fraction of ethylene for the whole polymer as a function of its molecular weight.

A polymerization kinetic model is needed to calculate $f_w^i(t)$. Most olefin polymerization kinetic curves with single-site coordination catalysts can be described with simple mechanisms, as the one listed in Table 4.4 [4], comprising site activation, monomer and comonomer propagation, and catalyst deactivation. For simplicity's sake, the model will assume that chain propagation depends only on monomer type, not on chain end type (Bernoullian statistics), since the use of higher order models (terminal, penultimate, or higher Markovian statistics) does not have any appreciable impact on the effect of RTD on polymer properties being illustrated in this section. In addition, chain transfer reactions are assumed to have no effect on the polymerization rate and do not need to be included in the model at this stage.

The polymerization rate, R_p, is given by the equation:

$$R_P = \frac{\mathrm{d}([M_1] + [M_2])}{\mathrm{d}t} = \left(k_{p1}[M_1] + k_{p2}[M_2]\right)\sum_{r=0}^{\infty}[P_n]$$
$$= \left(k_{p1}[M_1] + k_{p2}[M_2]\right)[Y_0] \tag{4.33}$$

where $[M_1]$ and $[M_2]$ are the concentrations of monomer and comonomer, respectively, at the active sites, $[P_n]$ is the molar concentration of living chains with length n, $[Y_0]$ is the total molar concentration of living chains in the reactor, and k_{p1} and k_{p2} are the propagation rate constants for monomer and comonomer, respectively.

After a few simple mathematical manipulations [5], Eq. 4.33 can be rewritten in the more convenient format below:

Table 4.4 Copolymerization mechanism for polymerization kinetics

Description	Chemical equation
Activation	$S_p + A \xrightarrow{k_a} P_0$
Propagation	$P_n + M_1 \xrightarrow{k_{p1}} P_{n+1}$
	$P_n + M_2 \xrightarrow{k_{p2}} P_{n+1}$
Deactivation	$P_n + M_2 \xrightarrow{k_d} C_D + D_n$

S_p, catalyst precursor; A, cocatalyst; P_0, active site; P_n, living polymer chain with length n; M_i, monomer of type i; C_D, deactivated active site; D_n, dead polymer chain of length n; k_a, activation rate constant; k_d, deactivation rate constant; k_{pi}, propagation rate constant for monomer type i

$$R_P = \widehat{k_p}[M]\frac{1 - e^{-k_a^*\left(1-\frac{k_d}{k_a^*}\right)t}}{1 - \frac{k_d}{k_a^*}} C_0 e^{-k_d t} \tag{4.34}$$

where $[M] = [M_1]+[M_2]$ is the total monomer concentration, $\widehat{k_p}$ is the pseudo-propagation rate constant for copolymerization [4] defined as:

$$\widehat{k_p} = f_{m,1}k_{p1} + \left(1 - f_{m,1}\right)k_{p2} \tag{4.35}$$

$f_{m,1}$ is the molar fraction of monomer type 1 in the reactor:

$$f_{m,1} = \frac{[M_1]}{[M_1] + [M_2]} \tag{4.36}$$

and $k_a^* = k_a[A]$ considering $[A]$ to remain relatively constant through the polymerization.

The parameters k_a^* (or k_a), $\widehat{k_p}$ (or k_{p1} and k_{p2}) and k_d are estimated by fitting experimental monomer uptake curves with Eq. 4.34 [5]. The value of $\widehat{k_p}$ in Eq. 4.35 depends on the comonomer composition in the reaction medium. Since semi-batch reactors need to be operated at constant comonomer concentration for acquisition of reliable polymerization kinetics data, the use of $\widehat{k_p}$ does not impose any limitation to the use of Eq. 4.34. When different monomer–comonomer ratios are used, distinct values for $\widehat{k_p}$ would be obtained following Eq. 4.35, provided that the model assumptions are adequate. If the terminal model for copolymerization needs to be used, similar expressions can be derived for R_P.

Equation 4.34 can be substituted into Eqs. 4.30 and 4.32 to predict the MWD and $\Phi_1 \times MW$ profiles of polyolefins made with two or more catalysts in a CSTR—with $E(t)$ given by Eq. 4.27—or a semi-batch reactor—with $E(t)$ given by Eq. 4.23.

Figure 4.22 compares the MWD and SCB-MW plots of an ethylene/1-hexene copolymer made with a supported dual-site catalyst in a semi-batch reactor and a CSTR. The average residence time of the CSTR was kept the same as the

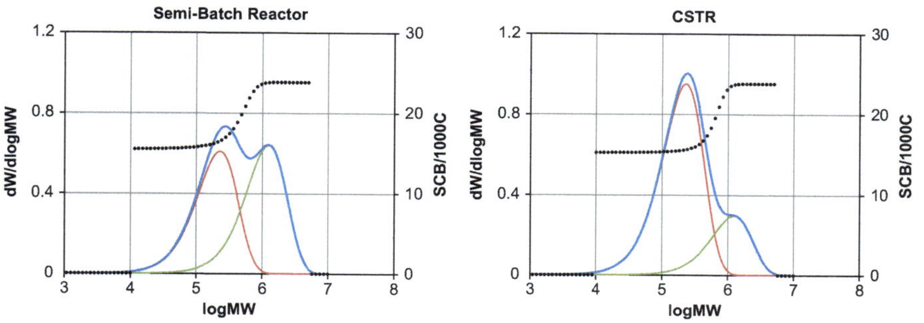

Fig. 4.22 SEC-IR plots of an ethylene/1-hexene copolymer made with a supported dual-site catalyst in a semi-batch reactor and a CSTR

polymerization time in the batch reactor, that is $t_R = t_p$. Comonomer composition across the MWD is reported as short chain branches per 1000 carbon atoms (SCB/1000 C), as usually done for polyolefins. The incorporation of 1-hexene in the polymer backbone generates a butyl SCB. The number of SCB/1000 C is calculated with the expression:

$$\mathrm{SCB}/1000\mathrm{C} = 1000\frac{1 - f_{m,1}}{6(1 - f_{m,1}) + 2f_{m,1}} \tag{4.37}$$

Figure 4.22 tells us that the product made in the semi-batch reactor is considerably different from the one made in the CSTR, even though the same polymerization conditions were used and the average residence time in the CSTR was kept equal to the polymerization time in the semi-batch reactor ($t_R = t_p$). Catalyst 1 makes polymer with lower M_n because C_1 has higher transfer to hydrogen and β-hydride elimination constants, and it also has lower propagation rate constants for M_1 and M_2. On the other hand, C_1 makes polymer with higher molar fraction of M_1 (lower SCB/1000 C), because it has a higher k_{p1}/k_{p2} ratio than C_2 (10 versus 6). More interestingly, the MWDs of polymers made in the CSTR and semi-batch reactor differ substantially because of their distinct RTDs and the different decay rates of the two catalysts. The fraction of polymer made by C_2 is higher in the CSTR because the activity of C_2 decays at a much lower rate than that of C_1. In general, the broader RTD of a CSTR will maximize the contribution of polymer made by the catalyst with slower decay rate. This example illustrates very effectively the importance of taking into account RTD effects when scaling up bench-scale results obtained in a semi-batch reactors to continuous pilot plant and industrial reactors.

Table 4.5 lists the average properties for the distributions depicted in Fig. 4.22. As expected, the properties of individual polymer populations made by each active site type in the dual catalyst remain the same, either when the catalyst is used in a semi-batch reactor or in a CSTR, but they differ for the whole polymer since the cumulative mass fractions of polymer made by each catalyst ($\overline{f_{w,\mathrm{SB}}^{1}}$) depend on the

Table 4.5 Average copolymer properties

	CSTR			Semi-batch reactor		
	C_1	C_2	Overall	C_1	C_2	Overall
M_n	113,000	620,000	140,000	113,000	620,000	195,000
M_w	226,000	1,240,000	467,000	226,000	1,240,000	745,000
PDI	2.0	2.0	3.3	2.0	2.0	3.8
$\overline{f_{m,1}}$	0.96	0.94	0.96	0.96	0.94	0.95
$\overline{f_w^1}$	0.76	0.23	1.00	0.48	0.51	1.00

RTD. Note the particularly large differences in M_n and M_w between the two modes of operation. Consequently, RTD has no effect on the microstructure of polymers made with a single-site catalyst, but may have an important effect in the case of dual- or multi-site catalysts.

It is illustrative to investigate how the MWD varies when a single CSTR is replaced by a series of CSTRs, keeping the same overall average residence time in the reactor train constant. Figure 4.23 shows, as expected from our previous analysis, that as the number of CSTRs in the reactor train increases, the MWD of the polyolefin approximates the one made in a semi-batch reactor shown in Fig. 4.22 since their RTDs start to converge.

Figure 4.24 depicts how the M_n of the same polymer is affected by the number of reactors in the train. Most of the change in M_n takes place when only a few reactors are introduced, with later changes becoming incremental and asymptotically approaching the M_n value for a semi-batch reactor. Thus, the introduction of only 2 or 3 reactors in series can substantially affect the properties of the polymer, even for the case when all reactors are operated in the same way and have the same average residence times.

This analysis could be extended to other process parameters such as reactor average residence time, polymerization temperature, monomer–comonomer ratio and, hydrogen concentrations [5]. It suffices to say that RTD differences could play an important role in all of these cases. We would like, however, to finish this section with a counter example to illustrate a case when RTD would have no appreciable effect on the polymer properties. Figure 4.25 shows the kinetic curves of two different single-site catalysts. Differently from the case analyzed above, these two site types have rather similar kinetic profiles characterized by fast activation rates and similar rate decay curves.

The MWD-SCB plots for a dual catalyst comprising equimolar amounts of these two site types are shown in Fig. 4.26. In contrast to what was observed when the polymerization kinetics of the two site types differed significantly, in the present case there are practically no differences in the SEC-IR plot of the polymer made in both reactor systems because the catalysts activate and deactivate following very similar partners. Therefore, shorter or longer residence times in the reactor do not affect the relative amounts of polymer made on each active site type and, consequently, the overall properties of the polymer are not subject to RTD effects.

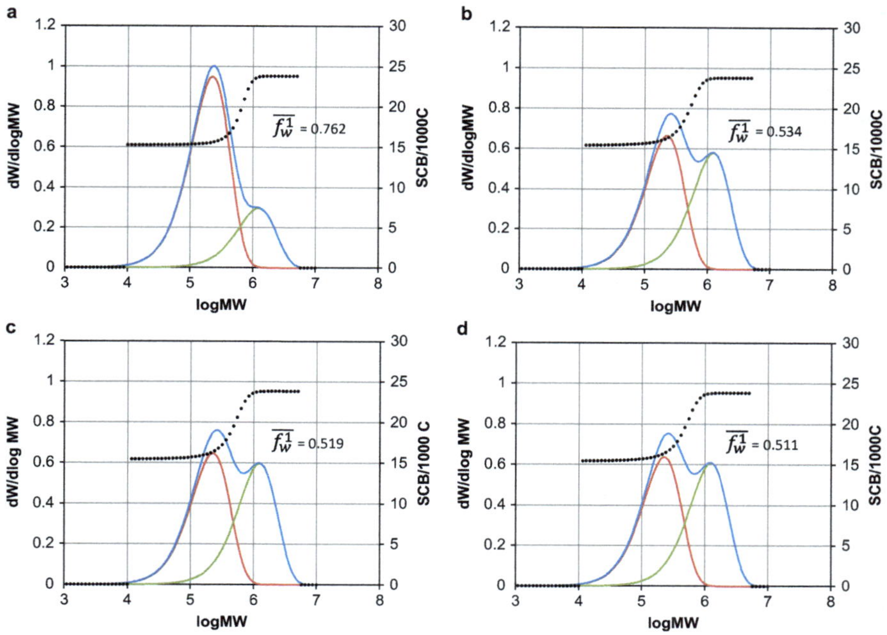

Fig. 4.23 SEC-IR plots of an ethylene–1-hexene copolymer made with n CSTRs in series: (**a**) $n = 1$, (**b**) $n = 2$, (**c**) $n = 3$, (**d**) $n = 4$

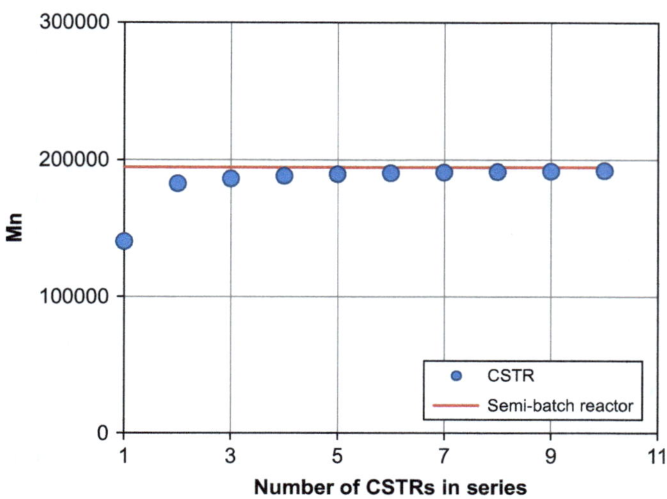

Fig. 4.24 Effect of using several CSTRs in series on polymer M_n

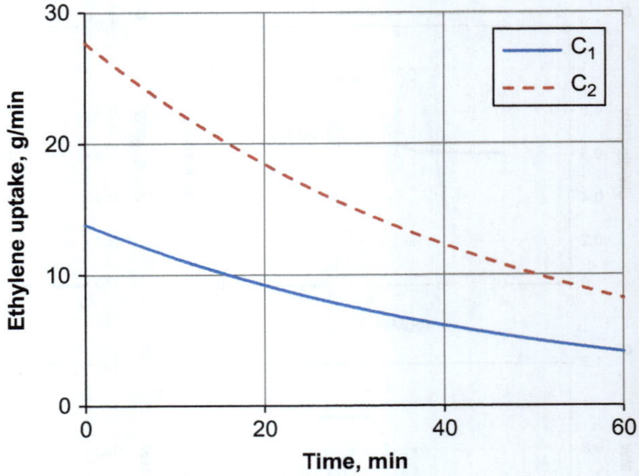

Fig. 4.25 Polymerization rates with two different single-site catalysts having fast activation and similar rate deactivation profiles

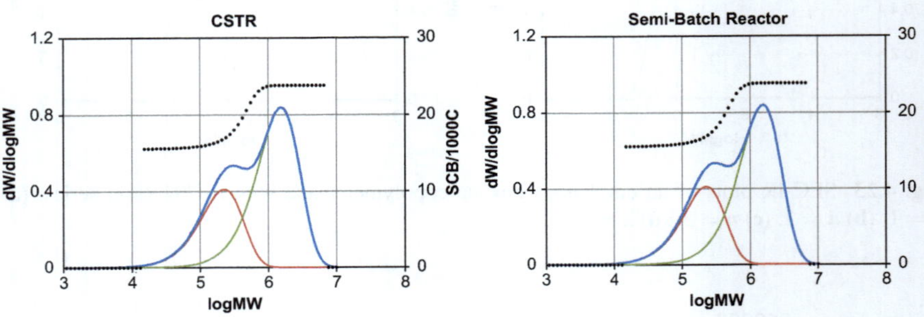

Fig. 4.26 SEC-IR plots of an ethylene–1-hexene copolymer made with the supported dual-site catalyst (Fig. 4.25) in a semi-batch reactor and in a CSTR

This last example emphasizes the important of understanding not only RTD effects but also the detailed polymerization kinetics of the catalysts under investigation. Both factors can have a significant impact on the polymer microstructure and on its final applications.

References

1. Brydson, J. A. (1999). *Plastics materials* (7th ed.). Oxford: Butterworth-Heinemann.
2. Odian, G. (2004). *Principles of polymerization* (4th ed.). New Jersey: Wiley Interscience.
3. Asua, J. M. (2008). Introduction to polymerization processes. In J. M. Asua (Ed.), *Polymer reaction engineering*. Oxford, UK: Blackwell Publishing.
4. Soares, J. B. P., & McKenna, T. (2012). *Polyolefin reaction engineering*. Manheim: Wiley-VCH.

5. Soares, J. B. P. (2014). The use of instantaneous distributions in polymerization reaction engineering. *Macromolecular Reaction Engineering, 8*, 235–259.
6. Flory, P. J. (1953). *Principles of polymer chemistry*. Ithaca, NY: Cornell University Press.
7. Kissin, Y. V. (1993). Ethylene polymerization kinetics with heterogeneous Ziegler-Natta catalysts. *Makromolekulare Chemie Macromolecular Symposia, 66*, 83.
8. Soares, J. B. P., & Hamielec, A. E. (1995). Deconvolution of chain-length distributions of linear polymers made by multiple-site catalysts. *Polymer, 11*, 2257–2263.
9. Alghyamah, A., & Soares, J. B. P. (2014). Fractionation of ethylene/1-octene copolymers by high temperature thermal gradient interaction chromatography. *Industrial and Engineering Chemistry Research, 53*(22), 9228–9235. https://doi.org/10.1021/ie403310d.
10. Cong, R., de Groot, A. W., Parrott, A., Yau, W., Hazlitt, L., Brown, R., Miller, M., & Zhou, Z. (2011). A new technique for characterizing comonomer distribution in polyolefins: High-temperature thermal gradient interaction chromatography (HT-TGIC). *Macromolecules, 44*, 3062.
11. Macko, T., Pasch, H., & Wang, Y. (2009a). Liquid chromatographic separation of olefin oligomers and its relation to separation of Polyolefins – An overview. *Macromolecular Symposia, 282*, 93.
12. Macko, T., Brüll, R., Alamo, R. G., Thomann, Y., & Grumel, V. (2009b). Separation of propene/1-alkene and ethylene/1-alkene copolymers by high-temperature adsorption liquid chromatography. *Polymer, 50*, 5443.
13. Ginzburg, A., Macko, T., Dolle, V., & Brüll, R. (2011). Characterization of polyolefins by comprehensive high-temperature two-dimensional liquid chromatography (HT 2D-LC). *European Polymer Journal, 47*, 319.
14. Ortin, A., Monrabal, B., & Sancho-Tello, J. (2007). Development of an automated cross-fractionation apparatus (TREF-GPC) for a full characterization of the bivariate distribution of polyolefins. *Macromolecular Symposia, 257*, 13–28.
15. Roy, A., Miller, M. D., Meunier, D. M., de Groot, A. W., Winniford, W. L., Van Damme, F. A., Pell, R. J., & Lyons, J. W. (2010). Development of comprehensive two-dimensional high temperature liquid chromatography × gel permeation chromatography for characterization of polyolefins. *Macromolecules, 43*, 3710.
16. Yau, W. W. (2007). Examples of using 3D-GPC-TREF for polyolefin fractionation. *Macromolecular Symposia, 257*, 29–45.
17. Al-Saleh, M. A., Soares, J. B. P., & Duever, T. A. (2010). Integrated deconvolution estimation model: A parameter estimation methodology for ethylene/α-olefin copolymers made with multiple-site-type catalysts. *Macromolecular Reaction Engineering, 4*, 578–590.
18. Al-Saleh, M. A., Soares, J. B. P., & Duever, T. A. (2011). The integrated deconvolution estimation model: Estimation of reactivity ratios per site type for ethylene/1-butene copolymers made with a heterogeneous Ziegler-Natta catalyst. *Macromolecular Reaction Engineering, 5*, 587–598.
19. Al-Saleh, M. A., Soares, J. B. P., & Duever, T. A. (2012). The integrated deconvolution estimation model: Effect of inter-laboratory ^{13}C NMR analysis on IDEM performance. *Macromolecular Reaction Engineering, 6*, 189–199.
20. Meyer, T., & Keurentjes, J. (2005). *Handbook of polymer reaction engineering*. Weinheim: Wiley-VCH.
21. Ray, W. H. (1986). Modelling of polymerization phenomena. *Berichte der Bunsengesellschaft für physikalische Chemie, 90*, 947.
22. Ray, W. H. (1991). Modelling of addition polymerization processes — Free radical, ionic, group transfer, and Ziegler-Natta kinetics. *Canadian Journal of Chemical Engineering, 69*, 626.
23. McKenna, T. F., & Soares, J. B. P. (2001). Single particle modelling for olefin polymerization on supported catalysts: A review and proposals for future developments. *Chemical Engineering Science, 56*, 3931–3949.
24. de Carvalho, A. B., Gloor, P. E., & Hamielec, A. E. (1989). A kinetic mathematical model for heterogeneous Ziegler-Natta copolymerization. *Polymer, 30*, 281–296.

25. de Carvalho, A. B., Gloor, P. E., & Hamielec, A. E. (1990). A kinetic model for heterogeneous Ziegler-Natta copolymerization. Part 2: Stereochemical sequence length distributions. *Polymer, 31*, 1294–1311.
26. Zacca, J. J., & Ray, H. W. (1993). Modelling of the liquid phase polymerization of olefins in loop reactors. *Chemical Engineering Science, 48*(22), 3743–3765.
27. Chadwick, J. C., Morini, G., Balbontin, G., Camurati, I., Heere, J. J. R., Mingozzi, I., & Testoni, F. (2001). Effects of internal and external donors on the regio and stereoselectivity of active species in $MgCl_2$-supported catalysts for propene polymerization. *Macromolecular Chemistry and Physics, 202*(10), 1995–2002.
28. Touloupidis, V. (2014). Catalytic olefin polymerization process modeling: Multi-scale approach and modeling guidelines for micro-scale/kinetic modeling. *Macromolecular Reaction Engineering, 8*, 508–527.
29. Dompazis, G., Kanellopoulos, V., Touloupides, V., & Kiparissides, C. (2008). Development of a multi-scale, multi-phase multi-zone dynamic model for the prediction of particle segregation in catalytic olefin polymerization FBRs. *Chemical Engineering Science, 63*, 4735–4753.
30. Hatzantonis, H., Yiannoulakis, H., Yiagopoulos, A., & Kiparissides, C. (2000). Recent developments in modeling gas-phase catalyzed olefin polymerization fluidized-bed reactors: The effect of bubble size variation on the reactor's performance. *Chemical Engineering Science, 55*, 3237.
31. Touloupides, V., Kanellopoulos, V., Pladis, P., Kiparissides, C., Mignon, D., & Van Grabezen, P. (2010). Modeling and simulation of an industrial slurry-phase catalytic olefin polymerization reactor series. *Chemical Engineering Science, 65*, 3208–3222.
32. Hutchinson, R. A., Chen, C. M., & Ray, W. H. (1992). Polymerization of olefins through heterogeneous catalysis X: Modelling of particle growth and morphology. *Journal of Applied Polymer Science, 44*, 1389–1414.
33. Soares, J. B. P., & Hamielec, A. E. (1996). Polymerization reaction engineering-metallocene catalysts. *Progress in Polymer Science, 21*, 651–706.
34. Fortuny, M., Nele, M., Melo, P. A., & Pinto, J. C. (2004). Deconvolution of molecular weight distributions using dynamic Flory-Schulz distributions. *Macromolecular Theory and Simulations, 13*, 355–364.
35. Anantawaraskul, S., Bongsontia, W., & Soares, J. B. P. (2011). Simultaneous deconvolution of molecular weight and chemical composition distribution of ethylene/1-olefin copolymers: Strategy validation and comparison. *Macromolecular Reaction Engineering, 8*, 549–562.
36. Soares, J. B. P., & Hamielec, A. E. (1995). Analyzing TREF data by Stockmayer's bivariate distribution. *Macromolecular Theory and Simulations, 4*, 305–324.
37. Khare, N. P., Seavey, K. C., Liu, Z. A., Ramanathan, S., Lingard, S., & Chen, C. (2002). Steady-state and dynamic modelling of commercial slurry high-density polyethylene (HDPE) process. *Industrial and Engineering Chemistry Research, 41*, 5601–5618.
38. Neto, A. G. M., Freitas, M. F., Nele, M., & Pinto, J. C. (2005). Modeling ethylene/1-butene copolymerizations in industrial slurry reactors. *Industrial and Engineering Chemistry Research, 44*, 2697–2715.
39. Touloupidis, V., Albrecht, A., & Soares, J. B. P. (2018). A methodology for estimating kinetic parameters and reactivity ratios of multi-site type catalysts using polymerization, fractionation, and spectroscopic techniques. *Macromolecular Reaction Engineering, 12*, 1700056.
40. Mehdiabadi, S., Soares, J. B. P., Bilbao, D., & Brinen, J. L. (2011). A polymerization kinetics comparison between a metallocene catalyst activated by tetrakis(pentafluorophenyl) borate and MAO for the polymerization of ethylene in a semi-batch solution reactor. *Macromolecular Reaction Engineering, 5*, 418–430.
41. Mehdiabadi, S., & Soares, J. B. P. (2012). Ethylene homopolymerization kinetics with a constrained geometry catalyst in a solution reactor. *Macromolecules, 45*, 1777–1791.
42. Chakravarti, S., & Ray, W. H. (2001). Kinetic study of olefin polymerization with a supported metallocene catalyst. II. Ethylene/1-hexene copolymerization in gas phase. *Journal of Applied Polymer Science, 80*, 1096–1119.

43. Shin, S. Y. A., Simon, L. C., Soares, J. B. P., Scholz, G., & McKenna, T. F. (2009). Gas-phase polymerization with transition metal catalysts supported on montmorillonite – A particle morphological study. *Macromolecular Symposia, 285*, 64–73.
44. Samson, J. J. C., Bosman, P. J., Wieckert, G., & Westerterp, K. R. (1999). Liquid-phase polymerization of propylene with a highly active Ziegler-Natta catalyst: Influence of hydrogen, cocatalyst, and electron donor on the reaction kinetics. *Journal of Polymer Science, Part A: Polymer Chemistry, 37*, 219–232.
45. Arriola, D. J., Carnahan, E. M., Hustad, P. D., Kuhlman, R. L., & Wenzel, T. T. (2006). Catalytic production of olefin block copolymers via chain shuttling polymerization. *Science, 312*, 714–719.
46. Zacca, J. J., Debling, J. A., & Ray, W. H. (1996). Reactor residence time distribution effects on the multistage polymerization of olefins – I. Basic principles and illustrative examples, polypropylene. *Chemical Engineering Science, 51*, 4859–4886.
47. Levenspiel, O. (1972). *Chemical reaction engineering* (2nd ed.). New York: John Wiley & Sons.
48. Dittrich, C. J., & Mutsers, S. M. P. (2007). On the residence time distribution in reactors with non-uniform velocity profiles: The horizontal stirred bed reactor for polypropylene production. *Chemical Engineering Science, 62*, 5777–5793.

Chapter 5
Industrial Multimodal Processes

Vasileios Kanellopoulos and Costas Kiparissides

5.1 Introduction

Polyolefins are the most widely used plastics today due to their low production cost and wide range of applications (packaging and other disposables, building and construction, agriculture, appliances, transportation, electrics and electronics, furniture, communication, automotive industry, etc.). It is well acknowledged that the degree of technological and scientific sophistication in relation to the catalytic polyolefin manufacturing has no equal among other synthetic polymer production processes. Presently, the total polyolefin world-production exceeds 130 million tons per year covering around 45% of the total plastic production (about 1.5 times the steel consumption in volume).

Polyethylene [including high-density polyethylene (HDPE), low-density polyethylene (LDPE), and linear low-density polyethylene (LLDPE)] and polypropylene (PP) cover 60 and 40% of the total polyolefins production, respectively. In the coming years, the annual POs world market growth is estimated to 4–6% [1, 2].

The diversity of available polyolefin grades has been the result of advances in both the development of novel catalytic systems [Ziegler–Natta (Z–N), metallocenes, etc.] and the use of new reactor configurations (e.g. multistage reactors) and operating conditions. Current polyolefin manufacturing processes take advantage of high-yield

V. Kanellopoulos (✉)
Borealis Polymers, Porvoo, Finland
e-mail: vasileios.kanellopoulos@borealisgroup.com

C. Kiparissides
Department of Chemical Engineering, Aristotle University of Thessaloniki, Thessaloniki, Greece

Centre for Research and Technology Hellas, Chemical Process and Energy Resources Institute, Thessaloniki, Greece
e-mail: costas.kiparissides@cperi.certh.gr

A. R. Albunia et al. (eds.), *Multimodal Polymers with Supported Catalysts*,
https://doi.org/10.1007/978-3-030-03476-4_5

catalysts and cascade-reactor technologies to produce polyolefins with desired molecular (bimodal molecular weight distribution, BMWD, copolymer composition distribution, CCD, etc.) and morphological properties (i.e. particle size distribution, PSD) at a low production cost. The various catalytic polyolefin processes can be broadly classified into solution-phase, slurry-phase, and gas-phase processes. Note that almost 70% of the total polyolefin production is carried out in gas- and slurry-phase catalytic olefin polymerization processes [3]. Each process has its advantages depending on the producer, their product goals and intellectual property considerations.

5.2 Properties and Applications

Polyolefins are typically produced in the form of pellets that are extruded, blow-moulded, or injection-moulded to fabricate products. Extrusion processes produce films, sheets, fibres, profiles, foams, and coatings. Blow moulding produces containers and parts. Injection moulding produces smaller containers and parts, while rotomoulding (or rotational moulding) produces parts in a huge variety of shapes and sizes.

PE and PP are often considered first for use in any application because of their excellent cost/performance value such as low density, easy recyclability, and processability. Polyolefins are easy to fabricate into useful products and have increasing design capability. A large number of products are made from them with targeted product applications matched to polymer composition and structure (see Fig. 5.1, PE products roadmap is depicted).

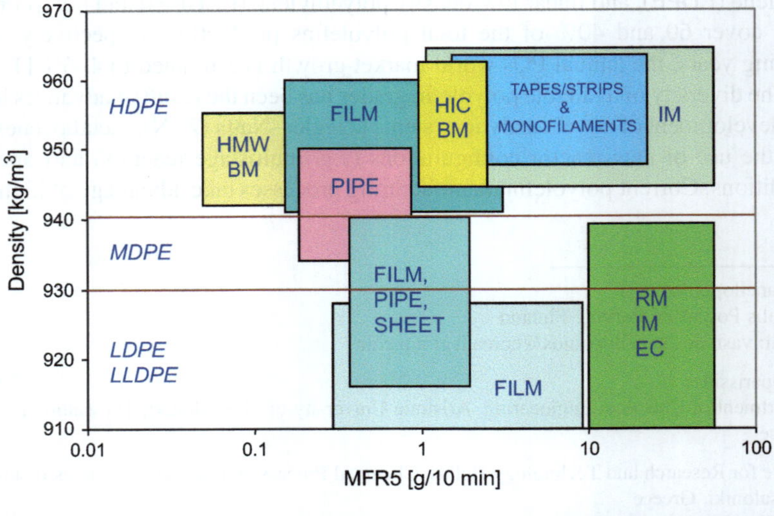

Fig. 5.1 PE products roadmap (*BM* blow moulding, *IM* injection moulded, *RM* rotational moulded, *EC* extrusion coating, *HIC* household and industrial chemicals)

Two important properties in polyolefins production are the density and the melt flow index (MI). It is a commercial practice to classify polymer resin grades based on the two properties instead of implicit properties such as molecular weight and composition. Polyethylene is generally divided into LDPE (0.91–0.93 g/ml) and HDPE (0.93–0.97 g/ml). LLDPE shares the same density range of LDPE but display many characteristics of HDPE. Sales specification of polymer resins are typically quoted to the third decimal value (in g/ml), and hence a very tight control is required.

In general, the density of polyethylene decreases with an increase in branch numbers; the more the branching, the lower the density. The crystallinity of PE decreases significantly with an increase in branch frequency and size. Hence, any physical properties related to crystallinity, such as stiffness and yield stress, will be affected by branching or chemical composition. Density decreases with an increase in molecular weight due to the inhibition of crystallization by longer molecular chains. The basic relationships between polymer properties and molecular structure can be summarized in Table 5.1 (case for PE). Molecular weight and molecular weight distribution affects almost the mechanical properties and directly affects the processability of the polymer. The degree of branching and its distribution change the mechanical and chemical properties of a polymer.

As illustrated in Table 5.1, end-use properties strongly depend on the polymer low-order and high-order molecular structures as well as their distribution (molecular weight distribution, etc.). For example, hardness of the polymer depends on polymer crystallinity that is determined by stereoregularity of polymer: a linear polymer with small side-groups is highly crystalline; a linear polymer with bulky but regular side-groups has low crystallinity; and a linear polymer with bulky and random side-groups is non-crystalline. Similarly, the morphological form of a high impact PP is often a key variable of end-use properties, and it depends on particle size distribution, polymer composition distribution, and processing history in the extruder. To determine the operating condition for each individual processing unit, it is indispensable to have deeper understandings about how the end user properties are affected by molecular structure, its distribution, and operating conditions of each processing units [4].

5.3 Polyolefin Reactor Technologies

5.3.1 The Evolution of Industrial Polyolefin Manufacturing Technologies

The first commercial polyethylene was produced under very high pressure (~3000 atm.) in a free radical reaction process which produces low-density polyethylene (LDPE) with narrow molecular weight distribution. The high operating pressure necessitates high capital investment (equipment to withstand the high pressure) and operating costs (power to compress reactants to the high operating pressure).

Table 5.1 Relationship between molecular structure and properties of PE [4]

		Molecular weight (mol. wt.)	Molecular weight distribution	Branching chemicals	Degree of branching	Degree of branching distribution	Long chain branching (LCB)
Mechanical and chemical property	Transparency	✓	✓	✓	✓	✓	
	Tensile strength	✓	✓	✓	✓	✓	✓
	Impact strength	✓	✓	✓	✓	✓	✓
	Rigidity				✓	✓	
	Heat resistance				✓	✓	
	Cold resistance	✓	✓	✓	✓	✓	
	Chemical resistance	✓	✓	✓	✓	✓	✓
	Heat seal	✓	✓	✓	✓	✓	
Processability	Bubble Stability	✓	✓				✓
	Draw-down	✓	✓				✓
	Extrusion torque	✓					✓

Table 5.2 Polymerization processes and reactor operating conditions

	Conventional high press. process	High-pressure, bulk process	Solution	Slurry	Gas phase
Reactor type	Tubular/autoclave	autoclave	CSTR	CSTR/loop	Fluidized/ stirred bed
Pressure (bar)	1200–3000	600–800	60–100	30–65	20–35
Temp (°C)	130–350	200–300	140–200	70–110	80–100
Kinetics	Free radical	Coordination	Coordination	Coordination	Coordination
Reac. medium	Monomer phase	Monomer phase	Solvent	Solid–liquid	Solid–gas

Variants of this technology licensed by ExxonMobil Chemical Co. and Basell Polyolefins continue to be used today with a worldwide capacity of over 6 MMtpy. This technology is also used to produce ethylene vinyl acetate (EVA) which is generally used in high clarity shrink-wrap sheets.

The subsequent technologies developed focused on achieving lower operating pressure (Table 5.2). With the development of catalyst to allow coordination polymerization, slurry and solution phase reactors were invented. These reactors operate at a lower pressure compared to the earlier high-pressure free radical process. The slurry-phase "loop reactor" by Chevron Phillips Chemical Co. operates at 40 atm. Development for low pressure polymerization processes also started the use of comonomers such as 1-butene and 1-hexene to produce linear-low density polyethylene (LLDPE) and high density polyethylene (HDPE). Another variant of Chevron Phillips technology developed by Mitsui Chemicals Inc. is named CX process, and it is used for production of HDPE and has a worldwide capacity of 4 MMtpy.

Each type of process has limitations on the product range that it can make. Figure 5.2 illustrates the capabilities of different processes in terms of density and melt index. Grades produced via the slurry process are limited to melt index up to 150 g/min. Solution processes have a wider window of operation in terms of densities and melt indexes. Gas-phase reactors have the widest operating window, but encounter difficulties in producing grades with very low densities due to stickiness and particle agglomeration.

5.3.2 *Gas-Phase Reactor Technology*

Gas-phase catalytic olefin polymerization is a highly attractive process since the polymerization is carried out in the absence of a liquid inventory. A wide range of polyolefin grades (i.e. MI: 0.1–100, density: 0.91–0.97 kg/m^3) can be produced by gas-phase polymerization. Gas-phase reactors commonly operate close to the dew point of the reaction mixture in order to achieve high monomer concentrations and thus high polymer yields. Due to the poor heat transfer characteristics in the gas-phase, particle overheating resulting in polymer melting should be taken into consideration.

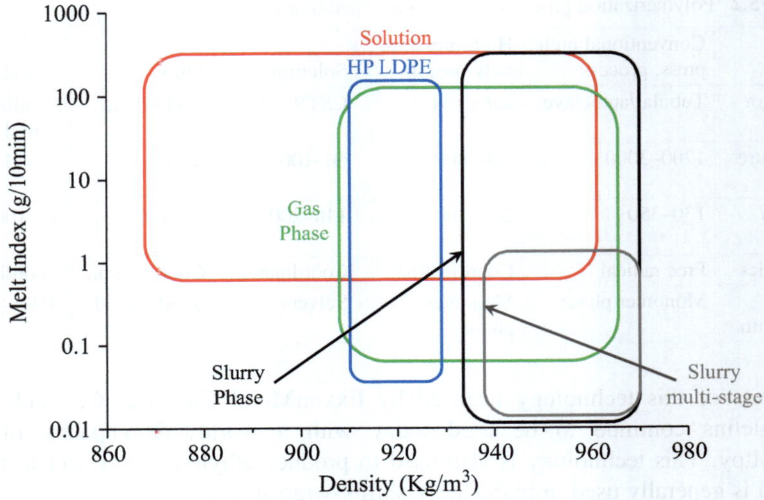

Fig. 5.2 Products made by different PE production technologies

The typical feature of this process is the fluidized bed reactor which widens at its top to reduce the gas velocity and entrainment of particles (an expanded dome reactor vessel-disengagement zone), a compressor, and a cooler. The earlier designs adopted the cooler before the compressor configuration. This is to cool down the reacting gas mixture, thus lowering the compression power required. The newer plants however adopted the cooler after compressor sequence.

In a catalytic fluidized bed reactor (FBR), see Fig. 5.3, catalyst particles in the size range of 20–80 μm are continuously fed into the reactor at a point above the gas distributor and react with the incoming fluidizing monomer(s) to form a broad distribution of polymer particles (e.g. 100–5000 μm) [5].

The particulate polyolefin is continuously withdrawn from the reactor at a point, preferably, close to the bottom of the bed through sequential operation of a pair of timed valves, defining a segregation zone. Subsequently, the particulate matter is transferred to a tank where nitrogen or other diluent is injected to remove residual monomers from the polymer (degassing unit). The polymerization heat is removed via the unreacted gaseous mixture exiting the bed, which is subsequently compressed, cooled (heat exchanger), and recycled back to the reactor.

Fouling of the cooler (polymer fines build-up in the exchanger tubes) often becomes the criterion for a shutdown. Industrial FBRs commonly operate at temperatures of 70–110 °C and pressures of 20–40 bar [6]. The superficial gas velocity is equal to 3–7 times the minimum fluidization velocity. The catalyst feed rate can vary from 0.01 to 0.5 g/s, depending on the catalyst activity and reactor capacity. Single-pass monomer conversion ranges from 2 to 5%, whereas the overall monomer conversion can be as high as 98%. The catalyst/polymer particles, during their stay in an FBR, can grow in size due to polymerization, can be entrained by the exiting

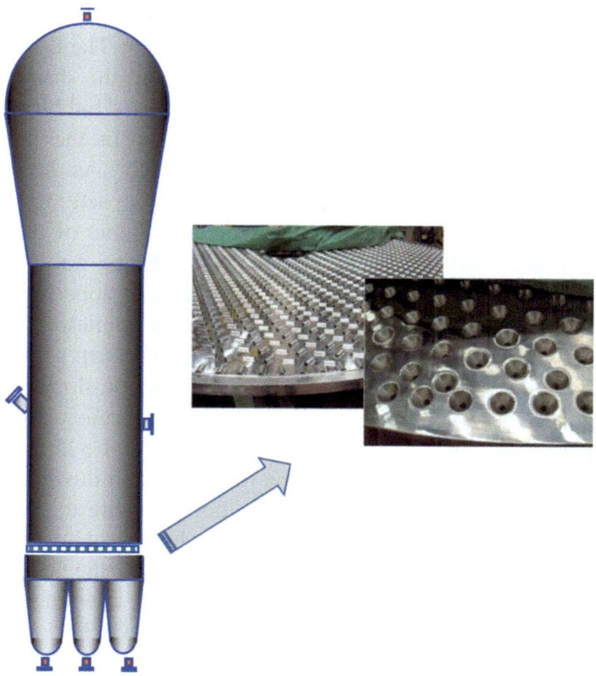

Fig. 5.3 Typical gas-phase reactor

gas stream, and can undergo particle agglomeration when the particles' temperature exceeds the polymer softening temperature [6].

The reaction zone has a height to diameter ratio ranging between 3 and 8. The disengagement zone has a diameter to height ratio of about 1–3. It is essential that the bed always contain polymer particles to prevent the formation of localized "hot spots" and to entrap and distribute the powdery catalyst. During start-up, the reaction zone is charged with a base of polymer particles before gas flow is initiated. Monomer (ethylene or propylene gas) is fed to the compressor inlet, whereas the α-olefin comonomer (1-butene or 1-hexene) is added to the reactor inlet. Fluidization is achieved by a high rate of gas recycle to and through the bed, typically on the order of about 50 times the rate of fresh gases. The pressure drop through the bed is typically on the order of 0.07 atm. (1 psi). A gas analyzer, positioned above the bed, determines the composition of the gas being recycled, and the composition of the fresh gas mixture is adjusted accordingly to maintain an essentially steady-state gaseous composition within the reaction zone.

The fluidized bed can maintain itself at an essentially constant temperature under steady-state conditions. To increase heat removal capacity and productivity, gas-phase reactors can be operated with an inlet gas temperature at the bottom of the fluidized bed which is below the dew point temperature of that gas. Recycle gas partly condenses in the external cooler, and the liquid droplets re-vaporize upon entry into the bed. It has been observed that droplets pass through the distributor as a mist and quickly wet the surface and pores of the polymer particles. The liquid vaporizes very quickly above the distributor. There is a dramatic change in the gas flow pattern in the gas feed region below the distributor. However, there is no apparent effect on bulk circulation patterns of bubbles and polymer particles above the gas distributor. The distribution plate at the bottom of a reactor plays an important role in the operation of the reactor. As the polymer particles are hot and possibly active, they must be prevented from settling to avoid agglomeration.

The reaction zone and disengagement zone of the reactor are connected by a transition section having sloped walls. During polymerization, some fine particles from the disengagement zone fall onto the sloped walls of the transition section. These fine particles build up during reactor operation. Since the fine particles contain active catalyst, they further polymerize and form solid sheets which can grow until they block the recycle gas flow or slide off the sloped walls into the polymerization zone, thus blocking the flow of gas and causing fusion of the polymerization particles. Thus, large chunks of polymer can be formed which can block the entire polymerization zone unless the reactor is shut down and the sheets are removed.

The initial stage of polymerization is where the risks of hot spots forming and grains bursting into fine particles are highest. Hot spots may lead to formation of agglomerates and to settling of polymer inside the fluidized bed. Furthermore, during polymerization small variations in the feed rates of catalyst, monomer, and comonomer or in the withdrawal rate of polymer will also cause an unexpected increase in the quantity of heat evolved by the polymerization. If the heat cannot be removed efficiently, these small variations can cause hot spots in the reaction bed and formation of agglomerates by melting polymer. Such variations can therefore make it difficult to obtain a polymer of consistent quality, in particular, of constant molecular weight and particle size. The aforementioned operability issues can be eliminated by adopting a pre-polymerization stage.

Pre-polymerization gives advantages in polymer particles size control, morphology and control of catalyst activity in the fluidized bed reactor. Pre-polymerization can be carried out in a liquid hydrocarbon medium or in a gas-phase stirred reactor at temperatures from 30 to 100 °C. Catalyst is introduced into the pre-polymerizer in the form of dry powder or in suspension in a liquid hydrocarbon or oil. Pre-polymerization is typically carried out under mild conditions and low degree of polymerization in order to separate the catalyst activation stage from the actual process (Table 5.3).

5.3.3 Slurry-Phase Ethylene Polymerization Reactors

Slurry-phase catalytic olefin polymerization is one of the most commonly employed processes for the production of polyolefins with MI values in the range of (0.01–150) and densities varying from 0.918 to 0.98 kg/m^3) [7]. Slurry-phase polymerization is commonly carried out in the presence of an inert diluent (propane, isobutene, hexane, heptane, etc.) in which the supported catalyst particles are suspended.

Table 5.3 Low pressure PE processes used by licenced technologies

Process	Technology	Features	Reactor configuration
Gas	Univation (UNIPOL I/PRODIGY)	• Gas-phase reactor • Bimodal catalyst • Simple process	
	Inneos (INNOVENE G)	• Cyclone removal of fines • Optimized disengagement • Clean loop process	
	LyondellBasell (SPHERILENE S)	• Avant Z catalysts • Direct catalyst feed • All PE market segments	
	Mitsui (EVOLUE)	• Long residence time • Extensive back mixing • Batch wise product removal	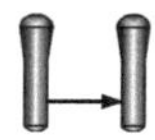
	Dow/Univation (Unipol II)	• Advance process control • Quick transition technology • Broad product portfolio	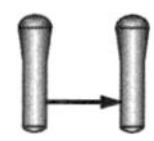
	Westlake (ENERGX)	• Similar to Unipol PE • Conventional gas-phase reactor • Selectively licencing	
Slurry tank	Mitsui Chemicals (MITSUI CX)	• Bimodal polymer grades • Autoclave-type vessels • Mixer for uniform mixing	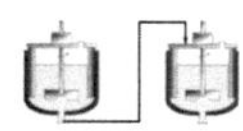
	LyondellBasell (HOSTALEN ACP)	• Trimodal polymer grades • Catalysts by LyondellBasell • Serial and parallel configuration	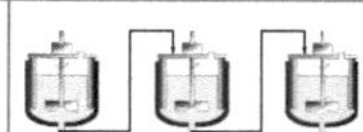
	Japan PE	• Formerly JPO/Nippon • CSTR type of reactors • Only few licences	
Slurry loop	Chevron Phillips (CPChem)	• Eight-leg loop reactor • Double loop configuration • Chromium, Z–N, metallocenes	
	Ineos Technologies (INNOVENE S)	• Formerly Solvey • Dual-loop bimodal technology • Low sensitivity to poisons	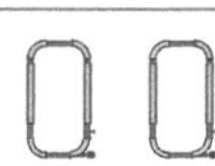

(continued)

Table 5.3 (continued)

Process	Technology	Features	Reactor configuration
	Japan PE	• Formerly JPO Showa Denko • Z–N, metallocenes • Few licences	
	Borealis (BORSTAR 3G)	• Increased homogeneity • Multimodal polymer design • Product wide range	
Solution	Nova Chemicals (ADVANCED SCLAIRTECH)	• Sufficient temperature control • Very fast grade transitions • High catalyst activity	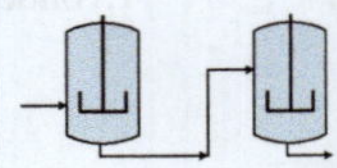
	Borealis (COMPACT)	• Formerly DSM/Stamicarbo • Wide range of PE specialties • Very short residence time	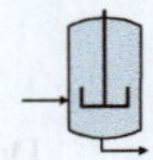
	Dow Chemicals (DOWLEX)	• Efficient heat transfer • Infuse catalyst technology • Low autoclave pressure	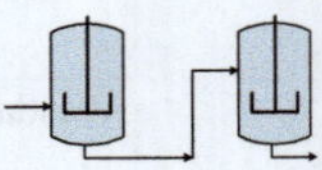
	ExxonMobil (EXXON)	• Parallel or/and autoclaves • Versatile process • Wide product range	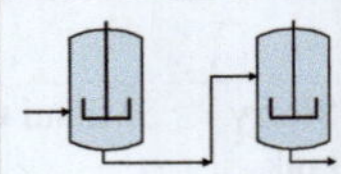

Multistage reactor configurations are commonly employed to control the molecular weight distribution of the polyolefins (i.e. unimodal, bimodal).

In the sixties, Phillips Petroleum Co. developed the first loop-reactor technology for the production of polyolefins in a slurry-phase. The slurry-phase loop-reactor technology has gained considerable acceptance and large quantities of polyethylene and its copolymers are annually produced via this process [8]. Nowadays, the continuous slurry-phase olefin polymerization in the presence of a heterogeneous Ziegler–Natta catalyst is one of the most commonly employed processes in the production of polyolefins, including HDPE, isotactic polypropylene (IPP) as well as their copolymers with higher olefins. The main reasons for the wide use of slurry-phase loop-reactor technology are (1) their simple design and operation, (2) their well-defined mixing conditions, (3) their excellent heat transfer capabilities, (4) the low power consumption requirements, and (5) the high monomer conversion rates [8].

Table 5.4 Reactor conditions of slurry PE technologies

Process	Reactor type	Reactor configuration	Diluent	*P* (bar)	*T* (°C)	Residence time
Mitsui	2 stirred autoclaves	Series or parallel	Hexane	<8	80–85	45 min per reactor
Basel (Hostalen Process)	2–3 stirred autoclaves	Series or parallel	Hexane	5–10	75–85	1–5 h per reactor
Equistar–Maruzen–Nissan	1–2 stirred autoclaves	Series or parallel	Hexane	10–14	75–90	45 min to 2 h per reactor
Chevron Phillips	Single loop	Multi-legged	Isobutane	30–40	85–100	1 h
Borealis Borstar	2–3 loops + FBR	Series	Propane	55–65	85–100	30–60 min in each loop, ~2 h in FBR
Innovene S	1–2 loops	Series	Isobutane	25–40	70–85	1 h

Since operating with liquid ethylene is not pragmatic, all slurry processes for polyethylene production employ diluents to dissolve the monomer, comonomer, and hydrogen. These were the first commercial processes for the production of polyolefins. The basic process consists of a series of CSTRs, with the polymerization taking place in a heterogeneous catalyst suspended in an inert diluent. Many variations of the slurry process were developed in the early 1970s. Since the activity of these early Ziegler–Natta catalysts was relatively low, a series of CSTRs were needed to push the reaction to completion. Catalyst residue removal, called de-ashing, was also necessary, greatly increasing the capital costs of the process. With the advent of high activity catalysts, the polymerization can now be completed in one or two reactors without de-ashing. Despite being the first process for ethylene polymerization, the slurry process is still economically viable and competitive these days [4]. The reactor conditions, reactor type, and configuration for the various PE slurry-phase olefin polymerization processes are illustrated in Table 5.4.

The slurry-phase reactors (i.e. autoclaves, loops) can be operated in series or in parallel, with the catalyst being fed only to the first reactor, along with ethylene and hydrogen. Typically, when a catalyst with a decay profile is used, low molecular weight homopolymer will be made in the first reactor. Since hydrogen is used to control the molecular weight and it can cause the polymerization rate to drop, it is added while the catalyst has its highest intrinsic activity. Higher molecular weight copolymers will be made in the subsequent reactors. In the second, and eventually third, reactor, it is usual to take advantage of the "comonomer-kick" (it is well known that small amounts of an α-olefin comonomer can increase the rate of polymerization of ethylene with respect to homopolymerization) to keep productivity up. This order of addition of catalyst, hydrogen, and comonomer is very common to most multistage reactor technologies, regardless of comonomer type or polymerization medium phase. Depending on the type of catalyst employed, the third reactor can be used to generate a very high or ultrahigh molecular weight component [4].

The slurry-phase processes for PE manufacture are also based on the loop reactor, which can operate at higher monomer and slurry concentrations than stirred autoclaves and allows for higher space-time yields. It should be mentioned that the residence time distribution (RTD) in loop reactors is also that of a CSTR, but instead of a stirrer, an axial pump is used to circulate the reactants around a pipe loop. Depending on the process, different diluents are used (propane, isobutane, *n*-hexane, etc.), see also Table 5.4. In such reactors, the fluid circulation pump is designed to provide high flow velocities of the reaction mixture (i.e. 5–7 m/s), resulting in well-mixed conditions [9]. Moreover, the high flow circulation rates ensures high heat transfer rates from the slurry phase to the coolant water, flowing in the reactor jacket and, at the same time, it limits reactor fouling caused by particle deposition on the internal surface of the reactor tube. Thus, the loop-reactor can operate at high-solids concentrations (e.g. larger than 30% w/w). This represents a clear advantage of the loop-reactor technology over the more conventional operation of stirred tanks where mixing and heat transfer limitations do appear at high-solids concentrations [10, 11].

5.3.4 Comparing Slurry and Gas-Phase Reactors: A Fundamental Approach

Usually, in a slurry-phase reactor, three phases can coexist, namely, a liquid phase (consisting of diluent, dissolved monomer, comonomer, and hydrogen), a polymer phase (i.e. polymer and sorbed quantities of diluent, monomer, comonomer, and hydrogen), and a gas phase (i.e. ethylene and hydrogen feed streams).

In Fig. 5.4a, b, the various external and internal resistances to the mass transfer of reaction species (monomer, hydrogen, etc.) from the gas or liquid phase to the polymer particle are depicted, respectively. In particular, in the slurry-phase olefin polymerization process, the gaseous reactants (ethylene, propylene, etc.) are first transferred from the bubble phase (where M_{i}^{*} denotes the concentration of "i" species in the bubble phase), to the continuous liquid phase, where they dissolved at a concentration equal to M_{i} (see Fig. 5.4a). Subsequently, the dissolved reactants are transferred to the external particle surface, through the liquid boundary layer surrounding the growing polymer particle, where $M_{\mathrm{i,S}}$ the dissolved species concentration at the external particle surface. $M_{\mathrm{i,R}}$ denotes the surface concentration of "i" absorbed species in the polymer phase at $\mathrm{r} = \mathrm{R}$, see Fig. 5.4a. On the other hand, in the gas-phase olefin polymerization there is only a single external resistance to the mass transfer of gaseous reactants from the bulk gas-phase to the polymer surface (see Fig. 5.4b) [12].

Consequently, in order to calculate the mass transfer rates of the various reactants from the bubble gas phase to the polymer phase in a slurry polymerization process, the solubilities of the reactive species in the different phases present in the slurry process (i.e. liquid and polymer) should be known. Moreover, the gas–liquid film mass transfer coefficient, k_{gl}, and the liquid–solid film, k_{ls}, mass transfer coefficient need to be calculated from available correlations.

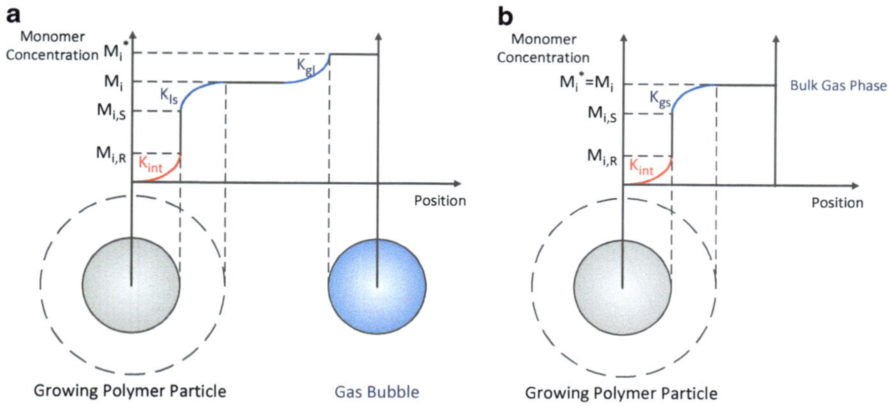

Fig. 5.4 (**a**) External (blue lines) and internal (red line) mass transfer resistances, at the particle level, in a slurry-phase process. (**b**) External (blue line) and internal (red line) mass transfer resistances, at the particle level, in a gas-phase process

The dynamic evolution of the polymerization rate and temperature in a single growing polymer particle will depend on the significance of external particle mass and heat transfer resistances. Figure 5.5 shows the dynamic evolution of the particle polymerization rate in both gas- and slurry-phase reactors at two different pressures (i.e. 20 and 30 bar). All other operating conditions as well as the molar feed composition were kept the same for both processes.

As can be seen from the results of Fig. 5.5, in the slurry-phase process the particle polymerization rate decreases due to significant external mass transfer particle limitations (monomer is transferred from the gas phase to the liquid and then to the growing polymer particles). That is, the rate of monomer transfer from the bulk slurry phase to the catalyst surface becomes the limiting step in the polymerization. It is important to notice that by changing the overall operating pressure from 20 bar (in the gas-phase process) to 30 bar (in the slurry-phase process), the particle polymerization rate attains the same steady state value (see black and blue lines in Fig. 5.5) after an initial transient period. On the other hand, in the gas-phase process particle overheating (i.e. accounting for the temperature difference between the particle's external surface temperature, T_s, and the temperature in the bulk gas phase, T_b) exhibits a higher deviation from the bulk phase temperature than the respective deviation observed in the slurry-phase process due to significant heat transfer limitations (see Fig. 5.6). It is apparent that external heat transfer limitations greatly affect the particle overheating, especially, in the case of highly active catalysts (e.g. catalysts exhibiting polymerization rates higher than 15,000 $g/g_{cat}/h$). On the other hand, no significant particle overheating is observed in the slurry-phase process due to the high heat transfer particle dissipation rate.

In Fig. 5.7, the final cumulative MWDs of polyethylene in gas-phase and slurry-phase reactors are illustrated. It should be noted that the MWD of polyethylene produced in a gas-phase reactor is shifted to larger chain lengths in comparison to

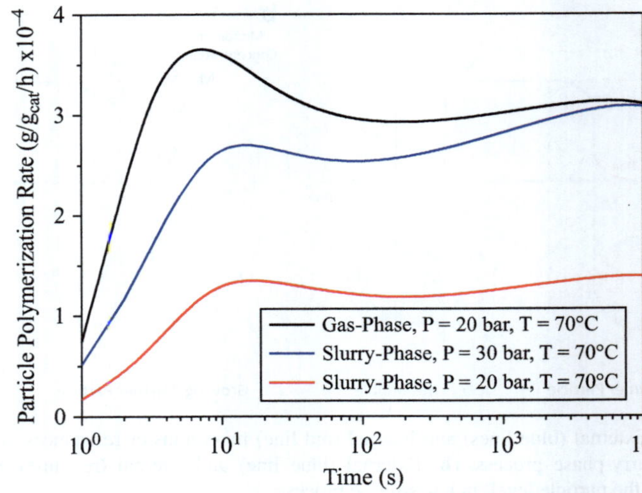

Fig. 5.5 Dynamic evolution of the particle polymerization rate in gas- and slurry-phase catalytic olefin polymerization reactors

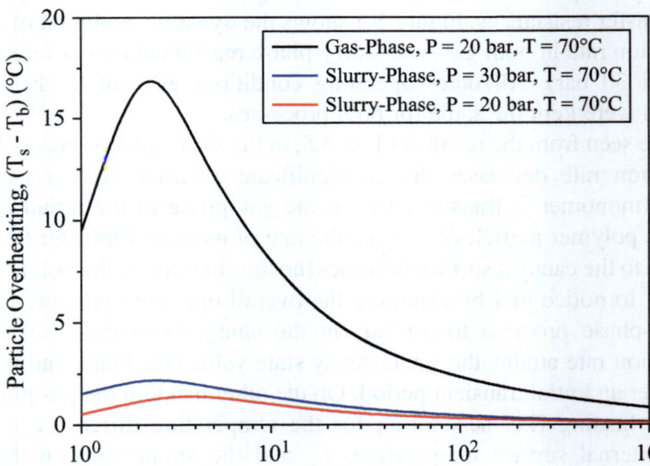

Fig. 5.6 Dynamic evolution of the particle overheating in gas- and slurry-phase catalytic olefin polymerization reactors

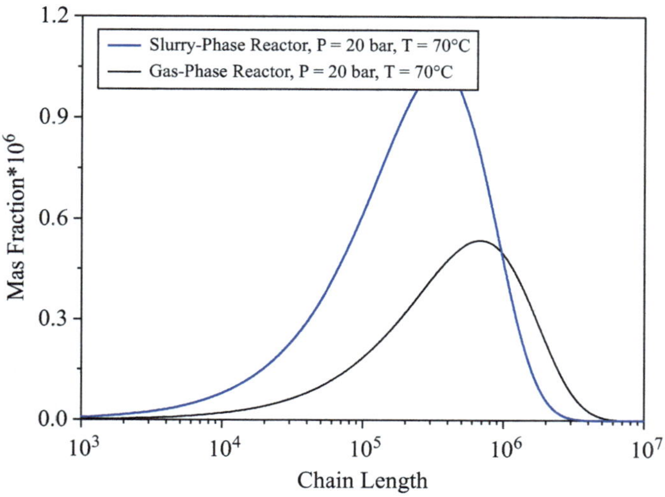

Fig. 5.7 Calculated MWDs of polyethylene produced in gas- and slurry-phase Z–N olefin polymerization reactors

the respective MWD of polymer produced in a slurry-phase loop reactor, under identical operating conditions. This can be explained by the fact that the particle polymerization rate in the gas-phase reactor will be higher than that exhibited in a slurry-phase reactor for the same operating conditions (pressure, temperature, feed composition, etc.).

In Fig. 5.8, the particle size distributions (PSDs) of particulate polymer produced in gas-phase and slurry-phase reactors are depicted at $T = 70$ °C and $P = 20$ bar. As can be seen, the polymer PSD in the slurry-phase reactor is shifted to smaller sizes in comparison to the respective PSD in the gas-phase reactor. Specifically, in the slurry-phase reactor the mean particle diameter is significant smaller than that obtained in the gas-phase reactor (i.e. 1450 vs. 1780 μm), due to the presence of significant monomer mass transfer limitations in the slurry-phase process. Thus, the polymer PSD is shifted to smaller sizes due to the lower particle growth rate. Moreover, the mass fraction of small-size particles (i.e. fines) in the slurry-phase reactor is substantially larger than that in the gas-phase reactor. It should be noted that the mass fraction of small-size particles in the slurry-phase reactor increases as the external monomer mass transfer limitations are significantly increased. However, as the monomer mass transfer resistance in the slurry-phase process decreases (i.e. the k_{ext} increases, $k_{ext} > 50$ m/s) the polymer PSDs in the slurry- and the gas-phase reactors become almost identical. It is evident from the previous analysis and simulation results that external mass and heat transfer resistances, at the particle level, can have a strong impact on the distributed polymer properties (i.e. PSD and MWD) in gas- and slurry-phase catalytic olefin polymerization reactors.

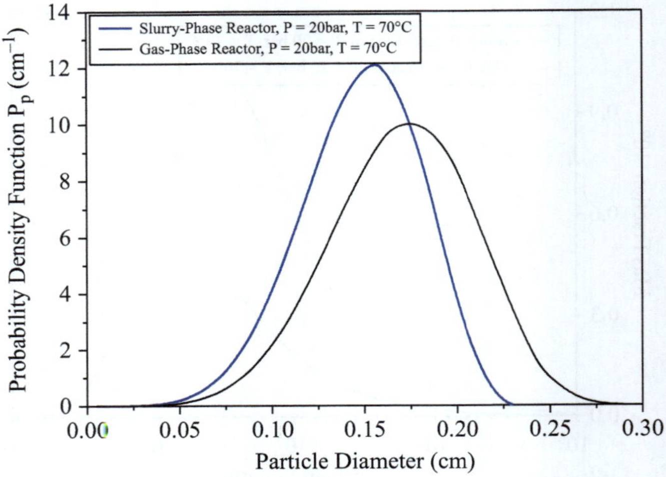

Fig. 5.8 PSDs of polyethylene produced in gas- and slurry-phase Z–N olefin polymerization reactors

5.4 Polyolefin Multimodal Processes

5.4.1 Introduction

The simplest polymerization processes have only one polymerization reactor. Even though a series of polyethylene or polypropylene grades can be made with a single reactor, certain products (those ones based on bimodal design) require the presence of two or more reactors in series (see Table 5.5). The use of multistage reactor configuration, despite increasing the plant capital cost, adds flexibility to the polyolefin production process, widens the product portfolio, and is becoming the standard in the industry nowadays. Impact polypropylene resins, for instance, need two reactors, one to make the isotactic polypropylene phase and another to produce the ethylene/propylene rubber components. Likewise, polyethylene grades also require the presence of multireactor configurations for producing polymer with broad MWD. In general, the distributed molecular properties of polyolefins, defined by the term polymer "chain microstructure", include the molecular weight distribution (MWD), the copolymer composition distribution (CCD), the short-chain branching distribution (SCBD), and the long-chain branching distribution (LCBD). The polymer chain microstructure to a large extent controls the rheological (see Fig. 5.9), the end-use properties and, thus, the application window of the produced polyolefins.

For example, high molecular weight polyolefins exhibit improved mechanical properties (e.g. toughness, strength, impact resistance and environmental stress cracking resistance, etc.). However, high molecular weight resins have higher melt viscosities and, therefore, it is more difficult to process them. On the other hand,

Table 5.5 Multimodal low pressure PE production processes

Configuration	Technology	Features	Reactor configuration
Loop-gas phase	BORSTAR	• Whole density range accessibility • Short transition times • Increased homogeneity	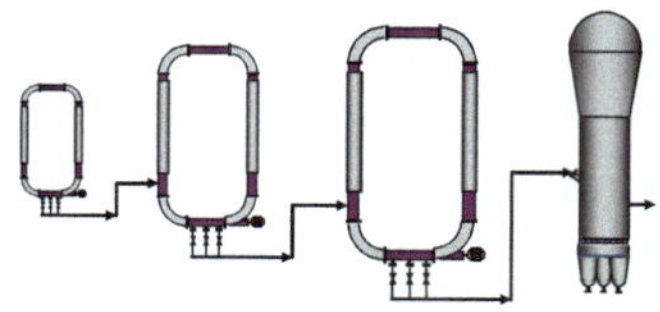
Gas phase–gas phase	UNPOL II	• Whole density range accessibility • Long transition times	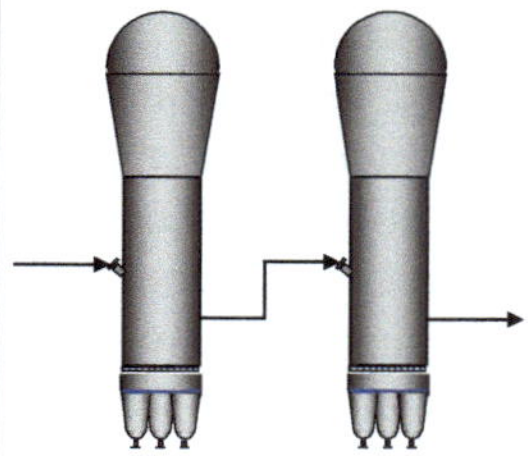
	EVOLUE		
	SPHERILENE C		
Double slurry loop	Fina (TOTAL PC)	• No LLDPE • Limited product design capabilities	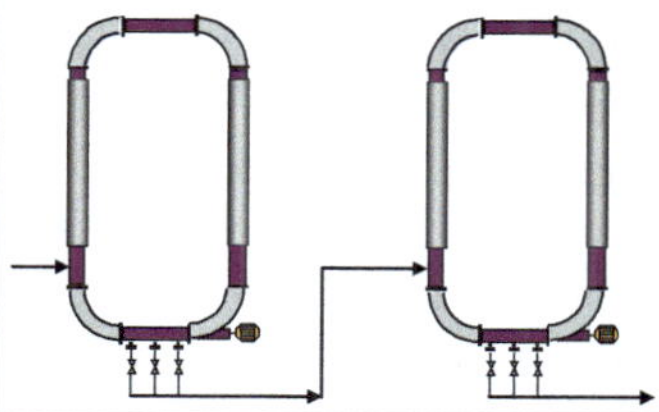
	Solvay (INEOS)		
	Showa Denko (JPE)		
Dual/triple slurry tank	HOSTALEN	• No LLDPE • High costs	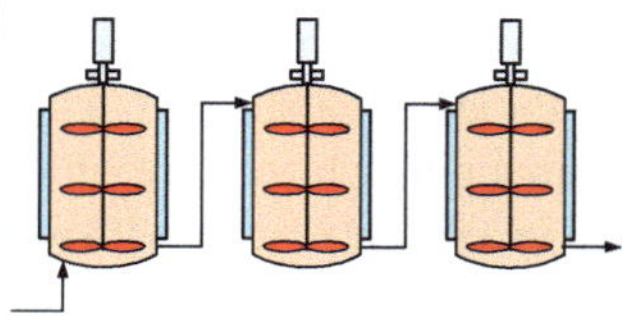
	MITSUI CX		
	EQUISTAR MARUZEN (NISSAN)		
	Nippon (JPE)		

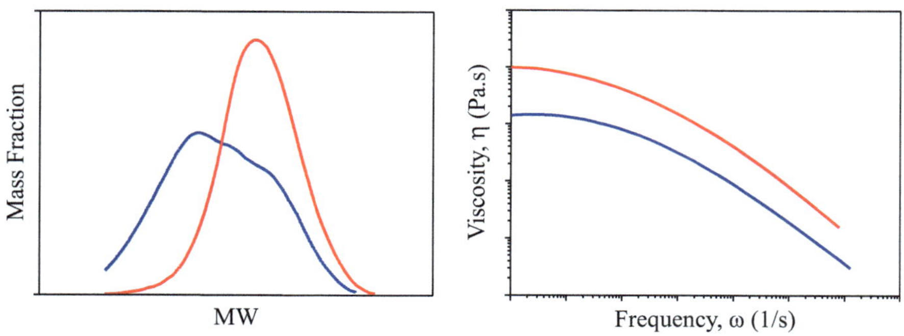

Fig. 5.9 Effect of MWD type on rheological properties [13]

polyolefins having a broad or bimodal MWD exhibit improved processability and, particularly, extrudability. Therefore, the accurate control of the molecular architecture of polyolefins in terms of the process operating conditions or/and catalyst design, is the most important issue in the production of polyolefins.

As it is depicted in Fig. 5.9, the molecular weight is as important as the density in terms of determining the final polymer properties. Often, a combination of properties will be required; for instance, high molecular weight products (i.e. low MFR) are stiffer and impart high tensile strength, whereas low molecular weight polymers are easier to process.

Therefore, many products will require a mixture of both long and short chains; it is very common to find polyethylene products with broad or multimodal MWDs. In the case of polypropylene, the final products can also be mixtures of homopolymers and random copolymers or of semi-crystalline homopolymers and amorphous elastomers. In general, there are two different approaches to make products with such mixtures of molecular weights and/or compositions: blending different products in an extruder or creating in-reactor product blends. The latter option is more effective than the former because it is almost impossible to achieve molecular-level mixing of two polymers (even if they have the same chemical structure) by mechanical mixing. Figure 5.10 illustrates the concept of different routes to obtaining in-reactor blends of different chain lengths. One can produce a broad, unimodal MWD that contains appropriate amounts of long and short chains to produce polyethylene with a decent balance of mechanical strength and processability, or one can use a multimodal MWD and tailor the averages and relative quantities of each mode to obtain the desired balance of properties. A unimodal MWD can be made in one reactor with a conventional catalyst at a set temperature and pressure. A multimodal MWD requires either a mixture of single-site catalysts (supported or in solution) that are used in a single reactor with a unique set of operating conditions or, more commonly,

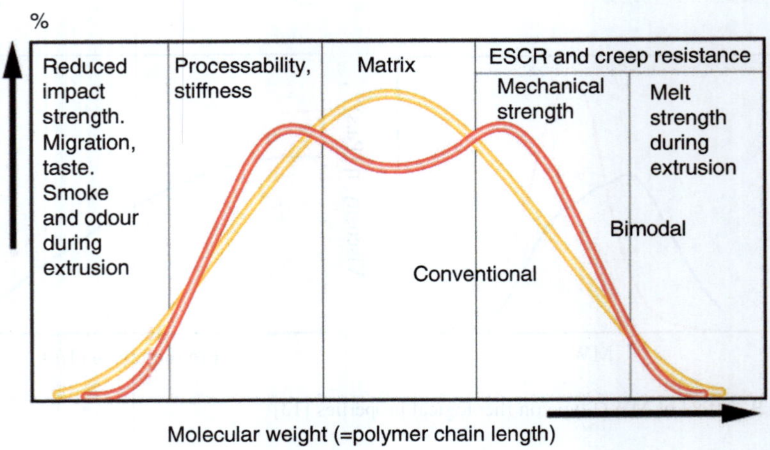

Fig. 5.10 Relationship between MWD and end-use properties—unimodal versus multimodal MWD and related properties affected by each fraction of MWD

one catalyst that it is exposed to different sets of conditions in a cascade of reactor zones, for instance, two or three reactors in series. The multiple reactor process will clearly offer far more fine control over the relative quantities and sizes of the polymer molecules in the final product than will the single-reactor option because of the increased number of degrees of operating freedom.

This chapter will be continued with a presentation of the most important multimodal processes for the production of polyolefins (both polyethylene and polypropylene) that are currently being licenced [14].

5.4.2 *Multimodal PE Processes*

The Split Loop Borstar Process

Background Borealis has not made its Borstar technology available for third-party licensing since November 2006. The technology had been available for license on a selective basis since a change in the majority ownership of Borealis in October 2005. Borealis continues to develop the Borstar process technology introducing a number of valuable process improvements including (1) redesign of catalyst feed system used for Borealis' proprietary high yield catalyst, (2) introduction of a continuous outlet from the loop reactor, (3) a higher pressure flash stage between the loop reactor and the gas-phase system aiming at energy consumption improvements, (4) use of hydrocyclone after the loop reactors, and (5) redesign of recovery area, improving diluent recovery, reducing energy consumption, and investment cost, operation of the gas-phase reactor in condensing mode.

Process Description As such, the Borstar process is based on two reactors in series to give the bimodal capability that is Borealis' route for achieving easy processing resins. The recently upgraded split loop Borstar technology from Borealis represents the third-generation Borstar technology and uses a combination of slurry loops (a pre-polymerization reactor and two loops) in series followed by an FBR to make a full range of polyethylene products. According to this process product portfolio can be largely extended while more homogeneous products with trimodal molecular weight distribution can be produced. As illustrated in Fig. 5.11, the split loop process consists of two loop reactors having different volumes, different solids concentration and operate under the same mean residence times. The polymerization in the slurry phase occurs in supercritical propane, followed by a flashing step and then at least one (possibly two) gas-phase reactor.

The slurry loop process for HDPE is well known and is the leading process for pipe and blow moulding grades; however, it has never been able to offer the full range of LLDPE grades, since the copolymer tends to soften at the operating temperature, dissolves in the diluent, and fouls the reactor walls. The slurry loop process, originally offered by Phillips, has been revolutionized by Borealis. The key difference between the Borstar process and a conventional slurry loop process is that

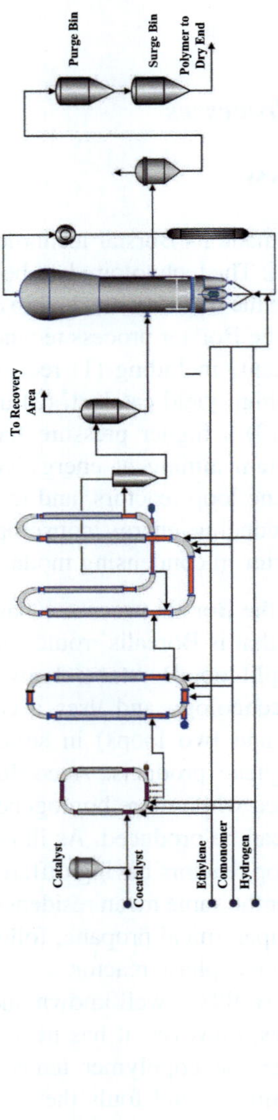

Fig. 5.11 The Borstar split loop process

Borstar uses liquid propane at a pressure above its critical point, whereas Phillips uses isobutane below its critical point.

The solubility of polyethylene at higher temperatures in supercritical propane is less than it is in isobutane, as shown in Fig. 5.12, so fouling of the reactor is avoided. Furthermore, in order to make a high melt index resin, a high hydrogen content is needed to cap the polymer chains, and this is possible in supercritical propane without bubbles forming.

Before catalyst enters into the loop reactors it is treated in a pre-polymerizer in order to improve the polymer particle morphology and give highly active catalysts time to reach an appropriate size for injection into the main loop reactors. This is necessary since the loop polymerization conditions are such that the local reaction rates are quite high, and fresh catalyst particles, especially the highly active ones used in this process, would not maintain their integrity without pre-polymerization. In Borstar split loop process, the catalyst slurried in propane and co-catalyst along with hydrogen and ethylene (after treating) are fed to a single loop pre-polymerizer. The pre-polymerization step ensures good morphology of the final polymer and makes up a few per cent of the total polymerization. No additional catalyst is required in the subsequent reactors (loops and gas-phase). This high activity catalyst results in ash levels of 1 ppm (LLDPE) to 2 ppm (HDPE) of titanium. The pre-polymer along with additional ethylene and hydrogen for molecular weight control are fed to the first loop reactor where the mean residence time is about 45 min. Subsequently, the material is continuously transferred in the second (larger in size, four-leg reactor) slurry loop reactor. Addition of comonomer (1-butene, or 1-hexene) as well as hydrogen (depending on the product design) is also fed to this loop reactor to get the lower density fractions.

The slurry discharge to the flash tank contains about 60 wt% polymer solids, but the low density of supercritical fluid makes it possible to operate at even higher (close to 70 wt%) solids concentration. The overheads from the flash tank are compressed and sent to the heavies scrubber column where a small amount of heavy components are separated and sent to off-sites for uses such as boiler fuel.

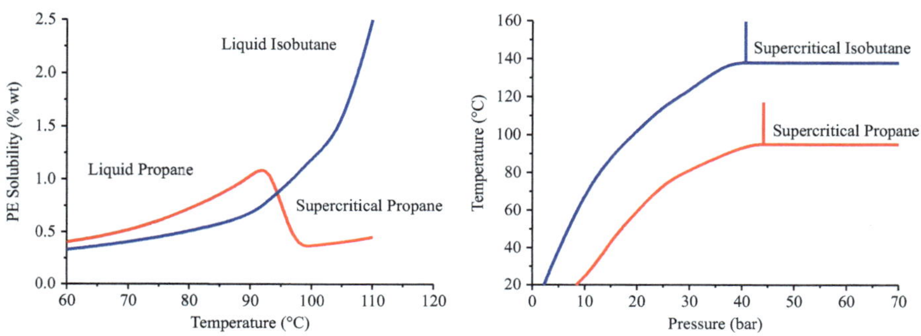

Fig. 5.12 (**a**) Polymer solubility and supercritical conditions. (**b**) Phase diagram of isobutane and propane

Meanwhile, the dried polymer from the flash tank is conveyed through a proprietary gas lock/conveying system to the gas-phase fluidized-bed reactor (FBR). In this reactor, the partial pressures of ethylene and comonomer are accurately controlled to give both the right composition of the polymer produced and to maintain a certain production split between the two reactors. Hydrogen is also used for molecular weight control. This polymerization reactor produces HMW resin fractions with MI typically below one and densities down to even 900 kg/m^3. This gives an overall LLDPE product output of densities down to 922 kg/m^3 (commercial) and 918 kg/m^3 (semi-commercial). This reactor operates at about 80 °C and 20 bar, and employs a bottom-mounted agitator that can enhance the mixing efficiency in the critical zone just above the distributor plate. A second FBR in this process would serve a function similar to that in the gas-phase processes mentioned above: allows one to make a copolymer phase with bimodal composition and/or MWDs.

Similar to gas-phase processes, hydrocarbons are vented from the top of the reactor and are compressed, cooled, and recycled to the reactor in order to remove the heat of polymerization. The use of propane as an inert component and the primary component in the atmosphere of the gas-phase reactor results in a productivity advantage. For example, the heat capacity is about three times that of a nitrogen atmosphere.

The polymer from the gas-phase reactor is discharged to the product degassing tank. The solids are sent to a purge bin in which the polymer is scrubbed with nitrogen to remove residual hydrocarbons and steam to deactivate the catalyst. The overheads from the purge bin are vented to the flare or optionally the small amounts of hydrocarbons are recovered in a membrane separation unit. The overheads from the product degassing tank are compressed and recycled to the gas-phase reactor. The polymer from the purge bin is conveyed to the downstream extrusion and pelletizing steps.

Key Features In Borstar split loop technology there is a flexibility to produce up to three different polymer qualities in each of the reactor of the configuration allowing novel product designs (see Fig. 5.13) with applications in pipe (e.g. trimodal concept for PE125, High temperature PE pipe, new HDPE Topcoat, trimodal material with higher melt strength), film and fibre (high toughness and low sealing temperature, multimodal material with good optical properties, high toughness linked with elasticity, etc.) and moulding (e.g. next generation injection moulding grades with enhanced modulus, etc.).

One of the main goals of having two loop reactors in series is the increase of polymer homogeneity in terms of balancing between the polymer produced in the loop reactors and in FBR. This is clearly depicted in Fig. 5.14 where it can be seen that the presence of the second loop reactor increases the probability of small sized particles which escaped from the first loop to stay in the second loop of the configuration, thus increasing their mean residence time in loop reactor conditions as well as producing more homogeneous product. It is obvious that the mass fraction of small-sized particles (fines) leaving earlier the loop reactor dramatically decreases when the split loop reactor configuration is employed. This has a tremendous effect on the final polymer properties, especially for the production of bimodal pipe grades since the formation of white spots can be eliminated.

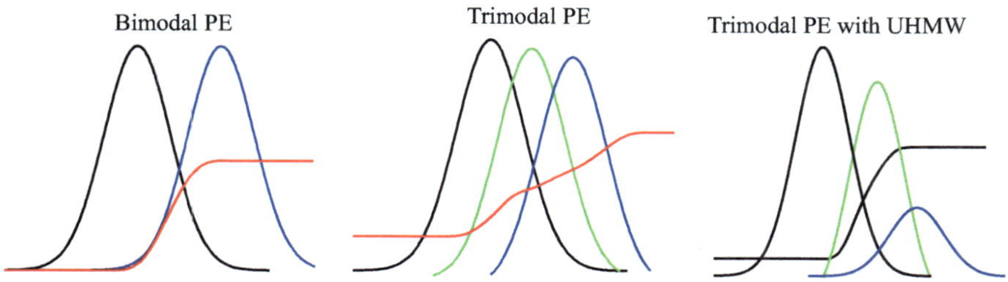

Fig. 5.13 Potential PE multimodal product designs to be produced by Borstar split loop process—Towards multimodal PE (MWD is depicted by black, green, and blue lines and CCD is depicted by red lines)

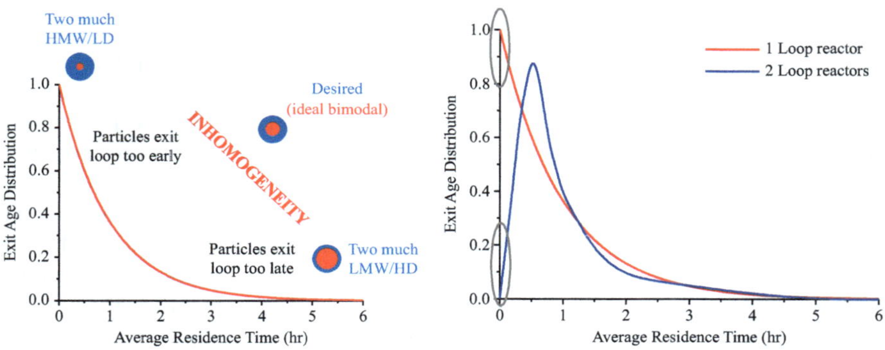

Fig. 5.14 Particles homogeneity in split loop reactor configuration

One of the striking features in Borstar loop reactors is the use of propane as diluent at elevated pressures (i.e. >55 bar) which increases polymer swelling, lowers polymer phase density, thus, enhances the solubility of reactants in the polymer phase (Fig. 5.15). This particular phenomenon has extremely big effect on the selection of the operating conditions, the solids concentration in the slurry loop reactors as well as in the determination of the pump power that is required to circulate the polymer fluff along the loop reactor.

The inter-reactor flash removes all propane, hydrogen, and comonomer used in the loops and allows the slurry- and gas-phase reactors to be run under independent conditions. This is a feature that is commonly found in the mixed-phase processes used in polypropylene production. Low molecular weight polymer is made in the loop, and higher molecular weight products in the FBR. The advantage of finishing in an FBR is that there one can add varying levels of comonomer without risking issues related to solubility (while the solubility of amorphous polyethylene is practically zero in supercritical propane). This makes the Borstar polyethylene process a true swing process capable of producing LLDPE with densities ranging from over 960 to 918 kg/m^3.

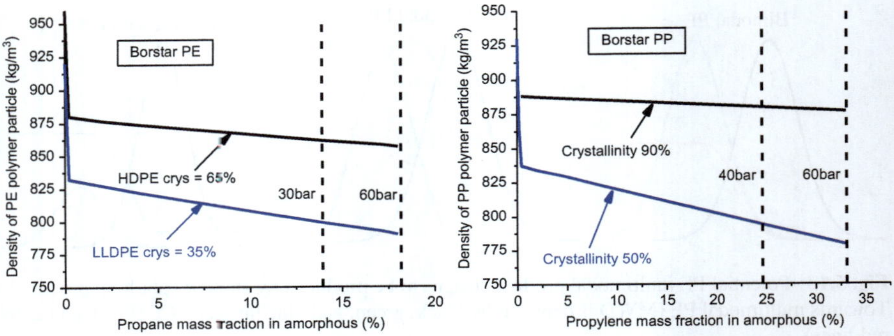

Fig. 5.15 Variation of PE and PP densities in Borstar loop conditions

During the operation of the slurry-phase olefin polymerization loop reactors the precise control of particle size and molecular weight distributions is important, but it is very difficult to realize in practice. During the operation of such a process the concentration of reactants, including monomer, one or more optional co-monomer (s), and hydrogen, varies and tends to decrease as the reactants are converted to polymer (i.e. reactants depletion phenomena). While reactants are depleted along the length of the loop reactor, fluctuations in temperature and reactants concentration occur along the reactor. As the length of the loop reactor increases, the concentrations of reactants tend to vary to a greater extent. The degree to which the concentration of reactants diminishes depends also on the polymerization kinetics which in turn is related to the employed catalyst system. The variation in concentration of a reactant is more pronounced when highly consumption rates are observed during the polymerization process. This contributes to non-uniform polymer properties resulting in poor product homogeneity.

The biggest difference in reactants concentration variations is generally observed between a point right before and a point right after a feeding point. The longer the reactor, the more important these concentration variations will be. As a result, polymer chains that are formed near the reactor inlet (as defined at the point at which the polymerization reaction commences) and polymer chains formed near the reactor outlet may exhibit different molecular properties.

A solution to the above issue is the utilization of multiple injection points. According to this principle, the multiple entries for feeding monomer, hydrogen or/and comonomer could be positioned spatially separated from each other on the reactor. Preferably, the multiple feeding entries of the reactants are positioned equidistantly along the reactor path in order to keep the hydrogen/monomer and comonomer/monomer ratios substantially constant along the reactor length. Alternatively, the reactants feeding entries may be provided at non-equidistant positions on the reactor. Particular optimized positioning sites of feeding points may be chosen as a function of operating conditions (i.e. temperature, pressure) and process parameters (hydrogen–monomer ratio, comonomer–monomer ratio, reactor pump activity, solids concentration distribution in the reactor, reactants flows, etc.).

In addition, due to spatial reactants concentration variations, reactor temperature may also vary throughout the reactor. Due to the consumption of monomer or/and comonomer in the loop, the reactor temperature may differ along the loop length. For example, in parts of the loop where less monomer is present, the polymerization rate will be reduced and the temperature will decrease. It should be noticed that fluctuating temperature throughout the reactor are absolutely detrimental with respect to polymer homogeneity in terms of composition. As a consequence of the different concentration of the reactants at various reactor locations, (co)polymers may have varying densities and particle size distributions throughout the reactor. A solution to the above challenging process operability and product quality inconsistency issues can be offered by splitting a loop reactor into two loop reactors of smaller volume/ shorter in length (split loop concept). By implementing such reactor configuration, a more even reactants spatial composition as well as a constant production rate along the reactors length can be achieved, and, as a consequence the polymer product homogeneity can be significantly improved.

The homogeneous particle size distribution (PSD) of the produced polymer in slurry-phase loop reactors is among the most critical issues towards its operability and performance. In this type of reactor, the broad catalyst (or/and pre-polymerized particles) size distribution in the feed stream in conjunction with the broad residence time distribution leads to the broadness of the PSD, which in turn results in decreased polymer heterogeneity. Broad PSD causes solids flowability problems due to the presence of fine particles and results in severe mass and heat transfer limitations due to the presence of large size particles. Moreover, when particles with broad PSD enter into a next polymerization reactor (especially in a gas phase) significant processability issues occur related to particle agglomeration, especially when copolymer grades are produced. It should be pointed out that PSD developments in gas-phase reactor dramatically affect the selected fluidization conditions and the hydrodynamic behaviour of the gas–solids mixture. Therefore, decreasing the polymer homogeneity in loop reactors can lead to a non-stabilized FBR operation in a configuration where slurry-phase loops is followed by a gas-phase reactor.

Product Portfolio Borstar PE produces a full density range of bimodal LLDPE/ HDPE (e.g. from 918 kg/m^3 for film to 960 kg/m^3 for injection and blow moulding grade resins) using its proprietary Ziegler–Natta based catalyst family. MI for bimodal products ranges from a high load melt index (HLMI at 21.6 kg) of 5 to an MI (at 2.16 kg) of above 10. Because of the claimed improvements in ESCR, mechanical strength, and processability, Borealis' current focus is on specialty (premium) polymers for pipe, blow moulding, wire and cable, and film applications across the LLDPE/HDPE range. Due to Borealis' market position in Europe, a strong target application is for pressure pipes.

The high-end fraction of the bimodal polymer gives it high melt strength and stiffness, good orientation, and good (low) swell properties. The low-end fraction gives extrudability without the tails that give taste and odour problems. As such, Borealis states that its HMW LLDPE film has extrudability that allows it to be used as a pure LLDPE resin on conventional LDPE lines. Additionally, it can be used in blends and/or in co-extrusion with all polyethylene resins. Borstar high strength LLDPE film has a non-blocking matt surface and is mainly used in pigmented applications or in co-extrusion, where the surface and optical properties are defined by the outer materials.

Advanced Cascade Process by Hostalen

Background and Key Features Hostalen ACP (Advanced Cascade Process) technology was made available for third-party licensing in 2009 and since then, more than 1.3 million tons per year of HDPE capacity has been licensed. Key recent developments relating to the Hostalen technology are as follows:

- LyondellBasell has developed the Hostalen ACP technology, which uses three stirred reactors in series, as an extension of their Hostalen bimodal technology to multimodal capability.
- Polymerization conditions can be separately adjusted for each reactor.
- The process uses a single family of Ziegler catalysts.
- Hostalen bimodal process can be expanded to Hostalen ACP.
- LyondellBasell has developed various bimodal and multimodal film, bottle, and other blow moulding grades based on Ziegler Avant Z509 and Avant Z501 catalysts.
- LyondellBasell's strategy for the Hostalen technology is to further develop it as part of its complete polyethylene technology portfolio for high performance HDPE products.
- LyondellBasell successfully introduced the new so-called multimodal HDPE products for film, blow moulding, and pressure pipe applications based on Hostalen ACP technology in Europe. Further polymers are under development for high performance film and blow moulding applications.

Process Description The Hostalen technology has been specialized in an ideal combination of catalyst and process to be able to generate proper, sophisticated, and advanced polymer structures (high molecular mass multimodal HDPE) specifically designed to match the needs of the more demanding and advanced application sectors, such as film, pipe, and blow moulding.

The Hostalen ACP process scheme is depicted in Fig. 5.16. Avant Z501 or Z509 catalysts, which are supplied by LyondellBasell, can be fed into the catalyst dilution vessel. After further dilution with hexane, the catalyst feed to the reactors is done through a catalyst-dosing device. When operated in series, catalyst is fed only to the first reactor.

For the Hostalen ACP process, the gas-feed mixture consisting of ethylene, comonomer, and hydrogen is continuously fed to the three reactors operated in series for multimodal HDPE production. The gas-feed mixtures to the three reactors for multimodal production are different; the first reactor is fed a gas-feed mixture consisting of ethylene and hydrogen, while ethylene and comonomer are fed to the following two reactors (the reactors operate in parallel for unimodal production).

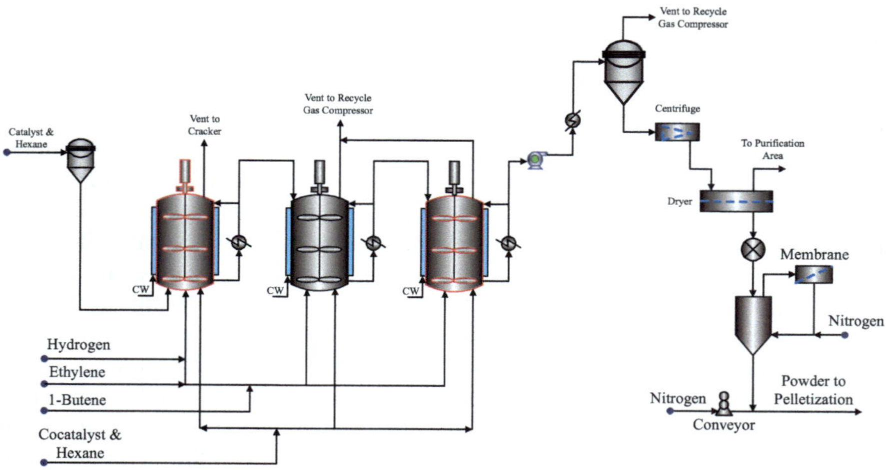

Fig. 5.16 Hostalen ACP by LyondellBasell

The catalyst is not sensitive to monomer impurities; ethylene purification is not required if the ethylene is supplied from a state-of-the-art cracker. Hexane solvent is simultaneously pumped to the reactors. Hexane is recovered downstream from the powder product and recycled back to the reactors. Co-catalyst is added to the recycle hexane stream.

The polymerization occurs at a pressure of 5–10 bar and a temperature of 75–90 °C; the reaction is exothermic and the heat of polymerization is removed via cooling water. Molecular weight, molecular weight distribution, and polymer density are controlled through catalyst concentration and variation of comonomer and hydrogen concentration and by running the reactors in series or parallel, while 1-butene is employed as the comonomer. Residence time is 0.7–2.5 h per reactor and catalyst yield can be as high as 50 $kgPE/g_{cat}$.

In the Hostalen ACP process, the polymer suspension flows from the reactors to the decanter feed tank (see Fig. 5.16). The vent gas from the decanter feed tank, as well as from the second and third reactors, is sent to further processing for comonomer and diluent recovery. The polymer slurry concentration is 200 kg/m^3 or 31 wt %. It is fed to a decanter centrifuge where the polyethylene powder is separated from the hexane mother liquor. The polyethylene powder is fed to a two-stage drying section and the hexane mother liquor is sent to a collection vessel.

In the drying section, residual hexane is stripped from the polyethylene powder with hot nitrogen in closed loops. The residual hexane concentration in the powder is less than 0.01 wt%. Dry powder is conveyed via a closed-loop nitrogen system to the pelletizing section. The average particle size is 140–250 μm. The small amount of vaporized hexane absorbed out of the nitrogen loops (not shown) is sent to the hexane collecting vessel. The majority of the hexane from the polymer slurry is directly recycled to the reactors, without purification. This enables reuse of contained co-catalyst and comonomer, cutting overall operation cost. A portion of the hexane stream (depending on the resin grade produced) is diverted to hexane distillation for removal of light ends and waxes from the mother liquor.

Product Portfolio The Hostalen ACP proprietary technology is based on LyondellBasell's own process know-how and is focused on premium products for film, pipe, blow moulding, tape, monofilament, and injection moulding applications. Products produced have a crystallinity of 60–80% and density ranging from 942 to 965 kg/m^3. The Hostalen process is based on a titanium Ziegler catalyst with improved morphology. The catalyst package consists of catalyst systems that produce a full range of materials, such as unimodal high density products with an MFI (at 2.16 g/10 min) range of 0.4–18 and a density range of 944–962 kg/m^3 and multimodal high molecular weight products with an MFI (at 2.16 g/10 min) range of 0.05–0.55 and a density range of 944–956 kg/m^3. Bimodal and more extensive multimodal products feature high stiffness combined with high toughness or high stress crack resistance when compared with unimodal products. For multimodal film grades, processing advantages include increased bubble stability and significant down gauging, in combination with increased mechanical properties. PE100 pressure pipe grades produced by LyondellBasell in the Hostalen ACP process have met the standards of the PE100+ Association.

The Spherilene C Process

Background and Key Features The Spherilene process for the production of polyethylene was developed and commercialized by Montell (prior to the formation of Basell) and its licensees as a cascade gas-phase technology beginning in 1994.

In 2005, Basell unveiled a modified Spherilene process that incorporated design aspects from their Lupotech G technology, to create a unified gas-phase technology that is licensed under the name Spherilene S. This process produces the standard slate of products normally expected for unimodal LLDPE and HDPE resins from a Ziegler-based gas-phase technology; for broad molecular weight HDPE resins, chromium catalysts are used.

For Spherilene C (multireactor configuration), catalyst development remains central to the process technology, with the development of a new series of Avant Z direct injection catalysts that require no pre-polymerization. The new catalysts continue to be based on a spherical magnesium chloride support and preserve the key features of the earlier Spherilene S catalyst system in terms of particle morphology control, high activity, and the ability to start production with an empty reactor. The catalyst chemistry and more stable support structure is said to provide expanded product properties and improved operational stability. The new catalysts also have enabled simplification of the process, eliminating certain catalyst preparation equipment and associated systems, thus decreasing the investment cost. LyondellBasell claims for the Spherilene C process that morphology development control is essential for excellent reactor operability (e.g. no reactor and exchanger fouling, no hot spots) and better resin uniformity (e.g. no "burnt or fused" particles that can cause non-homogeneity and gels in film).

For the development of Spherilene C, LyondellBasell's overall achieved objectives are as follows:

- To produce the range of LLDPE/HDPE (narrow and medium MWD), unimodal products with a single Avant catalyst family of Ziegler catalysts and to use its Reactor Granule.
- Technology to control particle morphology and incorporate multiple comonomers independently, while producing a highly consistent product.
- To improve heat transfer during polymerization for chromium HDPE resins with the use of propane as the inert.
- To achieve better economics through the possibility of monomer recovery, without any additional investment, by using propane as the inert.

A key advantage claimed by LyondellBasell is that the overall Spherilene C process achieves a high space-time yield without the use of any "condensing mode" of operation. The high productivity relates to an overall residence time of the polymerization steps, of about 2.5 h. Another process advantage is derived from the fact that a light hydrocarbon is used as the diluent instead of nitrogen in the system. This improves the heat transfer capability since it is proportional to the gas density, thus improving thermal stability at high polymer throughput (i.e. reducing chances of hot spots). Another factor is that the average particle size allows higher fluidization rates. In addition, using propane as the inert enables condensation of the recirculating gases without the need for a separate condensation-inducing agent.

Process Description The Spherilene C process is shown in the simplified flow sheet presented in Fig. 5.17. Feeding the supplied Avant Z catalyst via an "in process" activation step, the gas reactor can be started up from an empty condition since no seed resin is required. This allows for shorter start-up transitions and shutdown time while decreasing the complexity and logistics of operation.

While feeding the Avant C catalyst, a seed resin is required and this can be transferred from the powder silo with a transport system. Product withdrawn from the reactor is degassed using propane and treated with moist nitrogen to neutralize residual catalyst activity and sent to the powder silo for extrusion. The catalysts are injected directly into the polymerization reactor. The polymerization temperature and pressure depend on the molecular characterizations of the desired resin. Due to the combination of catalyst performance and reactor conditions, the product morphology is said to be excellent, with low fines, high bulk density, and minimal lump formation.

After the first gas-phase reactor the polymer particles are fed in the second gas-phase reactor. From the second reactor, polymer is continuously discharged from the bottom of the reactor to a disengaging bag filter, through a proprietary transfer device. The gas overhead is recompressed and recycled back to the reactors. The vapour from the top reactors exit is cooled to remove the heat of polymerization, mixed with the fresh monomer feeds, and fed to the reactors.

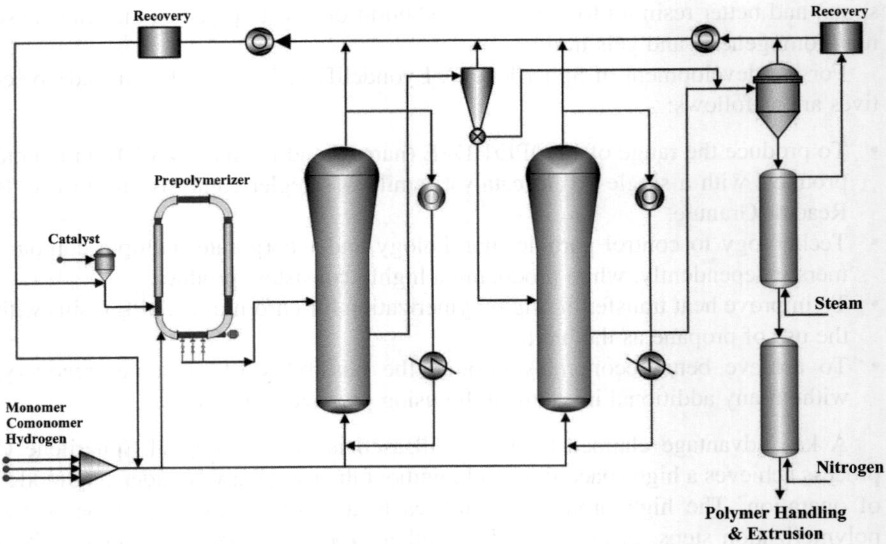

Fig. 5.17 Spherilene C Process by LyondellBasell

Depending on the product to be synthesized, the reactor process conditions and the polymer compositions vary (e.g. temperature from 70 to 90 °C, pressure: 15 to 30 bar, the gas-phase composition depends on the required product composition, the final product composition is 70–100 wt% ethylene and 0–30 wt% comonomer. From the second reactor bag filter (or even from the first, depending on the mode of operation), the polymer is sent to a degassing vessel for monomer stripping, which is accomplished with propane. By stripping the hydrocarbons with a hydrocarbon, LyondellBasell keeps the entire gas stream free of contamination and is thus able to recycle the stream back to the reactors with minimal cleaning and separation systems. The degassing is followed by catalyst/alkyl deactivation. The deactivation of catalyst residues and stripping of the dissolved hydrocarbons are obtained by means of contact between nitrogen, steam, and the polymer. The polymer is conveyed to the nitrogen degasser where the hydrocarbon stripping is completed. In a cost savings design aspect, the degassing silo serves a dual purpose as the intermediate silo prior to the extruder.

Product Portfolio Spherilene C and S (one reactor configuration in used) produces the standard slate of products normally expected for unimodal LLDPE and HDPE resins from a Ziegler-based gas-phase technology; for broad molecular weight HDPE resins, chromium catalysts are used. More specifically, from a product perspective, achieved objectives are as follows:

- To produce a wide product slate, from conventional 1-butene LLDPE grade for competitive resins (e.g. 1 MI/0.918 density) to 1-hexene HDPE.

- To extend the product slate with LLDPE and HDPE narrow and medium MWD specialty and conventional types with a single family of Avant Z catalysts.
- To enhance the value of the technology by producing MDPE grades for rotomoulding applications; these can be made with different comonomers based on the targeted application's need for price sensitivity.

With regard to the product range using a single catalyst family, the claimed density range demonstrated in the smaller scale plants is from 916 to 960 kg/m^3, with a melt index range of 0.01–100 g/10 min at 2.16 kg.

5.4.3 Multimodal PP Processes

In this family of processes, polymerization to make homopolymer and random copolymers with low ethylene content takes place in liquid propylene without the use of an inert diluent. This is a significant simplification over the traditional diluent slurry process, since propylene can be separated from the polymer by flashing and there is no need for the extensive diluent recovery system. Running in liquid monomer significantly enhances the reaction rate and productivity, but one typically uses at least two well back-mixed reactors in series in order to control the MWD (Table 5.6).

The Spheripol PP Process

Background and Key Features The Basell joint venture (formed October 1, 2000) included the polyolefins assets of Shell and BASF (i.e. Montell, Targor, and Elenac) incorporating the development and licensing activities for Spheripol. LyondellBasell Industries (LBI) was formed in December 2007 when Basell merged with Lyondell, creating the third largest independent chemical company in the world. Recent key developments for LyondellBasell's Spheripol polypropylene technology include the following:

- Higher pressure operation of the slurry loop, allowing higher hydrogen concentration and higher operating temperature in the loop.
- Z/N catalyst systems.
- New gas-phase reactor unit design (without agitator) and wider rubber capabilities.
- Downstream improvements to improve performance and generate maximum benefits from new catalyst properties.
- LyondellBasell launched the Avant family of catalysts in 2001.
- New catalyst families are in development for broad MWD, high isotacticity products, and "drop in" supported metallocene catalysts.

Table 5.6 PP multistage production technologies

Technology	Configuration	Features
Borealis (Borstar)	1 loop + 1 FBR + 1–2 FBR	– High heat transfer capability – High solids concentration, high conversion – Better randomness (comonomer distribution) – Higher isotacticity but not optimal comonomer distribution – Easy reactors production split control
Basell (Spheripol)	2 loops + 1–2 FBR	– Broad product window – Random copolymers are limited in comonomer content – Heterophasic PPs (iPP) are limited in rubber content – Full tailoring of matrix of heterophasic PP not possible
Basell (Spherizone)	1 MZCR + optionally 1–2 FBR	– Production of broad MWD polymer in one reactor – No residence time distribution issue – Enhanced homogeneity of intra particle composition profiles (enhances product properties) – Bimodality achieved in MW, comonomer content, and isotacticity – No higher comonomer content in low MFR phase possible – Very complex control – Temperature gradient and monomer depletion in downcomer
Mitsui (Hypol II)	2 CSTR + 1–2 FBR 1 loop + 1 FBR	– Designed for flexibility, own donor technology – Complete range of homophasic, random, and heterophasic PP with ZN and SSC catalysts (Wintec™) from very stiff to very soft (Tafmer™) – Elastomers/plastomers with SSC (Notio™)
Basell (Catalloy)	3 FBR	– Brittle catalyst, therefore special attention in pre-polymerization operation – High comonomer content polymers – Very low modulus PP is produced
Ineos (Innovene)	1 FBR + 1 FBR (horizontal stirred gas-phase reactors)	– Powder moves as plug flow—short transition times – Cost reduction by increasing reactor size – Broad MWD in two reactors results in enhanced stiffness, melt strength and processability – High comonomer products (Good impact/stiffness) – Mechanical reliability of long horizontal stirrer is an issue
NTH (Novolen)	1 FBR + 1 FBR	– Relatively small vessels, therefore rapid product change – Reactors can be run in series or in parallel – Very clean products due to pellet washing system

(continued)

Table 5.6 (continued)

Technology	Configuration	Features
		– High stiffness homoPP, 2400 MPa without filler – Good impact/stiffness of iPP – High rubber content (up to 50 wt%) – Very low extractables, and volatiles
Dow (Unipol PP)	1 FBR + 1 FBR	– Low cost process, with product tailoring limitations – No catalyst pre-polymerization – Limited improvements in stiffness, creep, melt strength in reactor made material
Sumitomo	1 FBR + 1 FBR	– High crystallinity catalyst – Licensed direct feed concept – No available elastomeric PP products
Dow Solution Process	1–2 CSTRs	– The most suitable process for elastomers – Very short time for grade change – Versify PP copolymers

- Basell and ExxonMobil signed a metallocene polypropylene (mPP) technology research and development agreement in 2001 under which they will cross license patents and know-how. LyondellBasell is responsible for licensing to third parties. LyondellBasell and ExxonMobil will separately manufacture, market, and sell mPP resins.
- Product developments have included high stiffness homopolymers and copolymers under the ADSTIF trademark and high MFR (MFR 50 and 100) impact copolymers. Ultra-high MFR 1200+ homopolymers have also been introduced for melt blown (MB) fibre applications. LyondellBasell has also developed butene-1 based random and heterophasic terpolymers for clarity applications.

Process Description The Spheripol process by Basell is shown in Fig. 5.18 for a two loop and a single gas-phase reactor configuration. The Spheripol process uses loop reactors to make propylene homopolymers. In this process, a small loop reactor is used to pre-polymerize the catalyst; the main polymerization, for homopolymer or random copolymer, takes place in one or two loop reactors, each with four to eight legs depending on production requirements (for production rates more than 450 kiloton/year two loops are employed). This gives the capability of producing bimodal resins to increase mechanical properties (e.g. for bi-axially oriented PP, BOPP), by using any conventional general purpose Z/N catalyst available on the market. The reactors are operated at 75–80 °C and 40–45 barg. This is a higher and wider operating window than in the past, allowing for better catalyst performance, giving, for example, polymers with higher crystallinity combined with high isotacticity. Catalyst activities are typically 40–50 kgPP/g_{cat}.

In each reactor, a pump circulates the reaction slurry mixture, containing 55 wt% solids, at high velocity to avoid solids settlement and improve heat transfer. Cooling water is circulated through the jacket for heat removal. The total reactor residence time is 1.4 h for impact copolymer production. For homopolymer production, the

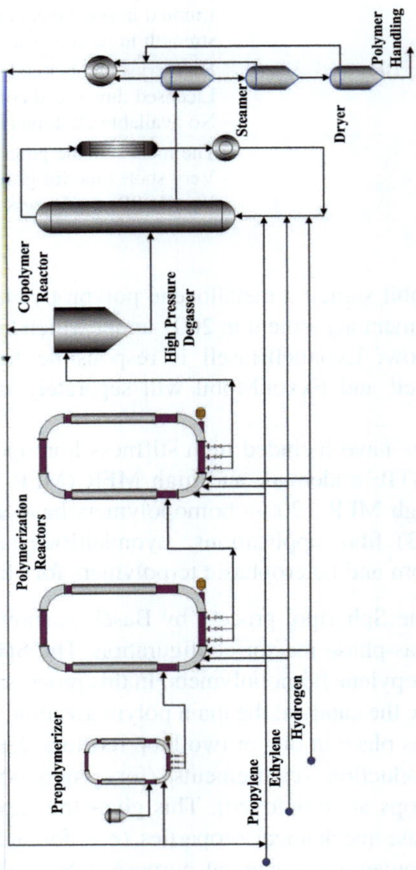

Fig. 5.18 Spheripol PP process

residence time is about 1 h. Product transition time may vary from 1 to 2 h (maximum) for homopolymer transitions, to a maximum of 6 h for the worst-case transition between butene-1 copolymer families and homopolymer at the end of a butene-1 copolymers campaign (due to butene-1 removal from the system).

Part of the circulating mixture is withdrawn from the reactor and flashed through a heated, jacketed pipe into a high pressure degasser, which is operated at about 18 barg, where the powder separates from the monomer. The monomer vapour leaving the flash tank is condensed against cooling water and is pumped back to the reactor. The polymer powder from the degasser, still containing some monomer, flows to a low-pressure flash filter or to a copolymer reactor if high impact copolymer is being produced. The low-pressure filter operates at 0.7 barg. The recovered vapours are compressed to 18 barg and are combined with the high-pressure monomer vapour for condensation and recycle to the reactor. For the production of impact copolymer, ethylene is separated and recycled.

In high impact copolymer production, the powder from the high-pressure degasser (with active polymerization catalyst) enters a gas-phase reactor where ethylene and additional propylene are added. Cooling is provided by means of a gas recirculation loop. This reactor is operated at 10–14 barg and at 70–80 °C. LyondellBasell operates two fluidized-bed, gas-phase copolymer reactors in series for the production of low blush copolymer (not shown). Levels of up to 35 wt% of rubber phase in the polymer are commercially achieved by using a single-stage copolymer reactor system, even if a two gas-phase reactor configuration can provide a better production rate for specialty impact copolymer products. Increasing the number of copolymer reactors allows one to obtain more homogeneous (in terms of rubber content) copolymers because of the narrowed RTD. Typically, residence times in the copolymer reactors are very small compared to the loops because of the high specific reaction rates of the copolymerization step.

The polymer from the low-pressure filter (with a monomer content of 1000–2000 ppm) is sent to a monomer stripping system for the removal of residual monomer. Here, steam strips all hydrocarbons, which are recovered, dried, and recycled back to the low pressure degassing step or made available to the battery limits for credit for propane purge when using polymer grade propylene. Wet polymer from the steamer is dried by a hot nitrogen closed loop. The residual monomer content in the polymer leaving the steaming and drying section is 3–5 ppm.

Product Portfolio LyondellBasell also produces enhanced polyolefin resins from a process known as CATALLOY. The CATALLOY process makes modified polyolefins by alloying a variety of monomers and comonomers in a series of gas-phase reactors. These are not included in this process evaluation. LyondellBasell offers a full range of products. The Spheripol process can produce polypropylene homopolymer with a commercial MFR range of 0.1–100 g/10 min (for automotive and appliance compounds) in pellets and up to 1860 as flakes, the latter suited to fibre applications (melt blown and spun bonded) with a narrow MWD and without the need for peroxide cracking (specialized non-pelleted material) and has produced

MFR of 2000 in the pilot plant. LyondellBasell is improving the clarity and impact properties of its (1-butene) random copolymers, which are finding application in flexible packaging including foodstuffs (ketchup and syrup bottles) and cosmetics.

In random copolymer production with Spheripol, LyondellBasell commercially produces resin with MFR from 0.1 (the benchmark for water pressure pipe) up to very high fluidity 75 g/10 min (for high transparency thin-wall injection moulding) with up to 4.5 wt% ethylene with a Z/N catalyst system suitable for food and medical applications. In addition, it manufactures terpolymer materials containing ethylene and butene, to compete with the higher ethylene content gas-phase products to achieve low heat-seal temperatures, as well as a new family of 1-butene modified random copolymers for film and soft fibre applications under the Clyrell trademark. LyondellBasell offers impact copolymers with MFR from 0.1 to 100 g/min in pellets, containing up to 25 wt% ethylene in the ethylene/propylene copolymer with an excellent balance of mechanical properties, and has developmental grades exceeding these levels of ethylene and performance.

Multi-zone Circulating Reactor (MZCR) Process Technology: Spherizone Process

Background and Key Features The major recent process technology development for polypropylene has been the introduction by LyondellBasell of Spherizone, a mainly gas-phase, non-metallocene polypropylene process featuring a new multi-zone circulating reactor (MZCR), featuring a homogenous two-phase polypropylene cascade reaction in one polymerization reactor. This process was commercialized by LyondellBasell in August 2002 and has been offered for license since 2003. Its introduction was the major recent process technology development for polypropylene which mainly consists of a gas-phase reactor that features a new multi-zone circulating reactor (MZCR), which is capable of producing a homogenous two-phase polypropylene in one reactor. The principle behind the MZCR is a reactor that offers two reaction zones with different process conditions, specifically temperature, hydrogen, and comonomer concentrations. Growing polymer circulates repeatedly between the two zones giving the prospect of enhanced homogeneity between the polymer phases. This technology is said to produce more uniform polymers with broader molecular weight distribution and advanced performance capabilities. A single reactor can be used to produce homopolymers (narrow to very broad MWD) and random copolymers with a better property balance than conventional polypropylenes, as well as new families of multiple phase propylene polymers including twin random copolymers or homo/random grades.

Process Description LyondellBasell developed the MZCR design by applying recirculation principles of catalytic cracker technology to their experience in single-zone polymerization reactors. A representative flow sheet of the process is shown in Fig. 5.19. It consists of two interconnecting reaction zones, a riser and a downer, separated at the top by a cyclone. As the polymer particles form, they

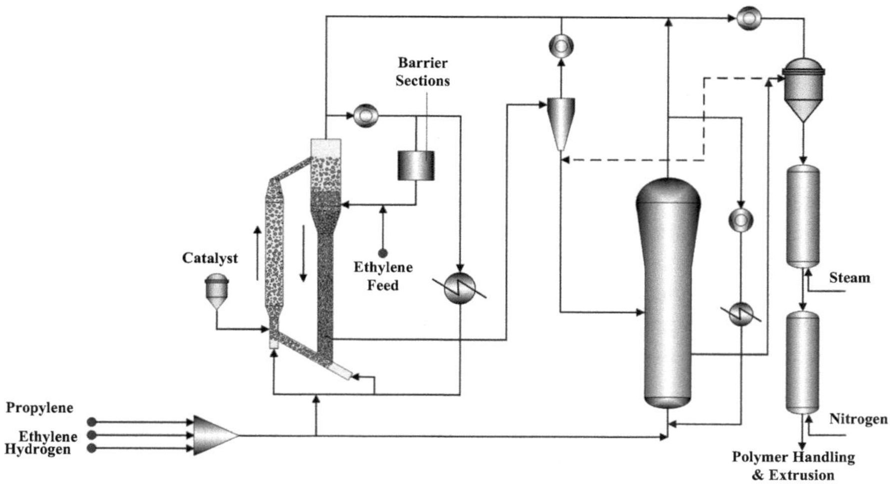

Fig. 5.19 Multi-Zone Circulating Reactor (MZCR) PP process technology

continuously circulate between the two interconnected zones, each with a distinct polymerization environment and fluid dynamic regime.

The barrier fluid, for instance propane near or below the dew point, is denser than the gas phase leaving the riser and is intended to help stop the entrainment of the lighter gases in the riser, in particular, hydrogen and/or ethylene, into the downer. Barrier fluid is fed in slight excess with respect to the theoretical interstitial volume of the downer in order to maintain a total downward flow of gas in this section of the reactor. The result of this separation of the particles and evacuation of the gas phase leaving the riser means that the two zones can be operated under very different polymerization conditions to produce polymers with different compositions in each zone. The riser usually operates with hydrogen concentration 2–4 orders of magnitude higher than that in the downer, meaning that it is possible to produce a resin with a bimodal MWD in a single continuous reactor using a standard olefin polymerization catalyst—something that is impossible with a standard gas-phase reactor. Along with other factors, the number of passes the particles make between the two zones determines the homogenization of the final polymer. The particles typically pass through the zones more than 40 times.

In the first zone (the riser), the gas velocity is high enough that this section is a fast fluidized bed. The particles are blown into the cyclone at the top of the reactor, which separates the particles from the gas phase of the riser. The particles then pass into the "downer" section that behaves like a moving packed bed where the particles move simply by the pull of gravity. According to Covezzi and Mei [15], the particle density in the downer is close to 90% of the bulk density of the final powder, whereas it is less than half that level in the riser. Product withdrawal is done at the bottom of the downer in order to reduce as much as possible the amount, and therefore cost, of recycling the gas phase. Much like the slurry loop reactor discussed below, the recycle ratios in the reactor are very high, so the particles move around with a residence time approaching that of a CSTR.

The uniformity and homogenization of the polymers produced give them higher melt strength, better impact performance, improved creep characteristics, and improved processing when compared with existing materials. LyondellBasell has suggested that in the future, more than one barrier fluid layer could be used in order to further improve the process.

The two reactor zones in one circulating system should give a cost advantage over existing polypropylene processes that utilize two or more separate reactors. LyondellBasell claims that energy consumption is significantly lower than for conventional processes.

Product Portfolio The process is claimed to produce resins with improved processability, better mechanical performance, and other superior technical characteristics, which should improve the products for traditional market applications such as biaxially oriented film (BOPP), packaging, fibres, consumer goods, and automotive industry. It is also believed that the new products will be suitable for new applications, replacing other materials. As with the Spheripol process, a gas-phase, fluid-bed reactor is added for impact copolymer production, which in the case of Spherizone, adds the capability of propylene ethylene rubber phase incorporation into the new homopolymer/random copolymer or twin random copolymer matrix structure coming from the MZCR polymerization step.

According to LyondellBasell, an increasing range of new propylene polymer families for various applications are currently available from commercial Spherizone plants and a substantial increase in polypropylene potential is being developed for special and large volume applications (e.g. new grades designed for non-woven soft textile applications in high spinning speed machines, clear blown film grades for monolayer or multilayer (side layer) structures, high clarity and high stiffness random copolymers with low temperature impact for the thermoforming sheet and blow moulding markets, pipe grades with outstanding stiffness (flexural modulus above 1900 MPa) and good impact performance, high stiffness grades specifically designed for high-barrier sheet applications, such as crystal clear cups and trays for packaging fruit and vegetables, with good dimensional stability in thinner gauges after hot-filling for microwave reheating and substitution of traditional glass and metal small packs, new homopolymer/random copolymer structures specifically designed for very high speed production of BOPP films, very high impact-stiffness balance copolymers for the automotive industry.

The Borstar PP Process

Background and Key Features Borealis' Borouge joint venture built two 400,000 ton per year lines in Abu Dhabi, which came on-stream in September 2010. The next phase of the project includes plans for two 480,000 ton per year lines that are expected to start up by early 2015.

The Borstar polypropylene process is based on the concept of using a dual reactor system where a slurry loop is followed by a gas-phase reactor. This twin reactor configuration is employed for homopolymer production and, with the addition of further downstream gas-phase reactors, impact copolymers can be produced. One or two downstream reactors are employed depending on the required rubber performance and content of the final polymer.

Borealis developed the polypropylene process from its successful polyethylene technology and, as in polyethylene, the loop reactor can be operated in supercritical mode, that is, above the critical point of liquid propylene (for polypropylene production, propylene replaces the propane diluent used for polyethylene production).

Borealis offers a proprietary Advanced Process Control System (BorAPC), which interfaces with conventional control systems and controls variables using non-linear predictive models. Claimed benefits include: higher production, more stable operation, improved product consistency, and faster transitions.

Process Description Conventional bulk phase polypropylene processes operate below the critical point of propylene. To avoid gas bubble formation due to inerts and lights (e.g. hydrogen and ethylene), polymerization takes place at 70–80 °C. The Borstar process can be operated in supercritical conditions, thus avoiding gas bubble formation. This allows for an extended operating window, allowing for higher temperature operation and higher hydrogen concentrations, which gives the opportunity for very high MFR products.

The Borstar process is based on a slurry loop/gas phase sequential reactor process. Simplified process flow sheets are shown in Fig. 5.20. For copolymer production, there are three reactors and a pre-polymerizer in series. The pre-polymerizer, loop reactor, and first gas-phase reactor are used to make homopolymer and random copolymer; the second gas-phase reactor is used for impact copolymer production.

Catalyst slurried in a carrier, co-catalyst and donor, hydrogen, and propylene (after treating) are fed to the pre-polymerizer. No additional catalyst is required in the subsequent reactors. The pre-polymerizer is a small loop reactor that operates at 50–60 barg and low temperature. The pre-polymer along with additional propylene and hydrogen for molecular weight control are fed to the slurry loop reactor. When producing random copolymers, ethylene is also fed to the loop reactor. The reaction takes place in bulk propylene. The loop reactor typically operates at 80–100 °C and about 50–60 barg. The slurry concentration is in the range of 40–60 wt%. The slurry from the loop reactor is fed directly to the first gas-phase reactor. Additional homopolymer (or random copolymer) is made in this gas-phase reactor. This is a fluidized bed reactor that typically operates at 80–100 °C and about 22–35 barg. In the homopolymer/random copolymer section, the loop reactor has a residence time of less than 1 h. The first gas-phase reactor has a 1.5–2 h residence time.

The homopolymer/random copolymer is fed to the separator for polymer degassing. This is a low pressure flash tank (LP) that is used to separate the polymer powder and hydrocarbon gases. The gas is fed to the recovery system and recycled to

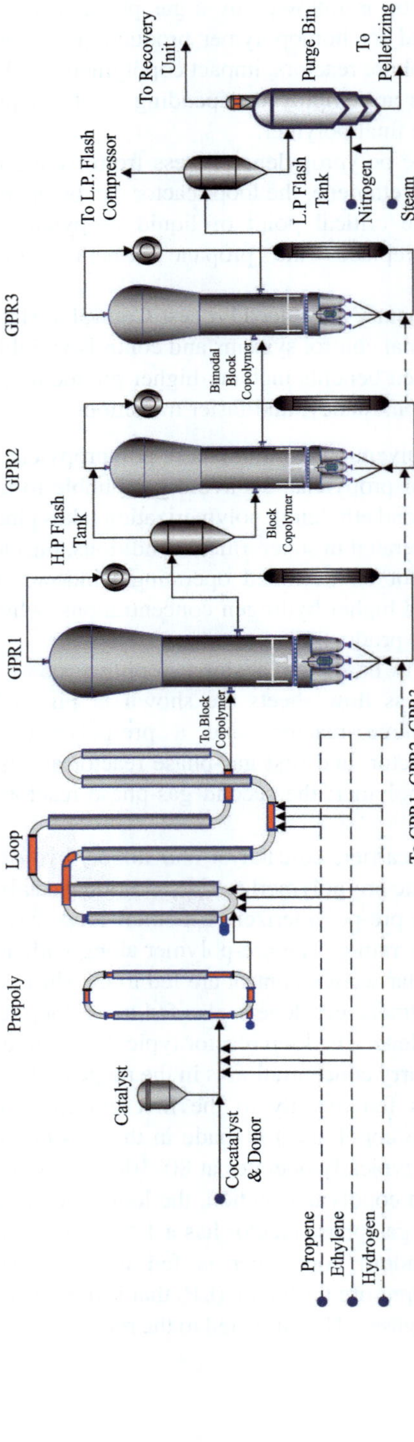

Fig. 5.20 Borstar PP process technology

the reactors. The polymer flows either to the second gas-phase reactor (if producing impact copolymer) or to the purge bin. The second gas-phase reactor is also a fluidized bed reactor that operates at lower conditions than the first gas-phase reactor, typically 75–90 °C and about 15–25 barg. The polymer from this reactor is also degassed and then flows by gravity to the purge bin. The polymer is scrubbed with nitrogen to remove residual hydrocarbons and steam to deactivate the catalyst. The polymer from the purge bin is conveyed to the downstream extrusion and pelletizing steps via a closed-loop nitrogen conveying system.

Product Portfolio Borealis' polypropylene products offer a broad application range covering conventional and advanced applications. The grade range offers:

- "Super-modality"—the ability to tailor tacticity distribution, molecular weight distribution and comonomer distribution between polymer fractions.
- Borstar nucleation technology (BNT) provides advantaged performance where stiffness and dimensional stability are important.
- Bormod PP with an excellent balance of impact strength and stiffness and offering significantly reduced cycle times for processors.
- A strong range of pipe grades.

The key application areas for Borstar PP include the following:

- Thin wall injection moulding applications, where improved stiffness and higher heat resistance than conventional grades offer the opportunity for lower weight products and faster cycle times with superior performance.
- Thermoforming, with similar benefits in terms of stiffness and heat resistance, plus improved transparency.
- Cast film with good mechanical and barrier properties.
- Fibre grades with unique bonding and tenacity properties.
- Injection moulding copolymer grades for crates, etc., with an excellent combination of stiffness and impact resistance.
- Automotive applications for interior and exterior applications.
- Pipe applications where the higher temperature resistance of polypropylene is an advantage compared with polyethylene, for example, in hot water pipes. The rigidity of the Borstar grades is an advantage for low-pressure pipe applications, such as sewage pipes, where polypropylene competes with PVC.

Future development priorities include high melt strength resins, soft polypropylene grades, high stiffness applications, terpolymers and low extractables, and low catalyst residue grades for medical applications.

The Hypol PP Process by Mitsui Chemicals

Background and Key Features Mitsui Chemicals and Idemitsu Kosan established a joint venture company, Prime Polymer, in April 2005 for their polyolefin businesses. Mitsui Chemicals owns 65% of the joint venture company, with Idemitsu Kosan owning the remaining 35%. Mitsui Chemicals retained ownership of the Hypol II process technology and is solely responsible for process development, licensing, and the catalyst business associated with Hypol II process technology.

The Hypol polypropylene process has been designed for a high degree of flexibility and operability across a wide product range. As such, Mitsui Chemicals' past philosophy was not to focus on having the lowest capital cost. However, given the competitive nature of polypropylene technology offerings, Mitsui Chemicals focused its engineering efforts on streamlining the process with respect to its design philosophy. As a result, Mitsui Chemicals introduced Hypol II, which has an overall investment cost reduction of about 15% without losing its flexibility and operability, making it very competitive with other leading bulk technologies. Principal areas of cost reduction are the following:

- High throughput by using a higher activity catalyst.
- High reactor efficiency by adopting a loop reactor for homopolymerization.
- High monomer yield by adopting an improved drying facility.
- Elimination of de-ashing and atactic polymer removal.
- Reduction in overall installation costs (e.g. streamlined layout and lower bulk costs, as steel, civil, and electrical)
- Mitsui Chemicals also continues its R&D in metallocene-based syndiotactic and isotactic polypropylene. The effort for isotactic development is substantially less than that for polyethylene.
- Mitsui Chemicals recently developed several types of sixth generation catalysts, based on phthalate and toluene-free technology. Mitsui claims that these catalysts are applicable to any bulk, slurry, or gas-phase process as a drop-in for environmentally unfriendly catalysts.
- Some grades of the new catalysts also offer a broader molecular weight distribution and a narrower comonomer content distribution of impact copolymer. These unique characteristics are said to greatly improve mechanical properties and widen the application range of polypropylene.
- A high crystallinity product, offering a good balance of impact strength and rigidity, can be produced by the incorporation of a proprietary electron donor component within the catalyst system.

Process Description The original Mitsui Hypol process is based on two stirred autoclave reactors in place of the loops, followed by two stirred fluidized-bed gas-phase reactors (see Fig. 5.21). The stirred autoclaves are replaced by loop reactors in the Hypol II process, giving higher throughput and further reducing capital costs.

The process flow sheet is based on the state-of-the-art Mitsui Chemicals process. According to this flow sheet, the plant consists of a single train made up of a catalyst feed system and a reactor system followed by monomer flash, recycling, stripping, and polymer finishing and handling (not shown). Mitsui Chemicals employs a series of two reactor systems for homopolymer and impact copolymer production. The first reactor system comprises two loop type reactors, while the second reactor system comprises of at least one fluidized-bed, gas-phase reactor. The bulk polymerization

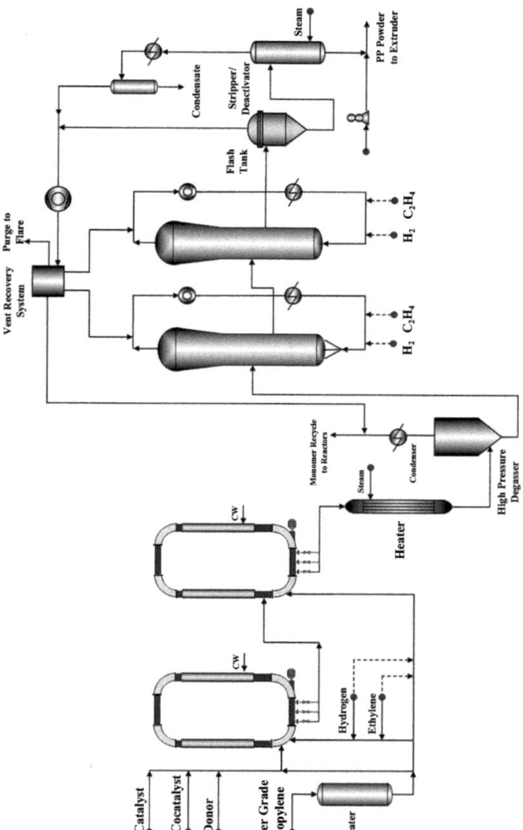

Fig. 5.21 Hypol II PP process technology

unit has a residence time of about 1–1.5 h and about 50% conversion per pass of propylene. The reactors operation depends on the product capability of the plant and the product being produced, homopolymer or impact copolymer.

The process operation is similar to LyondellBasell's Spheripol process. Although the specific design of the gas-phase reactor is highly proprietary, this design is based on a fluidized and scraped vessel with a 0.5–1.0 h residence time. With such a reactor design with wall scrapers it is possible to eliminate fouling and sheeting problems. It is claimed that this helps achieve higher rubber contents when making impact copolymers than is possible in a conventional FBR. Mitsui claims that spherical and uniformly sized particles ensure reliable operation and performance of the gas-phase reactor and finished system.

The polymer from the flash tank is purged with steam to recover the residual monomer and deactivate the remaining catalyst. Polymer is sent to the downstream powder storage and finishing sections similar to those previously discussed.

Product Portfolio Hypol II can produce typical homopolymer and impact copolymer grades. Moreover, Mitsui Chemicals Co. claims that the inherent flexibility of the gas-phase reactor allows for production of impact copolymers with an ethylene content of 25 wt% (40 wt% rubber fraction). Ethylene content as high as 30 wt%, corresponding to a 50 wt% rubber fraction, has also been demonstrated commercially. Mitsui Chemicals' HYPOL II process can produce a polymer with an MFR range of 0.30–80, with commercially demonstrated material in the 0.10–600 MFR range.

The Unipol II PP Process

Background and Key Features Dow Chemical took ownership of the UNIPOL gas-phase polypropylene technology as part of its acquisition of Union Carbide Corporation in April 2001, and is solely responsible for process development, licensing, and the catalyst business for the Unipol PP technology.

In the Ziegler–Natta field, Dow is targeting a new generation of catalysts with advanced polymer properties, expanded capability for making a broader range of products, improved post-reactor processability, higher productivity, and improved reactor operability. More specifically, according to Dow, the new generation of catalysts will bring significant improvement for advanced impact copolymer and random copolymer with reduced need for post-reactor peroxide cracking. The catalyst systems are designed to possess a kinetic profile suited to the gas-phase process, to provide for robust reactor operation.

The SHAC 300 series of PP catalysts are based on a proprietary precursor platform, produced in a new world scale facility, and provide improved polymer particle morphology. Specifically, the special precursor is said to provide enhanced sphericity and porosity with narrower particle size distribution. Resin bulk density and "flowability" are improved, which can lead to throughput increases of up to 30%. Higher ethylene content copolymers can be produced without encountering stickiness problems. The unique distribution of porosity in the catalysts derived from the special precursor plays a major role in the improvement of the properties of impact copolymer and random copolymer.

Dow has developed SHAC Catalyst Advanced Donor Technology (ADT) systems, which combine the process operability of the third generation catalysts with the improved product performance of the fourth generation catalysts. A portfolio of fully compatible catalyst and donor systems enable a user to cover a very broad range of product capabilities, including high crystallinity, high melt flow impact copolymers, high melt flow random copolymers, and high xylene soluble homopolymers. The use of SHAC ADT catalyst systems allows the resin producer to customize many aspects of the polymer produced, including melt flow, levels of oligomers, and molecular weight distribution. The recently developed members of the ADT and CONSISTA families also introduce additional benefits of polymer property enhancement.

The ADT and CONSISTA donors are offered only by Dow. They replace the traditional silanes that have historically had broad use within the industry. They have become the systems of choice for new Unipol PP Technology licensees. SHAC Catalyst ADT Systems have been introduced for all major product types. In addition to the product performance benefits that ADT provides, according to Dow, the technology has also proven to have an operational benefit due to the favourable kinetics that it brings to the catalyst system, making the operation of the Unipol PP reactors even more robust and bringing the reactor on-stream time to even higher levels.

Process Description Process flow sheets for polypropylene production using the Unipol PP gas-phase technology are presented in Fig. 5.22. The process closely parallels that of polyethylene production and hinges on Dow's fluidized-bed reactor. Major processing blocks of the plant include catalyst handling, raw material treatment for removal of trace catalyst poisons, polymerization, resin purging, integrated pelletization, and storage, packaging, and loading.

Dow can produce random copolymers by adding ethylene to the homopolymer reactor. A specialty line of propylene-1-butene random copolymers can also be produced, and such grades have been made at licensee plants. The production of high impact copolymers requires the installation of a second polymerization reactor that has no internal stirrer. Dow produces polypropylene in pellet form using either additives dry-blended into a side stream of powder from the production line or a direct feed of additives in neat or pre-blend form.

In the process, fresh propylene passes through molecular sieves to guard against trace quantities of water entering the gas-phase fluid bed reactor. Depending on the quality of feedstock, removal of other trace materials (oxygen, acetylene, sulphur-containing compounds, etc.) may be necessary. In the polymerization process, the reactor systems consist of vertical pressure vessels with expanded upper section for minimizing the carryover of polymer particles. The reactant gas stream is circulated through the bed and is cooled in an external heat exchanger, thus removing the exothermic reaction heat.

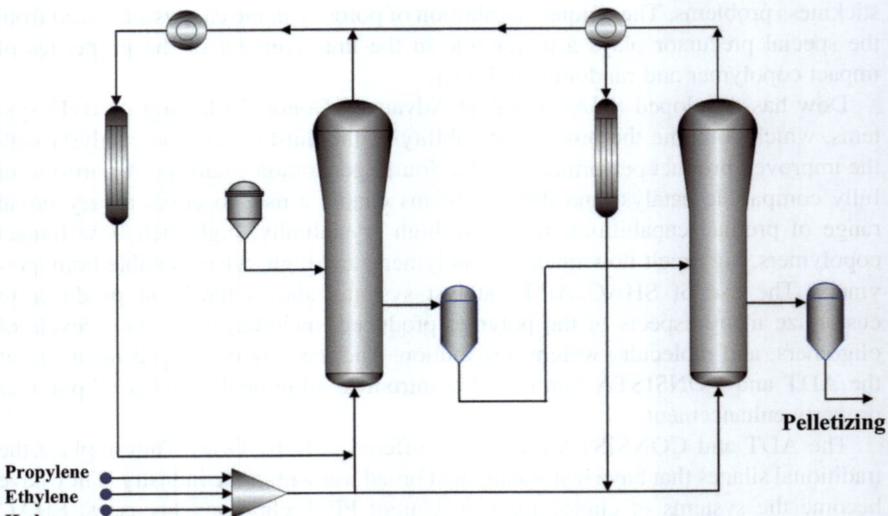

Fig. 5.22 Unipol II PP process technology by Dow Chemical Co.

There is some condensing of the propylene and propane (an inert hydrocarbon that comes in with the propylene feed) that enhances heat removal via vaporization of condensates. A blower in the recycle loop provides the pressure increase to overcome the difference in pressure in the loop. The reactors operate at around 34 bar and 70 °C or higher.

The fresh liquid polymer-grade propylene is combined with recycled reaction gases and enters the reactor beneath the gas distribution plate. The cycle gas stream provides the fluidization for the growing polymer bed. The titanium-based catalyst is added, as needed, as a slurry in mineral oil. The co-catalyst is metered to the reactor. Hydrogen is used in the reactor for molecular weight control. Also, an electron donor material that acts as a selectivity control agent is added.

The fluidized-bed reactors have average residence time of approximately 1.0 h each and extensive back mixing. Therefore, those systems are competitive with the liquid-phase processes in terms of transition times for type and grade changes (e.g. typically 3–4 h). Dow also claims that an increase in melt flow rate can be made more quickly than in a liquid-phase reactor because the equilibrium concentration of hydrogen can be increased more quickly in a gas-phase system versus a liquid-phase system.

In homopolymer production (typically one reactor is used), polymer is drawn off periodically and is separated from the gas in a series of high-pressure gas/solid separators. The separated gas is recycled back to the reactor.

Ethylene is added to the homopolymer reactor for random copolymer production (second reactor). The polymer from the gas/solid separator is delivered to a two-stage purge tower where residual monomer (2000–3000 ppm) is stripped first with recovered lights and finally with nitrogen and a small amount steam is injected to neutralize the catalyst residue. The powder is then gravity fed to the downstream extrusion steps. The vented material from the purge tower and reactors is sent to a small recovery operation to improve overall monomer efficiency. Inert gases and heavy gases, such as propane, are removed and propylene is recycled to the reactor. Ethylene and propylene are separated and recycled to impact copolymer production.

In impact copolymer production, the resin (containing active catalyst) from the second reactor discharge vessels is diverted to the transfer vessel, which feeds the fluidized bed copolymer reactor prior to entering the purge tower. Ethylene, propylene, and hydrogen are also fed to the reactor. The cooling and discharge systems are similar to the homopolymer reactor system, except that the gas recycle (cooling) loop operates 100% in the gas phase.

The dry end of a Unipol PP technology plant is typical of a polyolefin plant for which there are several extrusion systems available. The twin-screw extruder/gear pump is typically used in most polyolefin plants.

Product Portfolio Products containing up to 7 wt% ethylene have been commercially demonstrated in random copolymer production, and products containing over 12 wt% ethylene have been produced in the pilot plant. Random copolymers of propylene and 1-butene have been produced commercially. These resins have compositions of up to 15 wt% 1-butene and a melting point range of 131–142 °C and are offered in a melt flow index range of 3.4–12. Up to 21 wt% 1-butene has been demonstrated in the pilot plant. These materials have excellent optical properties of clarity, gloss, and are gel-free.

Unipol PP Technology plants produce impact copolymers with up to 19 wt% ethylene and nominal 38 wt% rubber fraction commercially. Melt flows from the reactor range from 0.2 to over 100 g/10 min, with less than 0.1 and over 130 g/10 min melt flow products demonstrated in the pilot plant. Controlled rheology and direct reactor grades have been produced commercially with melt flows greater than 120 g/10 min.

5.5 Overview

The commercial development of POs has been and is driven by technological developments based on an understanding first of the catalyst system and then of the process, the polymer molecular structure–property relationship which determine the product applications. Therefore, the innovation developments have been and are carefully planned, based on the adoption of an ideal model that starts from the proper understanding, management, and handling of the catalyst system and ends to the selection of the best process technology and product portfolio, the essential elements to all the consequent market success.

The continuous product property improvements have been driven by significant technology development which is still very much progressing. However, the most important is the scientifically and technologically equilibrated development aimed at reaching, via a proper combination of the catalyst and process, the best polymer structure–property design tailored to the achievement of specialty materials for specific applications. Thus, technology developments continue to redefine state-of-the-art products and processes in the polyolefins industry. Companies are focused on cost reduction and commercialization of new multimodal technologies aiming at manufacturing products for new applications. In the particular field of multimodal processes the technology continues to be a strategic differentiation as companies competing for a leadership position. These dynamics coupled with the industry's continued competitive pressures have increased the competition among technology licensers and producers. As a result, companies need to understand the global landscape as a prerequisite for understanding their own technical and commercial position. In the present book chapter, state-of-the-art polyolefin multimodal processes, most of which are available for license, have been presented and their key features in terms of process design and product portfolio range have been illustrated.

References

1. Mulhaupt, R. (2003). Catalytic polymerization and post polymerization catalysis fifty years after the discovery of Ziegler's catalysts. *Macromolecular Chemical Engineering, 204*, 289–327.
2. Galli, P., & Vecellio, G. (2004). Polyolefins: The most promising large-volume materials for the 21st century. *Journal of Polymer Science, 42*, 396–415.
3. CMAI. (2008). Global plastics and polymer market report, *123*, 1–66.
4. Soares, J. B. P., & McKenna, T. F. L. (2002). *Polyolefins reaction engineering*. Weinheim: Wiley-VCH.
5. Xie, T., McAuley, K. B., Hsu, J. C. C., & Bacon, D. W. (1994). Gas phase ethylene polymerization: Production processes, polymer properties, and reactor modelling. *Industrial and Engineering Chemistry Research, 33*, 449–479.
6. Dompazis, G., Kanellopoulos, V., Touloupides, V., & Kiparissides, C. (2008). Development of a multi-scale, multi-phase multi-zone dynamic model for the prediction of particle segregation in catalytic olefin polymerization FBRs. *Chemical Engineering Science, 63*, 4735–4753.
7. Fontes, C. H., & Mendes, M. J. (2005). Analysis of an industrial continuous slurry reactor for ethylene–butane copolymerization. *Polymer, 46*, 2922–2932.
8. Touloupides, V., Kanellopoulos, V., Pladis, P., Kiparissides, C., Mignon, D., & Van-Grambezen, P. (2010). Modeling and simulation of an industrial slurry-phase catalytic olefin polymerization reactor series. *Chemical Engineering Science, 65*, 3208–3222.
9. Zacca, J. J., & Ray, H. W. (1993). Modelling of the liquid phase polymerization of olefins in loop reactors. *Chemical Engineering Science, 48*, 3743–3765.
10. Ferrero, M. A., & Chiovetta, M. G. (1990). Preliminary design of a loop reactor for bulk propylene polymerization. *Polymer – Plastics Technology and Engineering, 29*, 263–283.
11. Drusco, G., & Rinaldi, R. (1984). Polypropylene-process selection criteria. *Hydrocarbon Processing, 63*, 113–117.
12. Kanellopoulos, V., & Kiparissides, C. (2019). Mathematical modelling and simulation of gas- and slurry-phase catalytic olefin polymerization reactors: A comparative study. *Macromolecular Reaction Engineering* (To be submitted).

13. Pladis, P., Kanellopoulos, V., Chatzidoukas, C., & Kiparissides, C. (2008). Effect of reaction conditions and catalyst design on the rheological properties of polyolefins produced in gas-phase olefin polymerization reactors. *Macromolecular Theory and Simulations, 17*, 478–487.
14. Polyolefins Planning Service Technology Review. (2011). *Nexant*. Retrieved from www.chemsystems
15. Covezzi, M., & Mei, G. (2001). The multizone circulating reactor technology. *Chemical Engineering Science, 56*, 4059–4067.

Chapter 6
Multimodal Polypropylenes: The Close Interplay Between Catalysts, Processes and Polymer Design

Christelle Grein

6.1 Introduction

Polypropylene (PP) is beside polyethylene the largest polymer—with an installed capacity of around 70 MMT in 2015—and one of the fastest expanding ones with average growth rates of 5% per year. Building on key-features like excellent chemical resistance, good thermal stability, low density, low cost and mechanical fit over a wide range of flows it is widely used in applications like pipes, films, fibres, capacitors, boxes, crates, tanks and automotive parts [1]. In the most generic offering, homopolymers (H-PP), random copolymers (R-PP) with ethylene contents up to 5–6 wt% and heterophasic copolymers with ethylene–propylene rubber contents up to around 30 wt% (B-PP) are available.

Extending its penetration further and moving its utilisation from a pure commodity to a specialty product goes via the introduction of new functionalities either by modifying the PP continuous phase itself or—in the case of B-PP—by tailoring the ethylene–propylene rubber (EPR) phase. One of the possibilities is to play with modalities, for example by broadening the molecular weight distribution (MWD) or by adjusting the chemical composition of the different phases. Operating (multi-reactor) processes and/or using adequate catalyst technologies are in this respect the key-enablers.

The aim of this chapter is:

1. To review the main processes capable to make multimodal PP;
2. To discuss the needed catalyst requirements as a function of reactor and product attributes;
3. To show how and why multimodality is of benefit in higher end applications.

C. Grein (✉)
SABIC, Sittard, The Netherlands
e-mail: christelle.grein@sabic.com

A. R. Albunia et al. (eds.), *Multimodal Polymers with Supported Catalysts*,
https://doi.org/10.1007/978-3-030-03476-4_6

There is no univocal definition of bimodality or multimodality based on MWD (defined as the ratio between the weight average molecular weight, Mw, and the number average molecular weight, Mn, determined by size exclusion chromatography—SEC) as a combination of a couple of factors:

1. PP—in its most common commercial form—is isotactic by design and manufactured by Ziegler–Natta (ZN) catalysts which are multi-sites by nature. A polydispersity of 2 (following a Schultz-Flory distribution) expected and seen for polymer chains of different lengths produced with single-site catalysts, is thus not met, usually exceeded this value by more than a factor 2;
2. MWD has evolved with the different generations of ZN catalysts as a result of the different chemistry applied [2]:
   (a) The nowadays widely used fourth generation of $MgCl_2$-supported $TiCl_4$ systems with alkylphthalates as internal donors and alkoxysilanes as external donors shows an MWD between 6.5 and 8 in bulk polymerisation (70 °C, 2 h) for an MFR of ca. 15 g/10 min (corresponding to the reported intrinsic viscosities of 2 dl/g);
   (b) Earlier blends based on benzoic acid esters as Lewis bases (external donors) are broader, exhibiting an MWD between 8 and 10 for same flow properties;
   (c) Fifth generation systems using 1,3-di-ethers as internal electron donor (silane as an external one) have a narrow MWD around 4–5, while succinate based catalysts show a broad MWD around 10–15;
3. Different processes, process conditions (e.g. temperature, polymerisation time), co-internal donors (type, concentration), catalyst (type, concentration), H2-feed (MFR) also play a role in the MWD evolution. Hence lower values of MWD than the ones indicated above are usually in people's minds. A possible explanation is the combination of intrinsic (e.g. catalyst/donor response) and extrinsic factors (process settings, process conditions, reactors in series including prepolymeriser)—in combination with one medium (liquid or gas) and with one specific reactor hardware design—in commercial conditions that (can) lead to a more complex picture—and thus to broader MWD—than under well-defined test polymerisation where the equilibrium and final MWD are reached within minutes.

In the following document, the terms monomodal and bimodal will be understood as fulfilling the descriptions below:

- Monomodal PPs are homopolymers or random copolymers showing a Gaussian MWD distribution and—when relevant—no difference in chemical composition for phases of same nature. Changes in the chemical composition can be traced back by Fourier transform infrared spectroscopy (FTIR), differential scanning calorimetry (DSC), temperature rising elution fractionation (TREF), SIST (step-wise isothermal segregation technique) or any other relevant analytical method;
- Bimodal PPs are homopolymers or random copolymers showing two different MWDs and/or two different chemical compositions produced in consecutive reactors or with a dual site catalyst/donor system;

Table 6.1 Possible combinations of multimodalities in homopolymer (H-PP), random copolymer (R-PP), heterophasic copolymer (B-PP) and random heterophasic copolymer (RB-PP)

	H-PP	R-PP	B-PP	RB-PP
MWD continuous phase	X	X	X	X
Comonomer distribution continuous phase		X		X
MWD dispersed phase			X	X
Comonomer distribution dispersed phase			X	X

– Materials with a broad MWD produced in two consecutive reactors, but not showing two distinct peaks or inflexion points in SEC traces, will also be considered as bimodal. The same is valid for cases where a broad MWD is achieved by specific catalyst/donor combinations in one reactor—though such materials are not in the focus of this paper.

For heterophasic systems, the above definition would still apply. The considered phases would be—as primary focus—the dispersed phases, besides the continuous one(s) (which can also be multimodal).

Note that besides SEC, rheology parameters like polydispersity index (PI, cross-over point of $G'(\omega)$ over $G''(\omega)$), shear thinning index (SHI, calculated by dividing the zero shear viscosity by a complex viscosity value obtained at a certain constant shear stress value G^*) or melt elasticity index indicate the broadness of the molecular weight distribution of the polymer. The higher these values the broader the MWD.

The different combinations of multimodalities are illustrated in Table 6.1. The most complex set-up can be obtained with random heterophasic PP copolymers (RB-PP) in a train of 3 or 4-reactors in series (excluding a prepolymerisation reactor) with bimodal matrix (in MWD and/or comonomer distribution and/or comonomer nature) and bimodal dispersed phase (in MWD and/or chemical composition for the rubber phase). While ethylene (C2) is usually the comonomer of choice in view of its availability, reactivity and performance, 1-butene (C4) and 1-hexene (C6) can occasionally also be used. Should this be the case, it will be explicitly highlighted. If not, the comonomer considered by default will be C2 for R-PP, B-PP and RB-PP. The coming considerations will be restricted to isotactic polypropylene, bimodal syndiotactic PP remaining predominantly a subject of interest for academia [3, 4].

6.2 An Overview of the Main PP Processes and Catalyst Requirements

Catalyst, processes and polymer design are the essential and non-dissociable pillars to tailor the end properties of materials. Their close and complex links will be presented in the coming paragraphs.

Multimodality can be achieved by adapting the process settings in different reaction zones of a polymerisation plant. The two main variables are in this respect

(1) the hydrogen (H2) concentration playing the role of chain termination agent and thus controlling as prime parameter the length of the polymeric chains (i.e. the higher the H2 feed, the higher the MFR) and (2) the comonomer concentration (in the continuous or dispersed phase). Other parameters like reactor temperature, reactor split and residence time play a secondary role, but can become a special key in fine-tuning product performance.

Their influence on MWD is nicely exemplified in a modelling study from Zacca based on original data from Hungenberg [5]—using two vertical stirred bed reactors in series to produce a bimodal ZN-PP [6]. The high molecular weight fraction is made in the first reactor, the low molecular weight fraction in the second one using a $TiCl_4/MgCl_2$ catalyst. The base case is illustrated by Fig. 6.1(i) and used as reference when varying different polymerisation parameters. Decrease of catalyst flow rate (i.e. increase of the residence time) in the first reactor (Fig. 6.1(ii)), decrease of bed level in the second reactor (Fig. 6.1(iii)) and selective poisoning of the second reactor (Fig. 6.1(iv)) were shown to boost the productivity towards the first reactor as a consequence of controlled reactivity shifts.

Multimodality via reactor design can theoretically be gained in every polymerisation train with two or more vessels (prepolymeriser excluded), the exceptions where bimodality (or at least broad molecular weight and/or chemical distribution) is claimed to be obtained in a single vessel being: (1) technologies containing in a one reactor system multiple reaction zones—coupled with adequate feeding points for , for example, hydrogen and/or comonomers (e.g. Spherizone™), (2) dual-site catalysts feed in one reactor and/or (3) a dynamical oscillatory hydrogen feed, from very high to very low.

Multiple reactor concepts are run in series for simplicity and economic reasons as a result of the advances in catalyst technologies in the last 20 years in terms of mileage and tuning of activity towards longer residence times.

The parallel way of operating is offered by for example Lummus in the Novolen® process via its proprietary Versatile Reactor Configuration (VRC™) gas-phase technology [7]. Homopolymers and random copolymers can be manufactured in cascade or in parallel in two identical reactors to achieve higher capacities for a single train and to produce bimodal grades. Optionally different catalyst systems may be used in each reactor if operated in parallel. A drawback of this mode of operation is, however, that two separate populations of powders are generated in each reactor that—if different enough from a molecular weight point of view—will be difficult to mix intimately in the extrusion step. This is all the more true that industrial extruder set directly after a polymerisation train are usually not designed with high shear elements. It follows that the probability to create gels (i.e. high molecular weight non-melted polymer fractions) is higher with such a process than with a sequential reaction mode [8].

In all commercial processes, the first reactor is dedicated to the production of homopolymers or random copolymers. The ethylene–propylene rubber is made in subsequent steps. Two base processes can be distinguished in practise:

- Liquid or bulk processes,
- Gas phase processes,

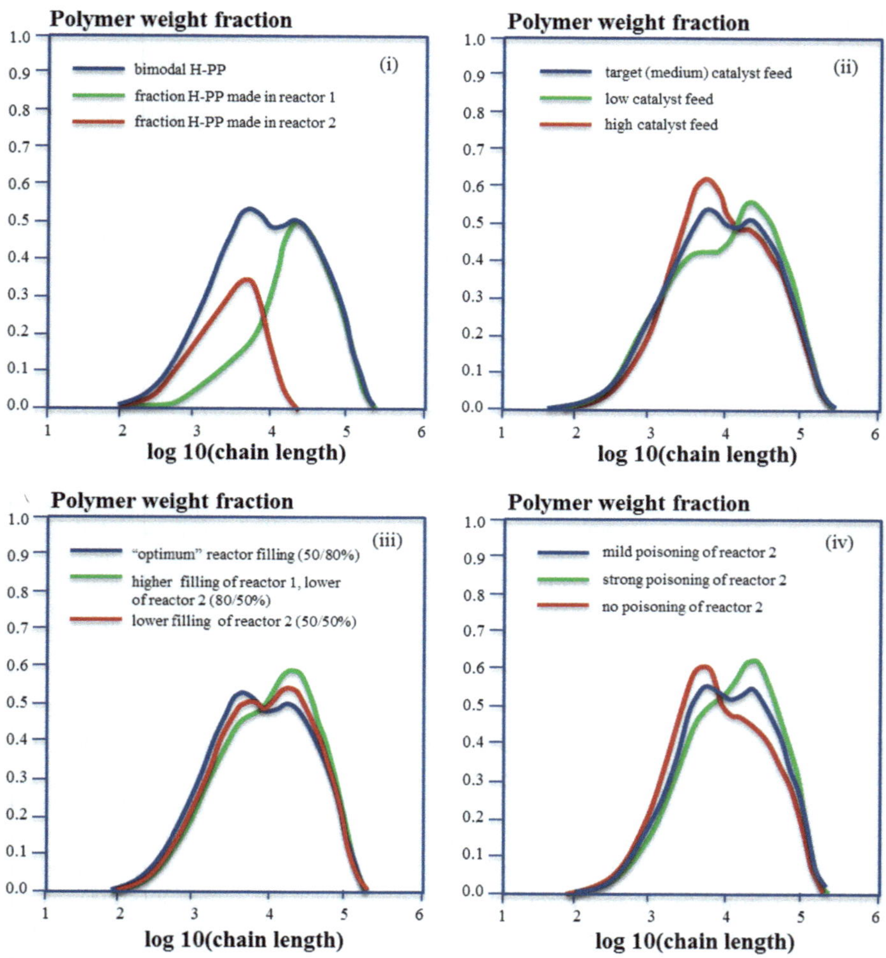

Fig. 6.1 (i) Example of bimodal PP produced in a two vertical stirred bed reactor process in series—used as base case for (ii), (iii) and (iv). Influence of (ii) catalyst flow rate, (iii) bed weight (split) and (iv) poisoning of the second reactor on the GPC curves. Figures adapted after an original paper of Zacca [6]

The bulk processes either consisting of loops (preferred and accounting for the production of 55% of all commercial polyolefins) or stirred-tanks reactors (outdated) are capable to make (bimodal) homopolymers and random copolymers up to comonomer concentrations of 5 wt%. Higher concentrations or the production of a rubber phase are not possible due to the high solubility of such elastomeric-like phases in the liquid monomer.

The gas phase processes are capable to make (bimodal) homopolymers or copolymers in a variety of comonomer concentrations. A two-reactor process is,

Table 6.2 Potential multimodal capabilities of main PP processes based on reactor design

	PrP	Polymerisation reactors					Potential bimodal capability	
	PrP	R1	R2	R3	R4	GP type	Continuous phase	Dispersed phase
Spheripol™	Y	LP	LP	GP	GP	FBR	X	X *if R4*
Hypol II	Y	LP	LP	GP	GP	FBR	X *if R2*	X *if R4*
ExxonMobil™	Y	LP	LP	GP	(GP)	FBR	X	(X *if R4*)
Catalloy™	Y	GP		GP	GP	FBR	X *if no DP in R3/R4*	X *if no CP in R3/R4*
Borstar®	Y	LP	GP	GP	GP	FBR	X	X *if R4*
Novolen®	N	GP		GP		VSBR	X *if no DP in R3*	
Unipol®	N	GP		GP		FBR	X *not developed*	
Innovene™	N	GP		GP		HSBR	X *if no DP in R3*	
Horizone	Y	GP		GP		HSBR	X *if no DP in R3*	
Spherizone™	Y	MZCR		GP	(GP)	FBR	X in MZCR	(X *if R4 or EPR in R1)*

Rx (with *x* from 1 to 4) stands for Reactor, *PrP Prepolymeriser*, *LP* Liquid phase, *GP* Gas phase, *FBR* Fluidised bed reactor, *VSBR* Vertical stirred bed reactor, *HSBR* Horizontal stirred bed reactor, *MZCR* Multi-zone circulating reactor, *CP* Continuous phase, *DP* Dispersed phase. The Spheripol™, Catalloy™ and Spherizone™ technologies are supplied by LyondellBasell, Hypol II by Mitsui Chemicals, ExxonMobil™ by ExxonMobil Technology Licensing, Borstar® by Borealis, Novolen® by Lummus Novolen Technology, Unipol® PP by Grace, Innovene™ by Ineos Technologies, Horizone by Japan Polypropylene Corp. (JPP)

however, generally used to manufacture heterophasic copolymers rather than bimodal H-PP or R-PP to maximise economics.

A high-level summary of the potential multimodal capabilities of the main PP processes is provided in Table 6.2. Most of them are modular, being offered in different reactor configurations (with a maximum of reactors represented in Table 6.2). Some of the processes show at a first glance the same (possible) reactor set-up; they differ, however, in some specificities of their equipment and features. Recent advances in reactor/process design—disclosed in dedicated patent applications—are not part of this overview.

Most of process licensors also provide catalysts—proof that both technologies go hand in hand and complement each other. A definition of the "Ideal Catalyst" was once given by Galli as "extreme target in the heterogeneous catalysis" [9]. A high performing catalyst should:

- Have very high activity and selectivity;
- Control the polymer microstructure (Mw and MWD chain shape, randomness, microtacticity);
- Control the polymer macrostructure (phase distribution and morphology, i.e. particle size, shape, porosity);
- Control the polymer properties (broad range of applications).

This definition is independent of the modality of the polymer and of the used process. Focus of the coming paragraphs will therefore be to describe the general attributes of catalysts valid for monomodal and multimodal systems. When relevant, additional catalyst requirements for use in multistage polymerisation will be highlighted in Sects. 6.3 and 6.4.

Control of the morphology is key to ensure high operability. Most desired are polymer particles with a regular shape, a narrow particle size distribution and the highest possible bulk density (considering the need or not to incorporate—high contents of—rubber). Spherical particles are preferred especially in bulk polymerisation as they allow good flowability, high packing and thus high solid content in the slurry, which in turn maximises the reactor throughput. Alcohol adduct and $Mg(OEt)_2$-based catalysts have therefore shown to be better choices than ball-milled catalysts—and are as a result in competition between each other in commercial environments [10]. As this structure is obtained via catalyst fragmentation and replication of the original catalyst morphology [11–14] the most desired shape of catalyst particles can be easily derived. Obviously a spherical shape is preferred over an elongated shape—this later being allowedly an exotic example (Fig. 6.2). Typical particle sizes are in between 10 and 50 μ, they grow 10–100 times during polymerisation. The pores of the PP particles actually behave as microreactors—characterised by their own kinetics, mass balance and energy balance—catalysed by the still active Ti or by other suitable chemical species. Smaller sizes are preferred for gas phase to prevent overheating and melting under high rates of polymerisation due to the poor heat transfer characteristics of the gas, larger sizes are used in bulk.

The determinant step in the process is the control of catalyst fragmentation—a very fast phenomenon [15–20]. Its uniformity—ensuring an even polymer growth rate—is guaranteed by additional catalyst features: (1) a controlled mechanical resistance to avoid abrasion, but to ensure support break-up, (2) a homogeneous distribution of the active centres and (3) an easy access to active sites by the monomer. High surface area and high porosity are also of benefit, even though compact catalysts have also be reported to undergo instant and uniform fragmentation at the beginning of the polymerisation—thus being capable to generate a spherical powder morphology—capable to incorporate comonomers—and high productivity [21, 22].

A tool often used to ensure proper fragmentation—an exothermic reaction—is the prepolymeriser (see Table 6.2). It allows reducing the polymerisation rates in the crucial initial stage thanks to mild polymerisation conditions. It fulfils three main functions:

- To prevent thermal overheating, like hot spots (that could deactivate the catalyst and create lumps) in a phase where intrinsic reaction rate is high and heat exchange with surrounding medium low;
- To avoid formation of fines;
- To optimise the catalyst activation/polymerisation phase in an environment where cocatalyst and external donor are already present.

Fig. 6.2 Polymer growth via catalyst fragmentation and replication [12]. Reproduced with permission

Unbearable asset in the case of fast starter catalysts, catalysts with a larger particle sizes and/or in liquid media that maximises the polymerisation rate due to high monomer concentration, it is usually not applied with catalysts showing a slow starter profile and/or in gaseous media—as the investment costs and complexity (e.g. stirring an additional reactor) are believed to overweight the benefits, all the more if access to prepolymerised commercial catalysts is given.

The terminology catalyst has been used previously, in its commercial acceptation, that is, consisting of (1) a support, (2) an active species and (3) an internal donor. In the case of a ZN-catalyst it comprises thus (1) an activated magnesium chloride ($MgCl_2$) support, of (2) titanium tetrachloride ($TiCl_4$) as active ingredient and of (3) an internal electron donor (often phthalate, diether or succinate based) as primary stereoregulator.

Talking about catalyst systems would actually be more complete as polymerisation could not take place without a cocatalyst for activation (usually triethylaluminium, i.e. TEAl). An external donor is also in most cases (if high stiffness is required) part of the concept to boost stereospecificity to its maximum (up to 99% isotactic pentads for the best performing systems). While the internal donor is suspected to influence the distribution of $TiCl_4$ active sites on the $MgCl_2$ support [12]—impacting therefore the MWD distribution—the external donor is also often used to broaden the attainable MFR range in polymerisation plants in view of the specific H_2-sensitivity associated with each external donor.

It has therefore been tempting to play with multiple external donors with different hydrogen responses in a single or multistage process to create extreme MFR splits. The feasibility of this approach has been demonstrated by Mitsui [23] where propyltriethoxysilane or vinyltriethoxysilane or beta-phenethylmethyldiethoxysilane were used in one reactor in combination with dicyclopentyldimethoxysilane (i.e. donor D) leading to an MWD up to eight while the base case with donor D only exhibited an MWD of four under the applied polymerisation conditions. Other studies with a mixture of dicyclopentyldimethoxysilane and isobutylisopropyldipentoxysilane were less successful: the behaviour of one donor was dominating the other one either because the complex that the strongest donor forms with the Ti-species and the external donor-complexed TEAl is more stable or because this complex is formed faster [5]. Lately the combination of internal donors (based on succinates or di-ethers) with different H2-responses and different intrinsic MWDs has been investigated and claimed to bring some advantages for film applications, especially for BOPP [24] or for fibres [25]. These pairs were the most recent advances of former mixtures based on for example phthalates and succinates [26]. However, all these systems are by nature quite inflexible as changing for example the hydrogen feed will impact the response of the system in a scaled proportional way.

Combining catalysts with different chemistry and H2-responses in series or using a dual-site catalyst in one or consecutive reactors have also been proposed as a path to lead to a bimodal molecular weight distribution [27–34]. Metallocene catalysts are often associated with ZN-ones to broaden the intrinsically narrow MWD of single-site fractions and to capitalise on their very high H2-sensitivity. However, one additional constraint in case of such ZN/metallocene combinations is the necessity to deactivate the active sites of the first fraction before proceeding with the polymerisation of the second part as TEAl—used to activate ZN-catalysts acts as a poison for metallocene catalysts. Alternatively methylalumoxane (MAO) as activator for the metallocene polymerisation is a poison for ZN-catalysts and leads to lower stereoregularities than expected. It follows that the combination of catalysts remains a niche and is discarded in commercial routine: operation in full-scale is far from being trivial, freedom in polymer design is limited (the MFR fractions of the different species cannot be manipulated separately neither can their split), and economic viability is questionable.

To conclude this section, it is worthwhile to recap and mention additional paramount features of a catalyst system besides controlled morphology, activity, stereoregularity, MWD, H2-response, that contribute to the suitability of a catalyst:

1. Its relative insensitivity vs. impurities;
2. Its C2 (or comonomer) response;
3. Its price (costs/ton PP) are there of prime importance.

From a very practical point of view, easy catalyst manufacturing, no issues in handling, in storage safety or in ageing, and easy feeding in the (pre)polymerisation vessel have also to be mentioned [35].

6.3 Tailoring the Continuous Phase: A Complex Undertake

6.3.1 Processes and Operating Conditions

Tailoring the continuous phase refers—for sake of simplicity—to the ability to run (very) different H2 and comonomer compositions in the first two reactors of a multistage process (R1 and R2)—after pre-activation or not of the catalyst. A summary of the combination possibilities of the individual vessels is shown in Table 6.2; the presence or not of a prepolymerisation section is also documented in view of its potential implications for catalyst design. Should the polymerisation train have enough reactors (three or more), trimodal set-ups can also be produced.

Typical—but not limited—operating conditions of R1 and R2 are reported in Table 6.3, they are widely unchanged over the last 30 years while the yearly volume produced in a single polymerisation train has grown from 100 kta to 500+ kta for most of the processes.

Amongst the main processes, Unipol™ is the only one, to the author's best knowledge, that does not actively offer bimodal (or at least broad) MWD capabilities, based on process design. Its additional small second reactor—which may complement its large first gas phase reactor—is indeed meant to make classical impact copolymers (improperly often called block copolymers, B-PP) rather than complex H-PP or R-PP [1].

The flexibility of other processes in terms of polymer design is linked to the physical nature of the polymerisation propylene medium (either liquid or gas phase) and to the separation capabilities of the hydrogen and/or comonomer between the first and second reactor (or polymerisation zones for Spherizone™, the so-called riser and downer).

Table 6.3 Typical operating conditions of R1 and R2 for the main PP processes

	Reactor 1 (R1)			Reactor 2 (R2)		
	Temp (°C)	Pressure (bar)	RT (min)	Temp (°C)	Pressure (bar)	RT (min)
Spheripol™	60–80	35–50	45–75	60–80	35–50	30–75
Hypol II	60–80	35–50	45–75	60–80	35–50	45–75
ExxonMobil™	60–80	35–50	45–75	60–80	35–50	45–75
Catalloy™	60–90	20–40	30–120	60–90	5–30	30–120
Borstar®	70–100	50–60	30–90	70–100	15–35	60–180
Novolen®	50–105	20–35	60–120	40–90	10–25	60–120
Unipol™	60–70	25–35	60–120			
Innovene™	60–85	22–30	60–120	60–85	22–30	60–120
Horizone	65–85	25–30	60–120	65–85	20–30	60–120
Spherizone™	70–90	25–30	60–120			

Information is mainly extracted from the references in brackets [1, 36–39]. *Temp.* Temperature, *RT* Residence Time

6.3.2 Bulk Polymerisation and Design of Continuous Phase

Bulk polymerisations are limited to a maximum C2 (as mainly used comonomer) incorporation of ca. 6 wt% which covers, however, a vast majority of the applications. The most prominent technology is Spheripol™ with its two loops in series, the Hypol II (where loops have replaced the stirred autoclaves of the Hypol I model) and the ExxonMobil™ PP process being the two other commercial examples.

The H2-content can theoretically be varied between its lower control limit and its highest solubility rate in liquid propylene at designed reactor pressure and given temperature. This playing field represents flowabilities going from fractional to three-digit melt flow rates (MFR, measured at 2.16 kg/230 °C) for the most advanced ZN-catalyst/donor systems. In practise, though, the full range is not used in view of (1) process stability issues, (2) catalyst yield deficits and (3) fine formation, not to mention the need for a very efficient flash system between both loops that is not standard equipment (the primary cyclone being normally placed after the second loop). Typical the low MFR fraction is polymerized in the first reactor to prevent hydrogen carry-over.

Some examples of broad MWD are referenced below, the first one having a rather theoretical character:

- Using a conventional titanium-based solid catalyst with di-isobutylphthalate as internal donor in a series of two autoclaves, one operated in the absence in H2, the other one with important quantities of H2 and extreme splits of up to 5:95 between both reactors. While the total MFR was around 30–60 g/10 min, the high molecular weight fraction exhibited intrinsic viscosities (IV) as high as 10 dl/g (decalin, 135 °C); Mw/Mn (MWD) up to 25 and Mz/Mw up to 10 could be achieved where Mz represents the z-average molecular weight [40];
- Using a specific diether (as internal donor) catalyst system in combination with C and D-donors and reactor splits of 65:35, MFR ratios of between 15 and 50 could be generated with different H2-feeds in two loops in series yielding in an end MFR of around 3 and MWD around 10–15 [41];
- Using a metallocene catalyst and reactor splits of 65:35, broad molecular weight low MFR pipe grades (Mw/Mn: 7 as from example 10) could be made with feeds of 60 ppm and 420 ppm H2 in two consecutive bulk polymerisers [42].

When it comes to the split between the two loops, a 70:30 ratio between both reactors is considered to be the maximum affordable economically, ratios around 50:50 are preferred. Indeed these reactors are of same size and have been built-in historically to increase productivity, narrow the powder size distribution and flatten out subsequent reactivity peaks in the GPR (as an undesirable result of highly active fines with a too short residence time in the bulk) rather than to allow extreme flexibility in design.

Comonomer splits—though possible—are in general not broad due to the important reactivity differences in both loops if a low C2/high C2 concentration is run: the delta in mileage between both reactors can truly lead to instabilities and therefore to run-aways.

From a product design envelope point of view, the Spheripol™ process is capable to make products with MWD of 15 and to incorporate in routine up to 4.5 wt% of ethylene in random copolymers acc. to LyondellBasell [43], more is feasible provided the right downstream settings. Promoted for bimodal purposes has thus been the Spherizone™ process in the last 10 years with claimed MWD capabilities up to 36 and C2 incorporation possibilities up to 7 wt% [43, 44].

6.3.3 Gas-Phase and Design of Continuous Phase

The loop look-alike like Spherizone™ process—originally described in EP0782587 [37]—is actually a gas phase process divided into two interconnected zones—a riser and a downer—separated by a cyclone. As there is no solubility limit in a gas phase environment, extreme hydrogen and comonomer contents might theoretically be run.

The two different zones can be operated at different concentrations of H2 and comonomer [45]—the low molecular weight fractions being produced in the riser that acts like a fast-fluidized bed and the high molecular weight part in the downer which can be described as a moving packed bed showing plug-flow characteristics. The separation of both phases is made via a barrier fluid introduced at the top of the reactor. Homogeneity of the material is excellent in view of the multiple paths of the powder through both sections—where alternating layers of each fraction are claimed to be produced—yielding in an onion-like structure and thus to an intimate intraparticle blend [46]. Unlike a polymer blend where each particle has its own molecular weight, the polymer produced in the successive riser/downer paths has several molecular weights in the same particle [47, 48]—a quite unique feature. The palette of potential grades is widespread from bimodal homopolymers [45] to bimodal random copolymers with up to high comonomer contents of 6–8 wt% [49] or terpolymers with C2 and C4 or C6 as classical comonomers for films [50, 51], fibres [52], pipes [53] or container applications [54].

The third PP process developed by LyondellBasell (or its former companies) is Catalloy™—actually the only one that has not been licensed. It is originally based on three mutually independent gas-phase reactors in series installed after a prepolymeriser. In principle, two or three of its reactors can be used to produce bimodal or trimodal H-PP or R-PP compositions in a polymerisation train primarily—but not exclusively—dedicated to softer compositions; ethylene and 1-butene can be fed in each reactor, including the prepolymeriser. Use is made of a catalyst with controlled porosity [55].

Examples of material designs of the continuous phase range from:

- Heterophasic PP/EPR systems with an (extreme) homopolymer split in terms of MFR used to make for example a polyolefin masterbatch for improved surface performance [56]. Interestingly the high flow homopolymer fraction is reported in the examples to be produced in the last GPR while (very) high Mw PP and EPR fractions are made in the first two reactors;

- To C2/C3-randoms (matrices) with achievable amounts of C2 theoretically sizeably higher than the ones obtained in bulk (where the solubility limit is around 5 wt%);
- To polypropylene/1-polybutene (C3/C4) grades with high comonomer content (up to 10 wt%) where high transparency, processability over a large(r) temperature window, and structural integrity in terms of thermomechanical stability are, along with a low cold extractable fraction, key—for, for example, stretch-blow moulding applications [57] or metallised films [58]. In WO2001/44367 for example [58], a material with 8.9 wt% of 1-butene was produced in 3 consecutive reactors containing individual fractions of 3 wt%, 30 wt% and 35 wt% of this comonomer;
- To C2/C3/C4 terpolymers (belonging to the Adsyl™ family)—with low melting temperatures (125–135 °C), low seal initiation temperature features (75–115 °C) in a stiffness range of 600–1100 MPa—used essentially as sealing layers in film applications [55].

Worthwhile to highlight in this section is the latest development of Innovene™ with its INstage technology that offers H2 staging in one single horizontal stirred gas-phase reactor with plug-flow behaviour and therefore the possibility to broaden the MWD of the continuous phase [59]. One of the key-features of the HSBR over other processes is indeed its narrow residence time distribution—linked to the nature of its vessel design: the HSBR can be described as a series of continuous stirred tank reactors (CSTR)—that even the characteristics of the powder throughout its reactor length and minimise the fraction of catalyst particles with very short or very long residence time. Provided discrete feed points and appropriate feed separation the ability to produce different well defined individual fractions in one reactor is possible. On top, being a process without a prepolymeriser, the spacing of the injection points for the catalyst and co-catalyst in the first HSBR has been assessed to be of particular importance to avoid overheating and thus lump formation if they are too close or under-activation if they are too far away [1].

This chapter would not be complete without mentioning the Novolen® technology with its vertical stirred bed reactor (VSBR) in series. Its versatile operation mode has already been commented before.

6.3.4 Hybrid Systems and Design of Continuous Phase

A hybrid system loop/GPR to run the continuous phase has been developed by Borealis with its proprietary Borstar® process. Both reactors are—by basic design—decoupled allowing potentially widespread hydrogen and comonomer settings. This is especially true with the possibility to run higher H2 and C2-contents in the first GPR and with the possibility to operate the sole loop either in sub-critical (liquid medium) or supercritical mode. It follows that the used catalyst should be designed to withstand higher operating temperature (95 °C) and to incorporate higher

C2-contents. For this later an adequate porosity plays a key role, especially when an additional elastomer phase is polymerised in subsequent reactors. An example of bimodal C2-spread is provided in EP2452959 [60] where a C2-content of around 4 wt% is produced in the loop and a C2-content of 8 wt% in the gas phase reactor, yielding in total C2-content close to 6 wt%.

Interesting to notice is a concept making use of the loop/GPR1/GPR2 in series to produce three independent homopolymer fractions of different molecular weights for high stiffness thin wall packaging applications with around 10 wt% EPR and an end MFR of around 30 g/10 min [61]. Claimed are homopolymers with MFR ranges per reactor of (1) 200–500 g/10 min in the loop, of (2) 30–200 g/10 min in the GPR1 and (3) of 0.03–5 g/10 min in GPR2—while the MWD stays preferably below 12. Examples in EP2727958 [61] show typical splits of 40:35:25, calculated MFRs as low as 0.07 g/10 min in the third reactor and the use of a classical fourth generation $MgCl_2$-supported catalyst system (with diethylphthalate based chemical as internal donor and diyclopentyldimethoxysilane as external donor, with or without use of polyvinylcyclohexane—PVCH—as in-situ nucleating agent).

A similar approach has been followed in the low MFR range for blow moulding [62] or for pipe applications [63, 64] both with ca. 10 wt% EPR using the same catalyst system as described before. Homopolymer flows ranging from 0.005 to 50 g/10 min in three reactors in series and more extreme splits (typically 35:50:15) have been produced. Moduli as high as 2000–2200 MPa were obtained as a result of the broad polydispersity and high stereoregularity of the continuous phase. Obviously the very low MFR fraction is beneficial for creep performance.

Variations of the above items propose the manufacturing of random copolymers for pipes in a 3-reactor configuration with MFRs ranging from 0.002 to 10 g/10 min and C2-contents up to 4 wt% to make random/random/homopolymer or random/random/random fractions where up to 2/3 of the polymer (the part with highest flow properties) is made in the second reactor [65] or to use a trimodal homopolymer as basis in a subsequent off-line compounding step [66].

6.3.5 Some Process Tricks to Create and Enhance Multimodality

Periodically varying process parameters—especially hydrogen—within a single reactor has occasionally been seen as tool to broaden the molecular weight distribution of PP [67–69]. For practical reasons, however, it remains a confidential area: the theoretical MWD between the two selected H2-feeds cannot be reached within a reasonable time frame [5], making it cumbersome to handle and limited in terms of achieved product performance.

Alternatively an oscillatory feed of C2 has also been reported to broaden the MWD [70]. Here too the translation to full-scale commercial operations is a challenge.

There are only few references dealing with the influence of temperature on the broadness of the MWD. EP2527593 [71] indicates a slight broadening of the molecular weight when producing a bimodal random (with a first fraction containing 1.5 wt% and a second one with ca. 6 wt% C2, the split between both reactors being 45:55) at temperatures of 80 °C in the loop and 90 °C in the GPR (vs. a comparative example run at 70 and 80 °C). Interesting to notice is, all other primary product properties being equal (melting temperature, crystallisation temperature, flexural modulus, puncture energy, room temperature notched impact strength, haze), the catalyst yield increases from 24 to 63 ton PP/kg catalyst between the inventive and comparative example. This feature might, however, also be (partially) ascribed to the use of a small C2-feed in the prepolymeriser of the inventive example with the used catalyst/donor combination. A deeper dive in the molecular structure of the resulting polymers may elucidate some of the open questions, for example, the ratio between the blockiness/randomness as a function of the C2-content of each individual fraction at a given temperature.

6.3.6 Unique Product Features Via Matrix Multimodality

One question that has not been answered so far is why multimodality is needed. The answer is straightforward: because it stretches the application window of multiple product lines. The impact on the different segments will be shown in the next sections. Worthwhile to keep in mind are the individual contributions of the different molecular weight fractions on the performance shown in Fig. 6.3 ranging from processability/cycle time for low Mw to creep/long term performance for high Mw.

Independently from the nature of the final product (H-PP, R-PP, B-PP or RB-PP), the three predominant reasons to broaden the molecular weight distribution of the continuous phase are [72]:

- To increase processability (via the introduction of a low molecular weight fraction)—this is valid for all applications [13, 73, 74] with the exception of fibre spinning where a narrow MWD (obtained in the reactor and/or with peroxide visbreaking in the extruder) is required to guarantee strand integrity at highest line speeds;
- To increase melt strength (via the introduction of a high molecular weight fraction), especially for following applications:
  - Pipes (MFR 0.1–0.5) to promote resistance to disentanglement as primary damage mechanism for fatigue types of loadings and thus enhance long-term properties [75–77].
  - Foam extrusion (MFR0.3–3) to ensure large cells and a uniform cell size distribution [78–80].
  - Extrusion (blow) moulding (MFR 1–5) to allow for production of large parts [81, 82].

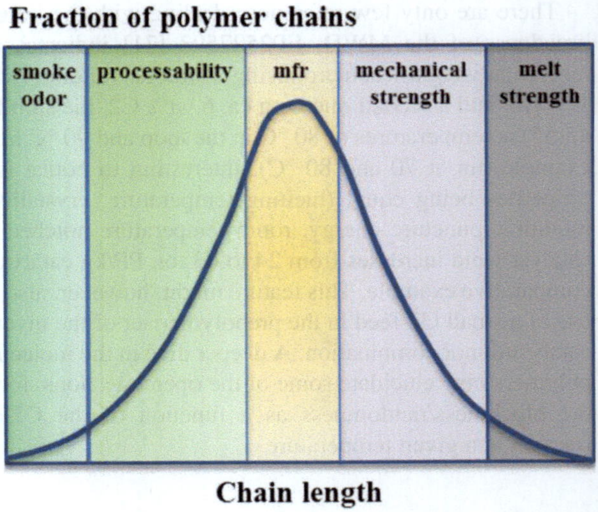

Fig. 6.3 Main contributions of the different fractions of polymer chains in terms of physical properties

- Extrusion coating (MFR 1–20) to accommodate high melt extension and promote orientation of the polymer melt while avoiding neck-in, draw resonance, edge weave and poor film quality [83].
- Film extrusion—especially the blown technology (MFR 1–4)—to ensure structural integrity (i.e. no bubble instability) and avoid thickness irregularities [84].
- (Stretched) Blow moulding (MFR 1–3) to avoid the phenomenon of draw-down (i.e. the stretching of the parison by its own weight) [85].
- BOPP and capacitor films (MFR 1–5) [41, 86].
- Sheet extrusion and thermoforming (MFR 1–5) to allow production of large parts and/or narrow wall thickness contribution, particularly at (sharper) corners [87].
- Non-oriented (cast) films (MFR 4–12) to reduce water vapour transmission rate [88].

• To increase the impact/stiffness balance, as a result of a broader MWD which accelerates nucleation rate and thus leads to products with higher crystallinity. This feature is of benefit especially for injection moulding applications and thicker parts obtained via extrusion.

High melt strength is usually achieved in a two-step process with the production of a very high molecular weight in one step and of an adequate Mw in the other one to reach the targeted end MFR. While polymerising in series with one single catalyst is the most straightforward choice [89, 90], consecutive combinations of catalysts in each reactor (ZN alone, metallocene alone or consecutive use of ZN and metallocene catalysts) have also been proposed as design tool [33, 91]. Taking the data of EP942013 [90] to illustrate this concept, an increase of the high molecular weight

Table 6.4 Characteristics of bimodal MFR 0.5 grades produced in a two-reactor batch slurry (BS), by melt mixing (MM) or by continuous slurry polymerisation (CS)

		IE1	IE2	IE3	CE1	CE2	CE3	CE4
		BS	BS	BS	BS	BS	MM	CS
		X-11	X-1	X-16	X-23	X-28	X-27	X-4
MFR-total	g/10 min	0.5	0.5	0.5	0.5	0.4	0.5	0.5
IV-R1	dl/g	8.9	8.7	9.4	3.9	10	9.4	9.1
Split	R1/R2	18/82	30/70	47/52	50/50	40/60	30/70	30/70
Gels	#	5	4	6	5	8	7850	718
Mw/Mn	–	8.3	11.5	9.0	5.5	38.2	10.3	10.2
Mz/Mw	–	3.7	4.1	3.7	3.1	4.0	4.1	4.3
Mw > 1500	kg/mol, %	9	14	30	7	22	14	15
MT	g	8.7	12.1	25.4	4.6	9.2	10.3	105
FlexMod	MPa	1813	1846	1945	1433	1550	1723	1920

R1 stands for first polymerisation step 1, R2 for second polymerisation step 2, IV-R1 is the intrinsic viscosity of the high molecular weight fraction (decalin, 135 °C) produced in R1, MT is the melt tension (measured at 230 °C at an extrusion speed of 15 mm/min and a rolling-up speed of 10 m/min). IE stands for inventive example, CE for comparative example. X-xx is the coding used in the patent for the illustrative examples [90]

fraction (described by its intrinsic viscosity IV-R1) leads to improved melt tension (MT) and stiffness (Flexural Modulus) at constant total MFR (Table 6.4). Comparing for example the inventive example 1 (IE1) with the comparative example 1 (CE1), a delta of 400 MPa in modulus can be seen as well as a double as high melt tension (8.7 vs. 4.6 g). Obviously Mw and thus Mw/Mn has to stay within reasonable limits to make the trick, an MWD of 38 being too much to boost stiffness as shown by CE2, a fact that is possibly linked to the intrinsic low crystallinity of very low MFR homopolymers (IV-R1 is 10 dl/g) and to a difficult compatibility between very low and very high Mw components. When competing with melt mixing (MM), the use of an in-reactor multistage process is also preferred in terms of improved gel performance and enhanced stiffness (compare CE3 with E1). Interestingly using a batch polymerization process is more favourable than a continuous mode in terms of material homogeneity (i.e. low gels) because all particles have the same residence time and composition (see CE4 vs. E2). This fact emphasis again the efforts made by the polyolefin producers to narrow the residence time distribution by using for example two loops or a CSTR-like reactor to manufacture the continuous phase and by working with catalysts with a narrow particle size distribution.

As an alternative to reactor design, chemical modification of the polymer in the extruder can also be used to achieve a bimodal molecular weight distribution and thus high melt strength [92]. Under appropriate conditions, the resulting material (thanks to its broad MWD and its long-chain branches) outperforms any reactor-made material. Considering the extruder as part of the polymerisation would, however, be too far-fetched and thus not discussed here. An introduction can be found in the enclosed references [93–96].

The advantages of having a high comonomer fraction and a low comonomer fraction in one product are less intuitive all the more that a narrow comonomer distribution—above a critical comonomer concentration—is preferred to optimise optics (the main driver for random copolymers). This is due to the tendency of higher comonomer fractions to segregate into a nanoscopic separate phase. Though, some benefits of splitting the comonomer distribution are highlighted below—often in combination with a bimodal or broad molecular weight distribution:

- To broaden the softening range (with for example a low seal initiation temperature) while maintaining a higher melting temperature, optimising the processing temperature, and minimising the amount of xylene cold solubles (XCS) at given comonomer contents (compared to bulk products)—which leads to improved performance in BOPP [86], thermoforming [87], heat sealing applications up to very soft films—including less blooming [97–99], ISBM [100] and thin wall packaging [71]. Benefits in oxygen permeation of coated articles have also been reported when using a bimodal C4 as a comonomer [101]. Combined with the right type of dispersed phase in a 3-or 4-reactor train, transparent soft compositions with low C6-solubles for, for example, extrusion blow-moulded articles have been obtained [102], building on a double matrix bimodality (i.e. different molecular weight and C2-content in two first reactors);
- To use the operation window of the continuous phase in a GPR/GPR train [103] or loop/GPR configuration for producing high comonomer (and/or terpolymers) grades to its full extent. This is the case in for example Borstar® where the solubility limitations in the bulk restrict the comonomer content to ca. 4 wt% in the loop [104]. Comonomers can be C2, C4 or C6 [105].
- To enrich one of the phases—the high molecular one as a general rule—with comonomer—while keeping the total comonomer content at a given level (for, for example, thermal purposes), in order to enhance for example creep properties in pipe applications [77]. EP0785954 highlights a precursor approach, where a double bimodality (MFR and C2-distribution) has been used to maximise the performance of C2/C3-based pipe grades (MFR 0.2–0.4 g/10 min, C2 of 3 wt%): a first very low MFR C2-richer phase was produced in the first reactor (MFR 0.02 g/10 min, C2-content around 3.5–4 wt%) in order to get excellent creep performance and a higher MFR fraction (20–40 g/10 min, C2-content around 2–3 wt%) was made in the second reactor to ensure good extrudability. Stiffness, slow crack growth properties measured under tensile load (ESCR type) and a longer time to failure at same hoop stress levels in standard pipe pressure tests have been also observed with this bimodal C2-distribution over the reactors. A rationale behind this outstanding performance is the non-linear incorporation of comonomer as function of the molecular weight in ZN-systems: shorter chains contain more comonomer than longer ones, but do not contribute to long-term properties. Enriching thus the high molecular weight fraction with comonomer guarantees a better structural integrity over time as creep resistance and slow-crack growth are improved this way [106]. A similar concept than for pipes can be applied as well to for blow moulding, film and fibres applications where good mechanical properties and improved comonomer distribution are needed.

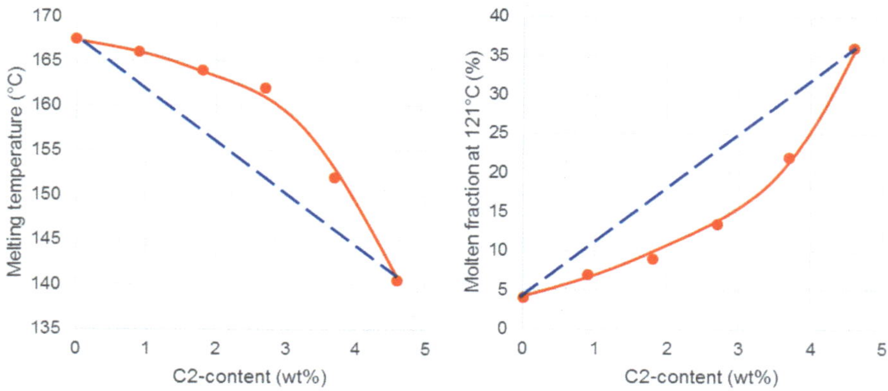

Fig. 6.4 Evolution of the melting temperature and molten fraction at 121 °C as a function of the C2-content of Homo/Random (H/R) systems (red dots) and of pure C2C3 randoms (blue dotted line, intrapolated) (data after [108])

Interesting to mention too is a family of materials where one phase is comonomer free and where the other one contains some comonomer. C2 is the most often used comonomer; working with C6 is less popular though still an area of investigation for high end sealing or optical applications [107]. Compared to a monomodal random system, it exhibits a higher melting temperature than the reference as can be seen in Fig. 6.4. Two designs routes can thus be explored):

- One for higher C2-contents and same melting temperature than the reference bringing advantages in stiffness, optics (haze, transparency, gloss) and processability (e.g. broader operation window) linked to a lower seal initiation temperature for cast and blown films), a lower stretching temperature for BOPP or (injection) stretched blow moulding [85, 87, 100, 109], a lower softening temperature for thermoforming [86], a better spinning performance for fibres [110].
- The other one at higher melting temperature and same C2-content than a reference, yielding in benefits in terms of heat stability—before and after a potential thermal treatment (like steam sterilisation)—and improving thus the overall performance when for example the impact strength and optical performance are of interest [108]. It can be combined or not with a bimodal matrix set-up. Should this be the case, benefits in processability and/or cycle time can also be derived, for example in thin wall packaging applications [30]. Improvements in rapid crack propagation, creep resistance and long-term pressure resistance in pipe applications have also been claimed [111].

As an example, Fig. 6.4 shows—as a function of the comonomer content—the expected evolution of the melting temperature of the C2/C3-random matrix of a model series of RB-PPs (containing similar types of rubbers) and that of its homologous systems based on homo/random as continuous phase (with a various split between both fractions).

Table 6.5 Normalised haze before and after steam sterilisation

Comonomer	Haze before steam sterilisation (normalised, series 1)			Haze after steam sterilisation (normalised, series 2)		
C2 (wt%)	Random (%)	Homo/random (%)	Delta (H/R-R) (%)	Random (%)	Homo/random (%)	Delta (H/R-R) (%)
0	0	0	0	100	100	0
0.9	20	0	20	80	40	40
1.8	40	0	40	60	5	55
2.7	60	23	37	40	0	40
3.7	80	62	18	20	0	20
4.6	100	100	0	0	0	0

The highest value (100%) represents the lowest haze in one series, the lowest value (0%) the highest haze. Values for the random series have been interpolated based on the two extreme C2-content of a series (i.e. 0 and 4.6 wt%). Data for the H/R-systems have been measured on 100 μ thin cast films

The positive deviation of the melting performance of the homo/random based compositions is believed to be linked the combined presence of highly crystalline PP domains and of that of disturbed C2-rich chains that do mix intimately together on a nano-scale while exhibiting still some discrete properties of their individual components, especially the increased nucleation ability offered by the PP homopolymer phase. It follows that the amount of crystals already molten at a certain temperature is lower in the case of a homo-random matrix than in the case of a pure random C2/C3 copolymer matrix. Such considerations can be used to approximate phenomena occurring during steam sterilisation, where obviously when plotting the fraction of the polymer molten at 121 °C of a DSC trace (first heat, 10 °C/min) as a function of the C2-content a very limited amount of the matrix is affected by changes in transparency with a homo/random matrix versus a pure random copolymer. The positive impact on haze retention of such a system after steam sterilisation can be derived from Table 6.5 by comparing normalised values for both random and H/R systems before and after treatment [108].

6.4 Tailoring the Dispersed Phase: A Polymer Design Challenge

6.4.1 Which Processes for Multimodal Rubbers?

Processes capable of tailoring the dispersed phase are less widespread than those capable to broaden the molecular weight or chemical distribution of the homogeneous phase. Three are worthwhile to be highlighted, the products made thereof constituting the large majority of the grades showing a multimodal (often bimodal) character of their rubbery phase:

- Spheripol™ from LyondellBasell in a four-reactor configuration (loop-loop-GPR-GPR)—where the dispersed phase can be produced in the last two reactors, in typical amounts from 20 to 40 wt%. Operating pressures (in GPRs) are up to 15 bars, temperatures in the range of 75–85 °C higher in the last reactor (up to 95 °C) when the reaction has to be boosted;
- Borstar® from borealis in a four-reactor set-up (loop-GPR-GPR-GPR) where the rubber is usually made in the two last GPRs, also in typical amounts from 20 to 40 wt% at pressures up to 20 bars and temperatures up to 95 °C;
- Catalloy™, a proprietary technology from LyondellBasell, consisting of three independent GPRs in series preceded by a prepolymeriser—capable to make 40 to 80 wt% of dispersed phase—a unique feature in the whole industry. Use is made of a catalyst with controlled porosity allowing the incorporation of high amounts of rubbery phase. In each phase of the (pre)polymerization, C2, C3 and 1-butene as monomers can be dosed. Temperature is typically in the range of 60–80 °C in each reactor, operating pressures are between 10 and 30 bars.

Interesting to mention is that though less common, making a bimodal rubber (including different types of comonomers) has also been shown to be possible by a Spherizone™ process where the downer of the first reactor is used as rubber maker, the second rubber phase being manufactured in the consecutive GPR [75].

While the limiting factors for Spheripol™ and Borstar® are the designed reactor size of the GPR along with residence time considerations of the catalyst, Catalloy™ has been primarily developed for (very) high levels of rubbers. The rubber production—for which different species in chemistry and/or molecular weight can be made—takes usually place in the second and third reactor (of comparable size than the first GPR), while homopolymer or random copolymer are normally produced in the first reactor in amounts from 20 to 60 wt% of the final composition. Commercial offering ranges from 20 MPa (Softell™) to 1200 MPa (Hifax™) in stiffness, a large part of the portfolio being between 80 and 550 MPa with the Adflex™ brand, that is, beyond the reach of conventional PP technologies.

Operating the first reactor as a GPR is most certainly the key of the Catalloy™ process to incorporate very large amounts of rubbery phase (up to 70 wt%). Indeed simulations (via a population balance modelling approach) of different processes run under industrial conditions (loop, VSBR, HSBR and FBR) showed that starting from a catalyst with the same characteristics in terms of structural design and activity, the porosity after the homopolymerisation stage is sizeably lower (by ca. 20%) for the liquid phase process than for the gas phase processes while the mean powder size follows an opposite trend [112]. This feature originates from the intrinsic higher reactivity and thus polymer yield in a liquid medium compared to a gas phase environment—which in turn reduces the pore volume available to incorporate a rubbery phase as well as the number of active sites available for copolymerisation. It follows that a higher amount of dispersed phase can be produced in the rubber reactor(s) when operating the first polymerisation stage in an FBR, VSBR or HSBR—all other parameters being kept equal [36].

6.4.2 Additional Catalyst Requirements

While the requirements described by Galli for the "Ideal Catalyst" [9] are still valid for multimodal rubber systems, their weight is somewhat shifted compared to manufacturing of bimodal homopolymer or random copolymers. Background is:

- The necessity to prevent stickiness when rubbery-like compositions are made;
- The need to keep an acceptable level of activity over multiple reactors, considering that the residence time can be between 6 and 8 h in a four-reactor configuration.

It follows that a catalyst based kinetic behaviour with a moderate initial activity followed by a slow decay is preferred (1) to make desirable particle morphology and (2) to control the reactor performance and conditions in the polymerization process [10]. Hence while a fast starter with high productivity can be used for homopolymers and random copolymers—provided a limited amount of fines is formed and provided the final powder morphology is right in terms of spherical and bulk density—a slower starter with a somewhat lower, but flatter, activity profile over the reactors will be selected for high rubber containing systems. A schematic view of the preferred activity profile of a catalyst for multistage polymerisation is provided in Fig. 6.5—along with the morphology of a PP powder with increasing rubber content.

Of particular concern is often the reactivity in the last reactor as even a highly engineered long-life catalyst is "dying out" after several hours in a polymerisation train. Here profit is taken of the comonomer reactivity, which is known to decrease with increasing the steric hindrance around the reactive double bond in following order: ethylene > propylene >1-butene > linear 1-olefins > branched 1-olefins [11]. Some practical considerations can be derived thereof:

1. C2 is the comonomer of choice for heterophasic C3-based systems, the 100 times higher reactivity of C2 versus C3 being used in the last reactor to boost the polymerisation. It is all the more evident that heterophasic copolymers with C3-based rubber with C4-, C6- and C8- as comonomers (based on ZN-catalyst systems) have not found their way in the market, with their reactivity being respectively 10, 30 and 60–100 times lower than C3 [11];
2. The longer the residence time, the more C2 will be incorporated in the rubber. This phenomenon is attributed to the presence of different active sites characterised by different oxidation states of the titanium atom. A simplified view accounts for two sites which relative ratios and reactivities vary with time. One site represents an oxidation stage of Ti^{3+}; it can produce a statistical copolymer, is predominant at the beginning of polymerisation, and will deactivate other time into Ti^{2+} as a result of the presence of the cocatalyst. Ti^{2+} on the other hand is less present and less active than Ti^{3+} when polymerisation starts, becomes more important with increasing residence time and favours the development of a more C2-rich copolymer, thus promoting a higher building of C2 into the rubbery phase with time [13, 36];

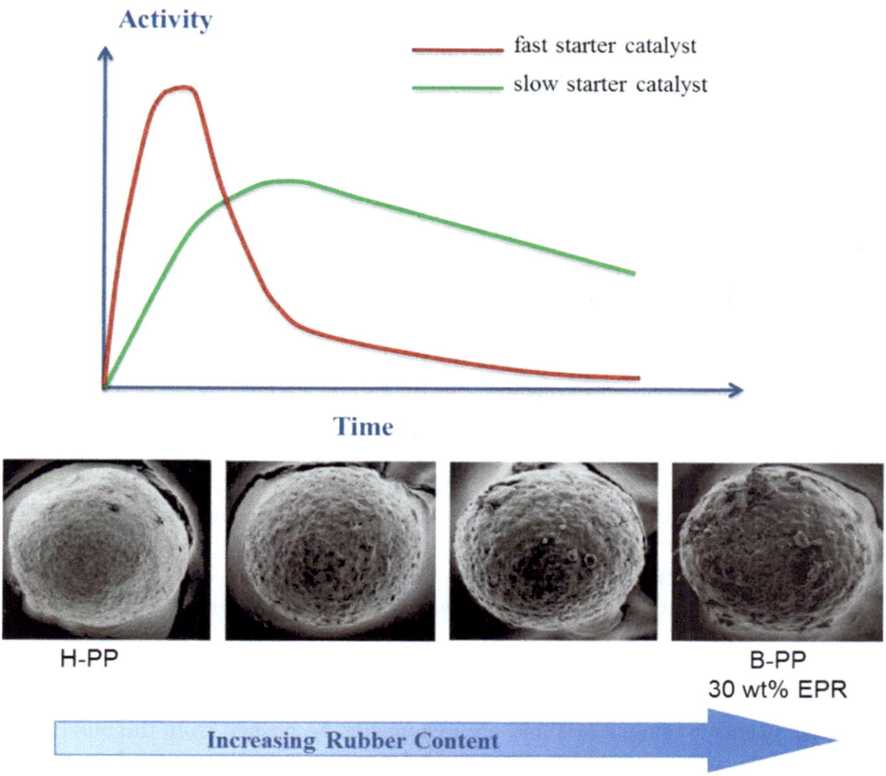

Fig. 6.5 Catalyst activity plotted against polymerisation time. In red: non-optimal profile for multimodal rubber incorporation. In green: preferred activity profile over a multistage process. The morphology of a PP powder with increasing rubber content is also provided

3. C2-rich rubbers are generally made in the last stage of the polymerisation (in extreme cases even HDPE-like substances). Not only it promotes reactivity and boosts catalyst lifetime, also secondary properties like low blushing or scratch resistance are improved ([113, 114] with a gas ratio of 0.97 in the last GPR, [115] with a gas ratio of 0.99 in the last GPR!, [116]). It constitutes thus an elegant alternative to the in-line compounding of HDPE on the polymerisation extruder should the equipment not be available.

An additional boost of the catalyst activity in the last reactor can be obtained via an increased temperature and/or pressure. Practically the pressure is, however, limited by the design of the process and an increase of temperature—provided the catalyst can withstand it—can lead to some undesirable stickiness. Moreover it has to be mentioned that changing the reactor conditions affects the respective contributions of the different feeds (H2, C2, C3), which in turn can impact the molecular weight, comonomer incorporation or rubber blockiness of the dispersed phase at for example given H2–C2 ratio. To increase the productivity of the gas phase reactors has also been proposed to split the feed of the external donor between the prepolymeriser and the slurry phase [117].

Working with a catalyst with the highest possible porosity (while minimising fines) is also of prime importance. Though yielding in some compromises in bulk density—and therefore in throughput, the incorporation of rubber inside the growing particles is an essential capability of a catalyst suited for a multistage rubber polymerisation as it prevents reactor fouling.

Figure 6.6 recaps the principle of fragmentation, replication and polymer growth in a schematic way—from the catalyst grain to the incorporation of rubber—based on the widely accepted double grain model with expanding microcore from the Ferrara school [13, 14]. Starting from the preferred spherical shape, the interstitial spaces between subparticles constitute the major contribution to the porosity of the catalyst macroparticles. The active centres are situated on the surface of the microparticles that are homogeneously dispersed within the catalyst subparticle.

The first step—activated by contacting monomer (C3) and additional chemical enablers—consists of a break-up of the catalyst grain in catalyst subparticles, a phenomenon which is extremely fast—almost instantaneous. These catalyst fragments—though discretely distributed in a uniform way—are kept together by the growing polymer (at the surface of catalyst microparticles) acting like a glue. They are dispersed in a uniform way, merge together and form larger polymer aggregates, the (pre)polymer subglobules, also called microcores. These subglobules tend to behave as individual polymeric flow units, that is, the catalyst microparticles undergo further fragmentation and are assumed to be pushed from the bulk to the surface of the polymer subglobules, where the reaction continues. By doing so the microcores form themselves agglomerates with interstitial spaces (or voids) which when polymerisation proceeds form polymer grains of increasing size (typically in the range of 100–1000 μ in diameter).

In case of a sequential polymerisation with a comonomer, the resulting phase is formed around the homopolymer subglobules. If the second monomer forms a crystalline material, it tends to stay where it grows; if a mixture of monomers, for example, C2 and C3, is used to generate a rubbery fraction, it migrates to the available space provided weak interglobular boundaries, and as the polymerization goes on, tends to fill them. Once full, the process becomes diffusion limited [118], sticky copolymer is made at the polymer surface, leading to stickiness issues and to blockage of the reactor.

This model explains the necessity to work with a high porosity catalyst, and thus with a high porosity (pre-)polymer grain to make a material with high rubber loadings (called PP/EPM heterophasic copolymer in Fig. 6.6) while a low porosity catalyst with high bulk density and compactness is sufficient for homopolymers. Particularly preferred is a catalyst having a large number of homogeneously dispersed and small pores (to optimise the filling ratio) and a combination of high inner and surface porosity to promote rubber incorporation while minimising friction of (stickier) particles thanks to their rough outer shell. Best performing catalysts have been reported to have porosity values—determined by mercury

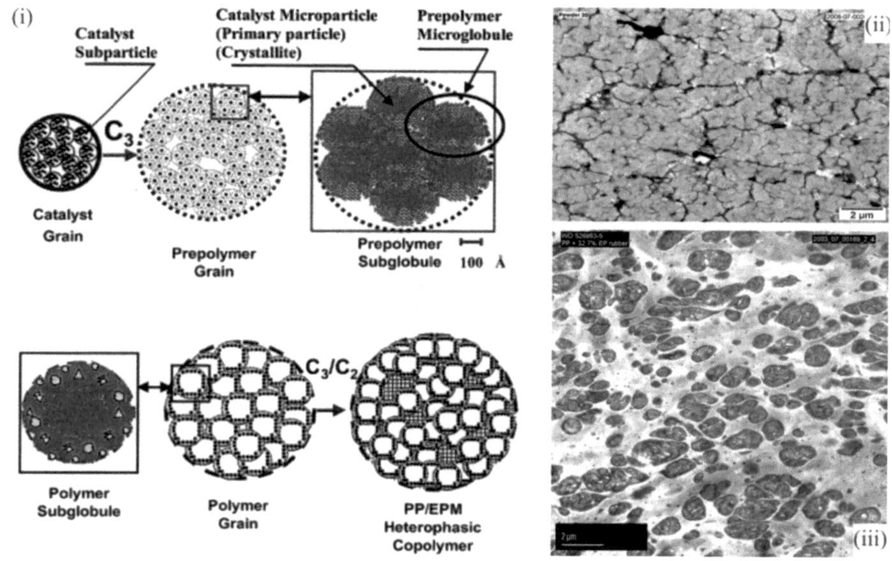

Fig. 6.6 Left hand side (i): Schematic view of the fragmentation, replication and polymer growth up to the formation of a heterophasic copolymer based on the double grain model with expanding microcore; top: detailed mechanism of prepolymer growth; bottom: expected morphology of PP and sequential PP/EPM copolymer grains. Reproduced with permission from [13]. Right-hand side: Microstructure observed via TEM (Transmission Electron Microscopy) (ii) in the polymer grain versus (iii) morphology observed in a PP/EPR after extrusion

porosimeter—around 50 vol%, pores ranging from 2 to 1000 nm, and mean pore radius around 100 nm [13, 14, 119].

Should the surface area of the catalyst be low, tailored introduction of inclusions that do not comprise catalytically active sites (like hollow voids) has been proposed [120] as well as use of a catalyst comprising a solid material without active sites, but with a high surface area up to 500 m^2/g and a particle size in the range of up to 100 nm [121, 122], this latter catalyst system being shown to allow the stickiness-free production of PP based polymers in a multistage polymerisation process with an XCS content of up to 50 wt%.

Important to notice is that the obtained morphology in the heterophasic polymer grain/powder are usually sizeably different from that of the system after extrusion. After mixing, the thermodynamically driven and stable induced phase separation between matrix and rubbery phase is the rule—while a kind of co-continuous structure is still visible in the powder (Fig. 6.6(ii) and (iii)). Worthwhile to notice too is that during extrusion there is no static situation, but a dynamic equilibrium with competing break-up and agglomeration mechanisms, yielding in a "metastable" continuous/dispersed phase structure.

In addition to the previously mentioned features, three secondary intrinsic catalyst requirements are also key to ensure a wide compositional playground in a multistage rubber reactor train:

- A broad H2-sensitivity, which allows to produce as well lower molecular weight than higher molecular weight copolymers in series. In this respect, design limitations will be essentially seen with catalysts with a good H2-response, the upper attainable limit of intrinsic viscosities being capped;
- The incorporation capability of comonomer in the rubber over a wide range, allowing extreme compositions to be run. Targeted are here—in routine—C2-contents (as most commonly used comonomer) between 25 and 65 wt%;
- The quality of the produced rubber in terms of sequential distribution of the comonomers. While a random distribution will lead to a more elastomeric-like substance—and its associated benefits, longer sequences (up to a critical level) will lead to a blocky composition which is preferred when the impact/stiffness balance has to be optimised.

6.4.3 Why Multimodal Rubbers?

Producing rubber in multiple reactors has obviously the advantage to maximise the amount of dispersed phase in a PP based heterophasic system. While the Catalloy™ process is the parade example for this feature with claimed EPR contents above 75 wt% [123], moving from one to two rubber reactors in for example the Spheripol™ or Borstar® process allows also to move the upper accessible limit rubber content from 25 to 40 wt%, independently of the modality of the rubber.

On top, a superior performance profile can be obtained with grades having multimodal rubbers compared to their monomodal homologous. Considering the most common system based on two ethylene–propylene rubbers produced in series, two primary factors can be varied: their molecular weight (or intrinsic viscosity of the rubber phase, IV/XCS) and their C3-content (C3/XCS). The possible evolution of some key properties of the individual fractions is provided in Fig. 6.7.

It is a result—at constant matrix MFR—of the complex interaction of the set rubber parameters (IV/XCS and C3/XCS) and of a third induced one, namely the EPR morphology—which evolution for monomodal fractions has been extensively reported elsewhere and which impact on primary (mechanics, flow) and secondary (surface) properties is crucial [125–129]. The maximum in performance for one given property is obtained with different IV/XCS and C3/XCS combinations and their respective splits. And as expected, none of the settings provides a maximum in a set of say seven individual properties, optimum settings have thus to be found. It is therefore tempting to include in one material produced in a multistage reactor train antagonistic features like low shrinkage (with low molecular weight C3-rich rubber) and good scratch resistance (with higher IV/XCS with a C2-rich rubber).

Figure 6.8 shows an example of a unique performance profile attainable with multimodal rubbers—here the optimisation of flow marks and scratch resistance as paramount secondary properties. All properties having been normalised, the enhanced design potential of multimodal components over single pure fractions is obvious [130].

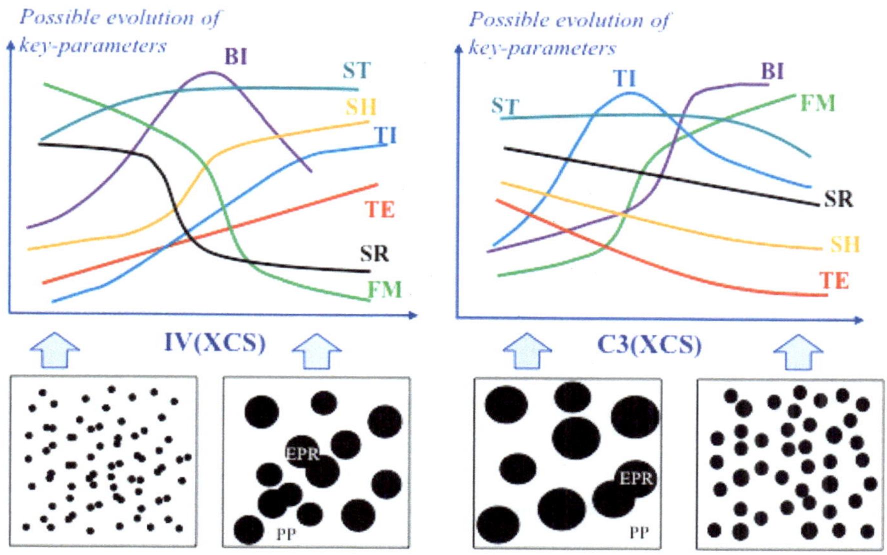

Fig. 6.7 Possible evolution of key-parameters as a function of the IV(XCS) and of the C3(XCS) of an Impact PP. A schematic view of the morphology of the PP/EPR as a function of the rubber characteristics is also provided. *IV* Intrinsic Viscosity, *XCS* Xylene Cold Soluble, *BI* Biaxial Impact (i.e. Impact Falling Weight), *ST* Stiffness, *SH* Shrinkage, *TI* Triaxial Impact (i.e. Charpy or Izod Notched Impact), *TE* coefficient of linear Thermal Expansion, *FM* Flow Marks, *SR* Scratch Resistance. Adapted after reference in bracket [124]

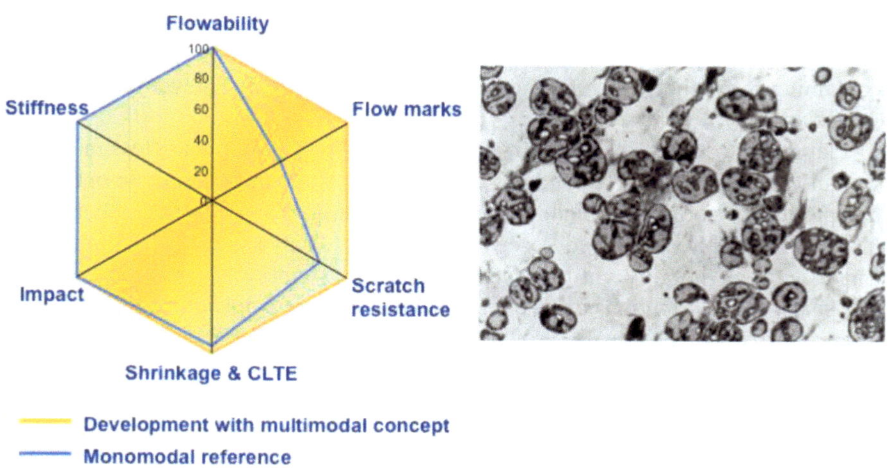

Fig. 6.8 Illustration of benefits of multimodal rubber—along with its morphology (20 × 10 μ)—versus monomodal rubber solution [130]

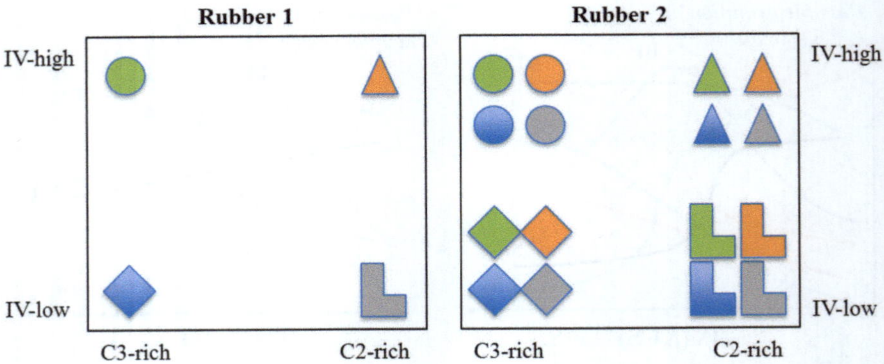

Fig. 6.9 Schematic view of variation possibilities in two consecutive reactors of molecular weight and C3-content of rubbery phase

A quick analysis of the theoretical possibilities offered by a bimodal rubber approach is highlighted in Fig. 6.9 limiting the considerations on C3-rich, C2-rich, IV-low, IV-high fractions when it comes to identify the rubber. Subtracting the four cases where monomodal rubber is produced (same colour, same symbol), 12 extreme combinations can be run in two consecutive reactors—and of course the whole spectrum in between. Interestingly this figure takes also into account the sequence in which each fraction is produced—where obviously some are less preferred (e.g. making a C2-rich rubber in the first rubber reactor and a C3-rich one in the last reactor) in view of the reactivity restrictions discussed in the above.

Several of these combinations deserve further attention. For the sake of simplicity the higher molecular weight fraction will be defined as high IV and the fraction containing the higher C3-content will be named as C3-rich whatever their absolute values are.

Playing with a high molecular weight C3-rich phase along with a low molecular weight C2-rich phase has turned out to be a successful concept from an operations and product performance point of view [131]. Manufacturing C2-rich rubbers in the last reactor improves indeed reactivity and thus catalyst yield and throughput. Enhanced melt stability via targeted comonomer assembly at controlled particles size shape and distribution can be gained following the principle that the C3-richer—more amorphous—phase will be at the interface matrix/rubber and that the C2-richer and more crystalline phase will reside within the particle building one or several discrete inclusions in the rubber particles. Several unique property combinations can be derived using the ratio of rubber produced in the last two reactors as additional polymer design tool:

- Improved mechanics at high flow for , for example, thin wall packaging [132]
- Superior surface appearance (i.e. less flow marks) at optimum mechanics for automotive applications [133, 134]
- Excellent scratch resistance or blushing with the use of a very C2-rich phase in the last reactor [113, 115, 116]

- Basis for transparent solutions with adequate stiffness/impact balance for use in (deep freeze) packaging with either a homopolymer or random copolymer [135, 136] or a homo/random polymer as matrix [137]
- Basis for (high flow) food compliant solutions with excellent impact stiffness balance and a low C6-extractable fraction [138].

Systems with a mix of C3-rich/low IV—C3-rich/high IV rubbers have also increasingly been used in the last years to offer out-of-reach profiles for monomodal rubbers. Use is made of the chemical compatibility between a C3-rich low molecular weight rubber fraction with the matrix to disperse a higher molecular weight fraction more homogeneously leading to an improved phase morphology and thus to an optimised impact/stiffness balance. The same considerations—illustrated conceptually in for example [139]—have shown to have some advantages too to stabilize the melt and thus to get a stable surface morphology upon injection moulding, yielding in a lower amount of surface defects (fish eyes). Additionally a very C3-rich rubber (getting closer to the PP matrix characteristics than to the ones of a classical rubber) shows a lower tendency to be dissolved in solvents (e.g. hexane or heptane) and constitutes thus a good basis for film or injection moulding (e.g. thin wall packaging) applications requiring low contents of extractables [140, 141] or excellent optical properties before and after a heat treatment for, for example, film or packaging applications [142]. Hence soft compositions with an MFR around four and of ca. 400 MPa (flexural modulus) based on a random matrix and 35 wt% of a very C3-rich rubber (C3/XCS: 80 wt%) with low IV/XCS (ca. 2 dl/g) could be reached in WO2013092615 yielding with a haze on 1 mm thin plaques before steam sterilisation (121 °C, 30 min) of 34% and after sterilisation of 49%, the comparative example with slightly poorer C3-content (around 70 wt%) showing a substantially worse optical performance with values of resp. 76 and 88%.

Soft grades with a (very) high content of rubber (35–80 wt%) and a C2/C3-random matrix are the last product family to be mentioned. The (bimodal) rubber is typically produced in two last reactors of a Borstar® (XCS 35–50 wt%) or Catalloy™ (XCS 35–80 wt%) process, typically consisting of C2/C3 rubber in the first cited process and of various combinations of C2/C3/1-butene as comonomers for the second one. Unless transparency is a key parameter [142–145], stickiness is prevented by producing (very) high EPR molecular weights in a fairly low flow matrix [146]. Use is often made of visbreaking to adjust the end MFR of the desired higher flow. In an illustrative example [123] soluble contents as high as 90 wt% could be achieved yielding with a stiffness as low as 20 MPa. The intrinsic viscosity of the first rubber was around 4 dl/g, that of the second rubber up to 6.5 dl/g. The split between both rubber reactors was close to 2:1 and their residence time in the range of 1:2—as process trick to circumvent the catalyst fatigue over time. In line with reactivity considerations the C2-richer phase was made in the last reactor (i.e. C2/XCS was around 25 wt% in the first rubber reactor, and close to 40 wt% in the last reactor). Specific catalyst features are also indicated: the surface area should be less than 200 m^2/g and its porosity above 0.2 ml/g.

Following trimodal homopolymer configurations, soft high rubber containing heterophasic PP/EPR have also been shown to be produced in a 4-reactor Borstar® polymerisation train, where the matrix was manufactured in the first reactor and were use was made of the three gas phase reactors to make three different EPR fractions as disclosed in EP2452975 [132]. Claims and examples show that apart from this trimodal rubber operation mode, the process is run in a classical way with claimed IV/XCS in between 2 and 4 dl/g and with the C2-content of the last rubber fraction being higher than that made in the penultimate reactor which in turn is also higher than in the first produced rubber (being in the range of 25–35 wt%). The EPR weight produced in each reactor differs too ranging from 40 to 80 wt% for the first rubber, to 10–35 wt% for the second one and to 5–30 wt% for the last ones. Total EPR weight is in between 39 and 46 wt% in the examples (claimed is 20–80 wt%). An additional difficulty in operations is linked to the high flow of the polymer homopolymer matrix (MFR $>$ 150 g/10 min), allowing finally to get compositions with a non-fractional MFR that can may be used further in compounding.

It would extent the frame of this review too much to comment the features of on-line compounding and reactive extrusion where the extruder attached to the polymerisation train can be used to further manipulate the polymer design of the material produced in the reactors. It has just to be noted that these tools offer an additional possibility to produce multimodal systems and thus offer unique high-end PP solutions.

6.5 Conclusions

Smart combinations of polymer fractions with different molecular weights and comonomer chemistry may lead to sophisticated high-end PP solutions. Designing them via a multistage polymerisation has shown to be an elegant, powerful and cost-effective way to reach novel property profiles.

Bimodal or even trimodal homopolymers broaden the standard monomodal product envelop—at given melt flow rate—in terms of stiffness, melt strength and processability. Multimodal random copolymers expand the softening range while maintaining good thermal stability or enrich the (high molecular) phase with comonomer to retard critical failure under for example fatigue conditions. Bimodal rubbers address multiple complex requirements by allowing the optimisation of both primary (impact, stiffness, softness) and secondary properties (e.g. surface aesthetics, organoleptics, transparency). Provided a polymerisation train with enough independent reactivity zones, tailor-made polymers with 4–5 discrete fractions—contributing actively to the end performance of a grade—can be produced in one single formulation. While the number of reactors (and their size) is thus an obvious prerequisite to manufacture high-end polymers, the used catalyst systems have to fulfil a bunch of complementary features to maximise process economics and freedom in polymer design. Being spherical, exhibiting a very high activity and selectivity over a given polymerisation time, having the targeted porosity, having a broad hydrogen response, allowing a qualitative and quantitative incorporation of comonomers are a few of the key attributes of a desired catalyst system.

Being an active research field for the last 15–20 years—playing with the golden triad "process configuration, catalyst optimisation and smart product design", multimodal polymers have by far not yet revealed their whole potential. It will therefore remain an area of interest for the next generations of scientists allowing innovative polyolefin producers to grow further the PP market by offering tailor-made solutions to demanding customers.

Acknowledgements This review is the result of multiple passionate discussions around polymer design and multimodality over the last 15 years. No list would be exhaustive enough to name all the people who have been involved in this fantastic adventure. Those belonging to my PO tribe will recognise themselves. Shall they all be thanked!

References

1. Gahleitner, M., & Paulik, C. (2014). Polypropylene. In M. Bohnet (Ed.), *Ullmann's encyclopedia of industrial chemistry* (pp. 1–42). Weinheim: Wiley.
2. Albizzati, E., Cecchin, G., Chadwick, J. C., Collina, G., Giannini, U., Morini, G., et al. (2005). Catalysts for polymerisation. In N. Pasquini (Ed.), *Polypropylene handbook* (2nd ed., pp. 16–18). Munich: Carl Hanser Verlag.
3. Rojo, E., Pena, B., Munoz, M. E., & Santamarıa, A. (2006). A study of the use of oscillatory flow to characterize isotactic and syndiotactic poly(propylene)s. *Macromolecular Chemistry and Physics, 207*, 1781–1788.
4. Winter, A., Dolle, V., & Spaleck, W. (1991). *Process for preparing syndiotactic polyolefins with large molecular weight distribution*. Hoechst, EP0516019.
5. Hungenberg, K. D. (1992). Strategies for PP-homo- and copolymers with broadening MWD in stirred bed gas-phase reactors, Dechema-Monographien. In *Proceedings of the 4th International Workshop on Polymer Reaction Engineering* (Vol. 127, pp. 502–509). Berlin: VCH Verlagsgesellschaft, Band.
6. Zacca, J. J., Debling, J. A., & Ray, W. H. (1997). Reactor residence time distribution effects on the multistage polymerisation of olefins—II. Polymer properties: Bimodal polypropylene and linear low-density polyethylene. *Chemical Engineering Science, 52*(12), 1941–1967.
7. Lummus. Retrieved from http://www.cbi.com/technologies/novolen-technology
8. Burkhardt, U., Atam, N., & Voigt, D. (1995). Processing of multimodal polymer materials, polypropylene (PP) and polyethylene (PE). In *Aufbereitung von Polymeren mit neuartigen Eigenschaften* (pp. 55–78) (German).
9. Galli, P. (1999). The reactor granule technology: The ultimate expansion of polypropylene properties? *Journal of Material Science – Pure Applied Chemistry, A36*(11), 1561–1586.
10. Dashti, A., Ahmad, A., Hiraoka, Y., Kim, S. K., Taniikeb, T., & Teranob, M. (2009). Kinetic and morphological study of a magnesium ethoxide-based Ziegler–Natta catalyst for propylene polymerization. *Polymer International, 58*, 40–45.
11. Albizzati, E., Giannini, U., Collina, G., Noristi, L., & Resconi, L. (1996). Catalysts and polymerisations. In E. P. Moore Jr. (Ed.), *Polypropylene handbook (polymerisation, characterisation, properties, processing and applications)* (pp. 11–98). New York: Hanser Publishers.
12. Bukatov, G. D., Sergeev, S. A., Zakharov, V. A., & Potapov, A. G. (2008). Supported titanium–magnesium catalysts for propylene polymerization. *Kinetics and Catalysis, 49*(6), 782–790.

13. Cecchin, G., Morini, G., & Pelliconi, A. (2001). Polypropene product innovation by reactor granule technology. *Macromolecular Symposium, 173*, 195–209.
14. Cecchin, G., Marchetti, E., & Baruzzi, G. (2001). On the mechanism of polypropene growth over $MgCl_2/TiCl_4$ catalyst systems. *Macromolecular Chemistry and Physics, 202*, 1987–1994.
15. Kittilsen, P., & Mc Kenna, T. F. (2001). Study of the kinetics, mass transfer, and particle morphology in the production of high impact polypropylene. *Journal of Applied Polymer Science, 82*, 1047–1060.
16. Noristi, L., Marchetti, E., Baruzzi, G., & Sgarzi, P. (1994). Investigation of the particle growth mechanism in propylene polymerisation with $MgCl_2$-supported Ziegler-Natta catalysts. *Journal of Polymer Science Part A: Polymer Chemistry, 32*, 3047–3059.
17. Pater, J. M. T., Weickert, G., Loos, J., & van Swaaij, W. P. M. (2001). High precision prepolymerization of propylene at extremely low reaction rates, kinetics and morphology. *Chemical Engineering and Science, 56*, 4107–4120.
18. Pater, J. M. T., Weickert, G., & van Swaaij, W. P. M. (2003). Polymerisation of liquid propylene with a fourth-generation Ziegler-Natta catalyst: Influence of temperature, hydrogen, monomer concentration, and prepolymerisation method on powder morphology. *Journal of Applied Polymer Science, 87*, 1421–1435.
19. Zheng, X., Smit, M., Chadwick, J. C., & Loos, J. (2005). Fragmentation behavior of silica-supported metallocene/MAO catalyst in the early stage of olefin polymerisation. *Macromolecules, 38*, 4673–4678.
20. Zheng, X., Pimplapure, M. S., Weickert, G., & Loos, J. (2006). Influence of porosity on the fragmentation of Ziegler-Natta catalyst in the early stages of propylene polymerization. *e-Polymers, 28*, 1–10.
21. Abboud, M., Denifl, P., & Reichert, K.-H. (2005). Study of the morphology and kinetics of novel Ziegler-Natta catalysts for propylene polymerisation. *Journal of Applied Polymer Science, 98*, 2191–2203.
22. Vestberg, T., Denifl, P., & Wilen, C.-E. (2008). Porous versus novel compact Ziegler-Natta catalyst particles and their fragmentation during the early stages of bulk propylene polymerisation. *Journal of Applied Polymer Science, 110*, 2021–2029.
23. Ishimura, N., Kioka, M., & Toyota, A. (1989). *Process for polymerising olefins and catalyst for polymerising olefins*. Mitsui, EP0385765.
24. Galvan, M., Pantaleoni, R., Fusco, O., Gaddi, B., Neumann, A., Mazzucco A., et al. (2011). *Propylene films and sheets*. LyondellBasell, WO20130004507.
25. Galvan, M., Neumann, A., Pantaleoni, R., Mazzucco, A., Fusco, O., Gaddi, B, et al. (2012). *Polypropylene fiber*. LyondellBasell, EP2682505.
26. Meesters, E., News, J., & Guidicini, A. (2002). *Highly stereoregular polypropylene with improved properties*. LyondellBasell, WO2004013193.
27. Ahn, T. O., Hong, S. C., Huh, W. S., Lee, Y. C., & Lee, D. H. (1999). Metallocene catalyst and its olefin polymerisation behaviour. *Polymer Engineering and Science, 39*, 1257–1264.
28. Ewen, A. J (1987). *Catalyst systems for producing polyolefins having a broad molecular weight distribution*. Fina, EP310734.
29. Galli, P., Collina, G., Szarzi, P., Baruzzi, G., & Marchetti, E. (1997). Combining Ziegler-Natta and metallocene catalysis: New heterophasic propylene copolymers from the novel multicatalyst reactor granule technology. *Journal of Applied Polymer Science, 66*, 1831–1837.
30. Kheirandish, S., Doshev, P., Jääskeläinen, P., Leskinen, P., Johnsen, G. K., & Ding, H. (2009). *High flow propylene composition*. Borealis, WO2011000557.
31. Lisovskii, A., Shuster, M., Gishvoliner, M., Lidor, G., & Eisen, M. S. (1998). Polymerization of propylene by mixed Ziegler-Natta and metallocene catalysts. *Applied Organometallic Chemistry, 12*, 401–408.
32. Lopez Moya, E., Van Grieken, R., Carrero, A., & Paredes, B. (2012). Bimodal poly(propylene) through binary metallocene catalytic systems as an alternative to melt blending. *Macromolecular Symposia, 321–322*(1), 46–52.

33. Mavridis, H., & Shankernaraynan, M. J. (2006). *Process for preparing high melt strength polypropylene*. Novolen, WO2007100436.
34. Tynys, A., Saarinen, T., Bartke, M., & Löfgren, B. (2007). Propylene polymerisations with novel heterogeneous combination of metallocene catalyst systems. *Polymer, 48*, 1893–1902.
35. Kulshreshtha, A. K., & Talapatra, S. (2005). Competitive new technologies in polyolefin synthesis, Section II E 10-11. In C. Vasile (Ed.), *Handbook of polyolefins* (2nd ed.). New York: Marcel Dekker Inc.
36. Debling, J. A., Zacca, J. J., & Ray, W. H. (1997). Reactor residence-time distribution effects on the multistage polymerisation of olefins – III. Multi-layered products: Impact polypropylene. *Chemical Engineering and Science, 52*(12), 1969–2001.
37. Govoni, G., Rinaldi, R., Covezzi, M., & Galli, P. (1995). *Process and apparatus for the gas-phase polymerization of alpha-olefins*. Montell, EP0782587.
38. Liebermann, R. B., & LeNoir, R. T. (1996). Manufacturing. In E. P. Moore Jr. (Ed.), *Polypropylene handbook (polymerisation, characterisation, properties, processing and applications)* (pp. 287–301). New York: Hanser Publishers.
39. Petrochemical Processes Handbook. (2010). Gulf Publishing.
40. Sakai, I., Takaoka, T., Moriya, S., Hashimoto, M., Nakagawa, H., & Yoshikawa, H. (1997) *Crystalline polypropylene composition*. Mitsui, EP0916701.
41. Standaert, A., & Gromada, J. (2007). *Process for the production of a bimodal polypropylene having low ash content*. Total, EP2070954.
42. Dolle, V., Fraaije, V., Terwyen, H., & Boehm, T. (1999). *High-molecular polypropylene with a broad distribution of the molecular weight and short isotactic sequence*. LyondellBasell, US6759500.
43. Ellis P. (2007, March). Basell's progress with Spherizone and recent catalyst developments. *The Metallocene and Single-Site Catalyst Monitor*, 4–12.
44. Syed, F. H., & Singh, B. B. (2003). *Spherizone – Basell's newest process technology platform* (pp. 1–7). CMR (Chemical Market Resources) Inc.
45. Govoni, G., & Covezzi, M. (1998). *Process and apparatus for the gas-phase polymerisation*. Basell, EP1012195.
46. Galli, P., Chadwick, J. C., Mei, G., & Vecellio, G. (1999). Multizone circulating reactor: The novel frontier of the polyolefin technology. In *SPO 99 – 9th International Business Forum on Specialty Polyolefins*, Houston, 12–13 October 1999.
47. Fernandes, F. A. N., & Lona, L. M. F. (2004). Multizone circulating reactor modelling for gas-phase polymerisation. I. Reactor modelling. *Journal of Applied Polymer Science, 93*, 1042–1052.
48. Fernandes, F. A. N., & Lona, L. M. F. (2004). Multizone circulating reactor modelling for gas-phase polymerisation. I. Reactor operating with gas barrier in the downer section. *Journal of Applied Polymer Science, 93*, 1053–1059.
49. Grazzi, M., & Cavalieri, C. (2006). *High shrink polypropylene films*. Basell, EP2099834.
50. Cavalieri, C., Pradella, F., & Galvan, M. (2010). *Propylene-based terpolymers for films*. Basell, EP2558508.
51. Marzolla, R., & Galvan, M. (2010). *Polypropylene-based terpolymers for films*. Basell, US20130165612.
52. Marzolla, R., & Galvan, M. (2011). *Propylene-based terpolymers for fibers*. Basell, WO2012143485.
53. Cavalieri, C., & Galvan, M. (2012). *Propylene-based terpolymers compositions for pipes*. WO2014068027.
54. Destro, M., Ciarafoni, M., Massari, P., Caputo, T., & Böhm, T. (2013). *Propylene based terpolymers*. Basell, EP2813438.
55. Schloemer, T. (2014). *Catalloy process resins – Expanding the polymer performance envelope (Adsyl, Adflex, Hifax, Softell)*. Proceedings from SPO Conference, Troy (Michigan, US), October 2014.

56. Washiyama, J., Minowa, K., Takenouchi, H., Nakajima, T., Yamauchi, M., Yokoyama, Y., et al. (2000). *Polyolefin masterbatch and composition suitable for injection moulding*. Basell, EP1328581.
57. Lesca, C., Medri, A., & Pelliconi, A. (1995). *Stretch blow-moulding process for the preparation of polypropylene containers*. Basell, EP0764514.
58. Beccarini, E., & Pelliconi, A. (1999). *Compositions of random copolymers of propene containing an alpha-olefin as comonomer*. Basell, WO200144367.
59. Rondelez, D. (2010). *A new face to the Innovene PP process*. Zürich: Maack Business Services (PEPP).
60. Pääviläinen, J., & Leskinen, P. (2010). *Process for producing random copolymers and their use*. Borealis, EP2452959.
61. Leskinen, P., Doshev, P., & Tranninger, C. (2010). *Heterophasic propylene copolymers with improved stiffness/impact/flowability balance*. Borealis, EP2727958.
62. Eriksson, E., Bergstra, M., Hätönen, J., & Follestad, A. (2010). *Multimodal polymer of propylene, composition containing the same and a process for manufacturing the same*. Borealis, EP2223943.
63. Bergstra, M., Leskinen, P., Malm, B., & Kock, C. (2010). *Heterophasic polypropylene with excellent creep resistance*. Borealis, EP2368937.
64. Doshev, P., Liu, Y., Malm, B., Schiesser, S., & Sundholm, T. (2010). *Heterophasic polypropylene with excellent mechanical properties*. Borealis, EP2368938.
65. Sundholm, T., Bergstra, M., Malm, B., Leskinen, P., & Kock, C. (2010). *Random propylene copolymers for pipes*. Borealis, WO2011104100.
66. Kniesel, C., Kock, C., & Doshev, P. (2011). *Heterophasic propylene copolymer with excellent stiffness and impact balance*. Borealis, WO2013041507.
67. Al-Haj, A., Stroomer, J., Betlem, B., Weickert, G., & Roffel, B. (2008). Molecular weight distribution broadening of polypropylene by periodic switching of hydrogen and catalyst additions. *Journal of Applied Polymer Science, 108*, 2446–2457.
68. Bergstra, M., & Leskinen, P. (2008). *A method to produce very stiff polypropylene*. Borealis, EP2174962.
69. Bernreitner, K., Hebesberger, E., Horill, T., & Pillinger, F. (2009). *Process for the production of polyolefins with broad molecular weight distribution*. Borealis, EP2393840.
70. Severn, J., Bergstra, M., Vestberg, T., Leskinen, P., Malm, B., Tuominen, O., et al. (2008). *Preparation of propylene copolymer with dynamically operated reactor*. Borealis, EP2174965.
71. Potter, E., Doshev, P., Kheirandish, S., Leskinen, P., & Gahleitner, M. (2011). *Random propylene copolymer with high stiffness and low haze*. Borealis, EP2527593.
72. Rätzsch, M., & Neißl, W. (1995). Bimodal polymer materials based on polypropylene (PP) and polyethylene (PE). In *Aufbereitung von Polymeren mit neuartigen Eigenschaften* (pp. 3–25) (German).
73. Abedi, S., & Hassanpour, N. (2006). Preparation of bimodal polypropylene in two-step polymerisation. *Journal of Applied Polymer Science, 101*, 1456–1462.
74. Felipe, S., Klimke, K., Doshev, P., Tran, A. T., Tynys, A., Obadal, M., et al. (2010). *High flowability long chain branched polypropylene*. Borealis, EP2386601.
75. Cavalieri, C., Galvan, M., & Tisi, F. (2010). *Polyolefin and composition for pipe systems and sheets*. Basell, US8709564.
76. Dolle, V., Chicote Carrion, E., & Terwyen, H. (1997). *Process for the manufacture of pipes from a propylene-ethylene copolymer*. LyondellBasell, EP0791609.
77. Jääskeläinen, P., Karbasi, A., Malm, B., & Pitkänen, P. (1994). *New polypropylene composition with broad MWD*. Borealis, EP0785954.
78. Nam, G. J., Yoo, J. H., & Lee, J. W. (2005). Effect of long-chain branches of polypropylene on rheological properties and foam-extrusion performance. *Journal of Applied Polymer Science, 96*, 1793–1800.

79. Park, C. B., & Cheung, L. K. (1997). A study of cell nucleation in the extrusion of polypropylene foams. *Polymer Engineering and Science, 37*, 1–10.
80. Tran, A. T., Klimke, K., Braun, H., & Filipe, S. (2011). *Polypropylene for foam and polypropylene foam*. US20140163126.
81. Bernreitner, K., & Doshev, P. (2010). *Polypropylene bottles*. Borealis, WO201113157.
82. Klimke, K., Doshev, P., Filipe, S., & Tran, T. A. (2009). *Heterophasic polypropylene with long chain branching*. Borealis, EP2319885.
83. Lee, C., Dong, D., Holland, C. S., Beren, J. R., & De Nicola A. J. (2011). *Controlled rheology formulations containing high melt strength polypropylene for extrusion coating*. Equistar, WO2012109449.
84. Giacobbe, J. M., & Pufka, S. (1996). *Polypropylene blown film*. LyondellBasell, EP0757069.
85. Pradella, F., Weiss, P.-A., Cavalieri, C., & Sartori, G. (2006). *Polypropylene compositions for stretched articles*. Basell, EP1963423.
86. Pelliconi, A., Lonardo, A., & Mei, G. (2000). *Bioriented polypropylene films*. LyondellBasell, EP1358252.
87. Pelliconi, A., Lonardo, A., & Mei, G. (2000). *Polyolefin sheets for thermoforming*. LyondellBasell, EP1377629.
88. Kirchberger, M., Jääskeläinen, P., & Gahleitner, M. (2001). *Non-oriented polypropylene film*. Borealis, EP1439957.
89. Huovinen, P., Harlin, A., Jääskeläinen, P., Karbasi, A., & Manner, N. (1997). *High melt strength polypropylene*. Borealis, WO1999016797.
90. Takaoka, T., Sakai, I., Nakagawa, H., Hashimoto, M., Matsuura, S., & Shinizu, T. (1997). *Polypropylene resin composition and use thereof*. Mitsui, EP942013.
91. Razavi, A. (2002). *Branched polypropylene*. Total, EP1422250.
92. White, D. A. (1995). *Bimodalization of polymer molecular weight distribution*. Exxon, EP0828761.
93. He, C., Costeux, S., Wood-Adams, P., & Dealy, J. M. (2003). Molecular structure of high melt strength polypropylene and its application to polymer design. *Polymer, 44*, 7181–7188.
94. Rätzsch, M. (1999). Reaction mechanism to long-chain branched PP. *Journal of Macromolecular Science, Pure Applied Chemistry, 36*, 1759–1769.
95. Borsig, E., van Duin, M., & Gotsis, A. D. (2008). Picchioni, Long chain branching on linear polypropylene by solid state reactions. *European Polymer Journal, 44*, 200–212.
96. Parent, J. S., Sengupta, S. S., Kaufman, M., & Chaudhary, B. I. (2008). Coagent-induced transformations of polypropylene microstructure: Evolution of bimodal architectures and cross-linked nano-particles. *Polymer, 49*, 3884–3891.
97. Kirchberger, M., Jääskeläinen, P., Pitkänen, P., & Gahleitner, M. (2003). *Process for preparing a polymer film comprising a propylene random copolymer*. Borealis, EP1270651.
98. Maier, R.-D., Galiatsatos, V., Roessner, D. A., & Setty, S. P. (2004). *Polypropylene resin composition*. Lummus Novolen, EP1620505.
99. Vestberg, T., Malm, B., & Jääkeläinen, P. (2005). *Heterophasic polymer composition and process for its preparation*. Borealis, EP1681315.
100. Ommundsen, E., Pääviläinen, J., Wacholder, M., Gahleitner, M., Oysead, H., & Johansen, G. M. (2005). *Multimodal propylene polymer composition*. Borealis, EP1963422.
101. Oysaed, H., Johansen, G. M., Jääskelainen, P., & Wacholder, M. (2008). *Polypropylene-based resin composition and molded article thereof*. Borealis, EP2121830.
102. Gahleitner, M., Klimke, K., Bernreitner, K., & Sandholzer, M. (2012). *Random propylene copolymer for bottles with good optical properties and low hexane content*. Borealis EP2733175.
103. De Palo, R., Pelliconi, A., & Huovinen, P. (2001). *Pipe systems of polypropylene compositions*. Basell, EP1448704.
104. Jääskeläinen, P., Pitkänen, P., Tuominen, O., Hafner, N., Gahleitner, M., & Töltsch, W. (2001). *Propylene random copolymer and process for the production thereof*. Borealis, EP1270628.

105. Pääviläinen, J., Doshev, P., & Reichelt, K. (2011). *Propylene/1-hexene copolymer composition with low sealing temperature*. Borealis, US20130203931.
106. Deblieck, R. A. C., van Beek, D. J. M., Remerie, K., & Ward, I. M. (2011). Failure mechanisms in polyolefins: The role of crazing, shear yielding and the entanglement network. *Polymer, 52*, 2979–2990.
107. Pääviläinen, J., Doshev, P., Reichelt, K., Leskinen, P., & Miettinen, H. (2010). *Propylene/1-hexene copolymer composition with broad sealing window*. Borealis, EP2386603.
108. Grein, C., & Schedenig, T. (2007). *Sterilisable and tough impact polypropylene composition*. Borealis, EP2176340.
109. Oysaed, H., Jaaskelainen, P., Ommundsen, E., Johansen, G. M., & Wacholder, M. (2007). Borealis, EP2083987.
110. Sartori, F., & Sartori, G. (2002). *Polypropylene fibres suitable for thermally bonded non-woven fabrics*. Basell, WO2004029341.
111. Ek, C.-G, Malm, B., & Nymark, A. (1996). *Method for preparing a multilayer pipe*. Borealis, EP0883769.
112. Zacca, J. J., Debling, J. A., & Ray, W. H. (1996). Reactor residence time distribution effects on the multistage polymerisation of olefins – I. Basic principles and illustrative examples, polypropylene. *Chemical Engineering Science, 51*(21), 4859–4886.
113. Cavalieri, C., Galvan, M., Pantaleoni, R., Beccarini, E., Tisi, F., & Rothhöft, W. (2010). *Polyolefinic compositions for injection-moulded drainage systems*. Basell, EP2516545.
114. Schwager, H., & Kerth, J. (1991). *Propylen-Ethylen-Copolymerisate aus drei verschiedenen Bestandteilen*. BASF, EP0515879.
115. Massari, P., Ciarafoni, M., & News, J. (2005). *Polyolefinic compositions having good whitening resistance*. Basell, WO2006125720.
116. Hagen, E., Torgersen, U., Rohaczek, G., & Iveland, A. (2002). *Polyolefin with improved scratch resistance and process for producing the same*. Borealis, EP1516000.
117. Pääviläinen, J., Leskinen, P., Hakola, S., Sandholzer, D., & Grein, C. (2010). *Improved process for polymerising propylene*. Borealis, EP2452956.
118. Debling, J. A., & Ray, W. H. (1995). Heat and mass transfer effects in multistage polymerisation processes: Impact polypropylene. *Indian Engineering and Chemical Research, 34*, 3466–3480.
119. Sacchetti, M., Govoni, G., & Ciarrocchi, A. (1990). *Components and catalysts for the polymerisation of olefins*. Montell, EP0395083.
120. Leinonen, T., & Denifl, P. (2005). *Catalyst particles*. Borealis, EP1803743.
121. Denifl, P., Leinonen, T., Haikarainen, A., & Vestberg, T. (2007). *Catalyst with low surface area*. Borealis, EP2065405.
122. Haikarainen, A., Denifl, P., & Leinonen, T. (2009). *Preparation of precipitated ZN PP catalysts with internal pore structure using nanoparticles*. Borealis, EP2251361.
123. Pelliconi, A., Garagnani, E., & Villanuena, A. (2001). *Soft polyolefin compositions*. Basell, EP2258766.
124. Daumerie, M., & Gahleitner, M. (2013). *Pushing the boundaries of polyolefin applications through innovations in catalysts and process*. SPE International Polyolefin Conference 2013, Houston, TX.
125. Doshev, P., Lohse, G., Henning, S., Krumova, M., Heuvelsland, A., Michler, G., et al. (2006). Phase interactions and structure evolution of heterophasic ethylene-propylene copolymers as a function of system composition. *Journal of Applied Polymer Science, 101*, 2825–2837.
126. Grein, C., Bernreitner, K., Hauer, A., Gahleitner, M., & Neißl, W. (2003). Impact modified isotactic polypropylene with controlled rubber intrinsic viscosities: Some new aspects about morphology and fracture. *Journal of Applied Polymer Science, 87*, 1702–1712.
127. Grestenberger, G., Machl, D., & Grein, C. (2009). *High flow PP/EPR blends for the automotive industry: Basic correlations between EPR composition and application properties*. Annual Technical Conference – Society of Plastics Engineers 67th (pp. 1054–1057).

128. Grestenberger, G., Potter, G. D., & Grein, C. (2014). Polypropylene/ethylene-propylene rubber (PP/EPR) blends for the automotive industry: Basic correlations between EPR-design and shrinkage. *eXPRESS Polymer Letters, 8*(4), 282–292.
129. Machl, D., Grein, C., & Bernreitner, K. (2006). *Multiphase PP blends for automotive interior: Composition effects on surface structure and scratch resistance*. Annual Technical Conference – Society of Plastics Engineers 64th (pp. 41–44).
130. Daumerie, M. (2010). *Driving progress in polyolefins*. Zürich: Maack Business Services (PEPP 2010).
131. Remerie, K., & Groenewold, J. (2012). Morphology formation in polypropylene impact copolymers under static melt conditions: A simulation study. *Journal of Applied Polymer Science, 125*(1), 212–223.
132. Grein, C., & Leskinen, P. (2010). *Improved process for producing heterophasic propylene copolymers*. Borealis, EP2452957.
133. Grein, C., Bernreitner, K., & Berger, F. (2004). *Novel propylene polymer compositions*. Borealis, EP1769029.
134. Grein, C., & Bernreitner, K. (2008). *Thermoplastic polyolefins with high flowability and excellent surface quality produced by a multistage process*. Borealis, EP2294129.
135. Kock, C., Grein, C., & Prades, F. (2010). *High flow and stiff polymer material with good transparency and impact properties*. Borealis, EP2431416.
136. Wolfschwenger, J., Grein, C., Bernreitner, K., Gahleitner, M., & Kamfjord, T. (2004). *Novel propylene polymer compositions*. Borealis, EP1833909.
137. Doshev, P., Kona, B., Jääskeläinen, P., & Malm, B. (2009). *Heterophasic polypropylene with improved balance between stiffness and transparency*. Borealis, EP2516125.
138. Kock, C., Grein, C., & Potter, E. (2011). *Heterophasic polyolefin composition having improved flowability and impact strength*. Borealis, WO2012116718.
139. Kanzaki, S., Shimojo, M., & Tsuji, M. (2001). *Polypropylene resin composition*. Sumitomo, US6518363.
140. Doshev, P., Kona, B., Bernreitner, K., & Grein, C. (2008). *High flowable heterophasic polypropylene*. Borealis, WO2010040492.
141. Sandholzer, M., Bernreitner, K., Doshev, P., & Potter, G. (2010). *Propylene polymer compositions having superior hexane extractables/impact balance*. Borealis, EP2397517.
142. Gahleitner, M., Klimke, K., Bernreitner, K., Sandholzer, M., & Horill, T. (2011). *Process for the preparation of a heterophasic propylene copolymer*. Borealis, WO2013092615.
143. Collina, G., Pelliconi, A., & Garagnani, E. (1996). *High transparency and high flexibility elastoplastic polyolefin compositions*. Basell, EP0846134.
144. Gahleitner, M., Bernreitner, K., Sandholzer, M., & Horill, T. (2011). *Soft propylene copolymer*. Borealis, EP2557096.
145. Gahleitner, M., Sandholzer, M., Bernreitner, K., & Leskinen, P. (2011). *Unoriented film*. Borealis, EP2546298.
146. Sandholzer, M., Bernreitner, K., Klimke, K., & Gahleitner, M. (2011). *Propylene copolymer for injection molded articles or films*. Borealis, WO2013092620.

Chapter 7
Bimodal Polyethylene: Controlling Polymer Properties by Molecular Design

Christian Paulik, Gunnar Spiegel, and Dusan Jeremic

7.1 Introduction

Polyethylene is one of the most widely used polymers, and it can be found in various industrial applications. The annual production and consumption of polyethylene is currently higher than 100 million tons worldwide, or about 40% of the consumption of all thermoplastic materials [1].

Polyethylene is used in a vast variety of applications that range from artificial human body implants over large diameter pipes for gas and oil transportation to films used for manufacturing food or other packaging. Material performance requirements needed for successful implementation and use of various polyethylene types obviously substantially differ depending on the application. While packaging film needs to be flexible, tough, and possibly transparent, material used for manufacturing pipes needs to sustain high pressure over a very long period of time. The possibly only material feature that is needed in at least almost all materials regardless the application is good processability of the polymer. That means fast and low energy demanding conversion, manufacturing of the actual objects.

Although many applications demand use of multi-material structures where polyethylene is combined with metal or other polymeric materials in order to achieve advanced performance, raising concern for environment and need for recyclability dictates intensive development and increased use of mono-material solutions. This chapter therefore discusses primary material performance that does not take into

C. Paulik (✉) · G. Spiegel
Institute for Chemical Technology of Organic Materials, Johannes Kepler University, Linz, Austria
e-mail: christian.paulik@jku.at

D. Jeremic
Borealis Polyolefine GmbH, Linz, Austria
e-mail: dusan.jeremic@borealisgroup.com

A. R. Albunia et al. (eds.), *Multimodal Polymers with Supported Catalysts*,
https://doi.org/10.1007/978-3-030-03476-4_7

account possible combinations in multilayered constructions where polyethylene is combined with non-polyethylene materials.

The success story of polyolefins and polyethylene in particular basically started with the discovery of high-pressure ethylene polymerization at ICI in the UK in 1933. This accidental discovery [2] initiated the focused development and production of low-density polyethylene (LDPE). The commercial breakthrough started with the development of catalyst systems, which enabled the polymerization of olefins at mild process conditions but good control of the molecular architecture. In 1953 Karl Ziegler at the Max-Planck-Institut für Kohlenforschung in Germany discovered that with a combination of transition metal halides (e.g., titanium chloride) with aluminum alkyls (e.g., triethylaluminum) one could polymerize ethylene to a linear, high molecular weight, and crystalline polymer of high-density (high-density polyethylene, HDPE) [3]. At about the same time Robert Banks and Paul Hogan at Phillips Petroleum Company developed chromium based catalysts, which also afforded the production of HDPE [4]. Metal organic catalysts for the polymerization of olefins were first described by Giulio Natta in the early 1950s [5] but only the discovery of the activating effect of methylaluminoxane (MAO) by Walter Kaminsky and Hansjörg Sinn in 1976 [6] lead to an industrial interest in these catalysts. Since then metallocene and late transition metal catalysts (sometimes called post-metallocenes) are back in the focus of economic interest and academic research.

The catalyst development also triggered the devolvement of applied production processes. Modern processes offer the possibility of a broad range of different PE products. By the high-pressure process LDPEs and copolymers of ethylene with more polar co-monomers (e.g., vinyl acetate) can be obtained. Low-pressure plants are able to produce materials in a broad density range and can operate in solution, slurry, or gas-phase conditions. For the production of bimodal polyethylene a combination of different reaction regimes is possible.

However, the success story of plastics is best illustrated by numbers: In 1950 approximately 0.35 million tons of plastics were produced in Europe, whereas 62 years later already 57 million tons were produced [1].

As can be clearly seen in Fig. 7.1 the PE family accounts for the most widely used polymer worldwide. The reason for the success of this class of materials is in its ability to be tailored for many different applications, including grocery bags, containers, toys, adhesives, home appliances, engineering plastics, automotive parts, medical applications, and prosthetic implants.

The classification of polyethylene is still based on two parameter that could be easily measured in the 1950s in industrial environments: the rheological parameter called melt index, reflecting the average molecular weight of the resin, and the density of the polymer [7]. According to the American Society of Testing and Materials (ASTM D1248-05 and ASTM-D3350), all polyethylene materials are divided into various classifications. The commonly used commercial classification is given in Table 7.1.

Commercial grades of polyethylene resins are most often copolymers of ethylene, with varying fractions of an α-olefin co-monomer. The most commonly used α-olefins are 1-butene, 1-hexene, and 1-octene [8]. By means of the co-monomer

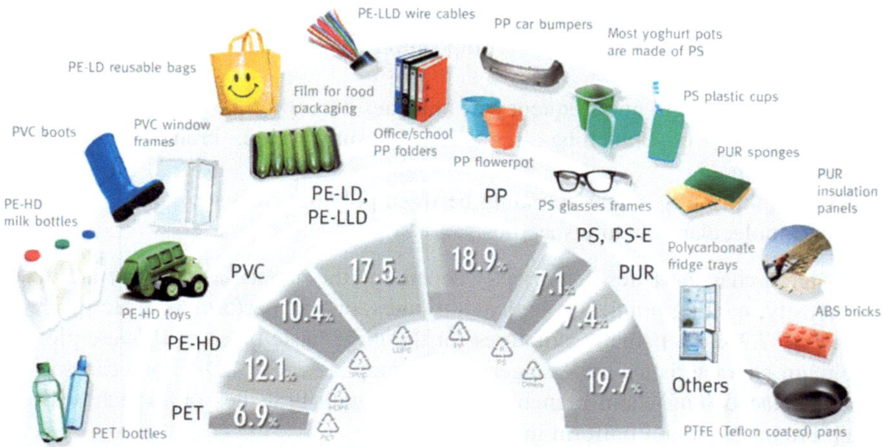

Fig. 7.1 Fields of application and product examples for different kinds of plastic with their market share and recycle code [1]

Table 7.1 Commercial classification of polyethylene resins

		α-Olefin content [mol%]	Crystallinity degree [%]	Density [g/cm^3]
Resins of high density	HDPE	0 to <0.5	65–60	0.960–0.941
Resins of ultrahigh molecular weight	UHMW HDPE			0.935–0.930
Resin of medium density	MDPE	1–2	55–45	0.940–0.926
Resin of low density	LLDPE	2.5–3.5	45–30	0.925–0.915
Resin of very low density	VLDPE	>4	<25	<0.915
Low density PE (high pressure process)	LDPE	0	45–55	0.910–0.940

the density and the crystallinity and in consequence the physical properties of the PEs are altered.

7.2 Molecular Structure and Properties of PE

Primary material performance depends on a few factors with the primary structure of polymer molecules and the composition of somewhat different polymer chains in the material as the most significant ones.

Under polyethylene polymer chain molecular structure, we predominantly understand the following attributes:

- Average molecular weight—length of the polymer chain.
- Molecular weight distribution—compositions of polymer chains with different molecular weights.
- Branching—length and frequency of branches occurring the polymer chain.
- Homogeneity of branching—regularity of substitution, branches, and their distribution.
- Cross-linking—chemical bounding between polymer chains.
- Other molecular structure specifics.

Material characteristics that directly depend on the molecular structure are material density, melting point, rheological attributes, optical performance, etc.

In Fig. 7.2 the schematic structures for the three basic types of PEs are plotted. The main characteristics are exaggerated for emphasis. For LDPE one can see the random placed long chain branching structure, with branches on branches. Short chain branches are not uniform in length.

Linear low density PE (LLDPE) has branches of uniform length originating from the co-monomer, randomly distributed along the polymer chain.

HDPE, high density PE, is basically free of both long and short branching. However, very small amounts of co-monomer may be added deliberately to achieve specific product targets.

Certain molecular structure features are most often a direct consequence of the polymerization process, that is, type of polymerization and the initiator that leads to bounding monomers into a polymer chain. Radical polymerization that occurs in the high-pressure process results in the polyethylene chains that are highly branched (LDPE), with relatively long side chains. These polyethylene chains do not crystallize well and therefore is density of the material low.

Polyethylene that is produced in the low-pressure process, product of the coordinative polymerization initiated by a transition metal containing catalyst, is mostly linear with little or no branching. These polymer chains can readily crystallize and therefore is the density of the material high (HDPE).

Since a few physical performance attributes depend on the crystallinity of the polyethylene, Du Pont (Canada) [9] introduced and Union Carbide later developed so LLDPE, in the late 1950s. LLDPE is copolymer of ethylene with other α-olefins such as 1-butene or 1-hexene. Due to alkyl residuals of the co-monomer built in the backbone of the polymer chain, the crystallization of the material is somewhat disturbed and it can be controlled by the frequency of branching or by the amount of co-monomer that is incorporated in the polymer.

Average molecular weight or length of the polymer chains obtained from the polymerization can also be controlled in the manufacturing process. This is done by addition of a so-called chain transfer agent, most often hydrogen that interrupts propagation of the polymer chain maintaining functionality of the catalyst's active site. Concentration of the chain transfer agent determines average molecular weight of the polymer made by the given catalyst.

Polyethylene crystallizes in "lamellas" (Fig. 7.3a); the thicker the layer, the higher is crystallinity as is the melting point. Total crystallization cannot be achieved even by unbranched chains (e.g., HDPE), since it is hindered by its high viscosity [10].

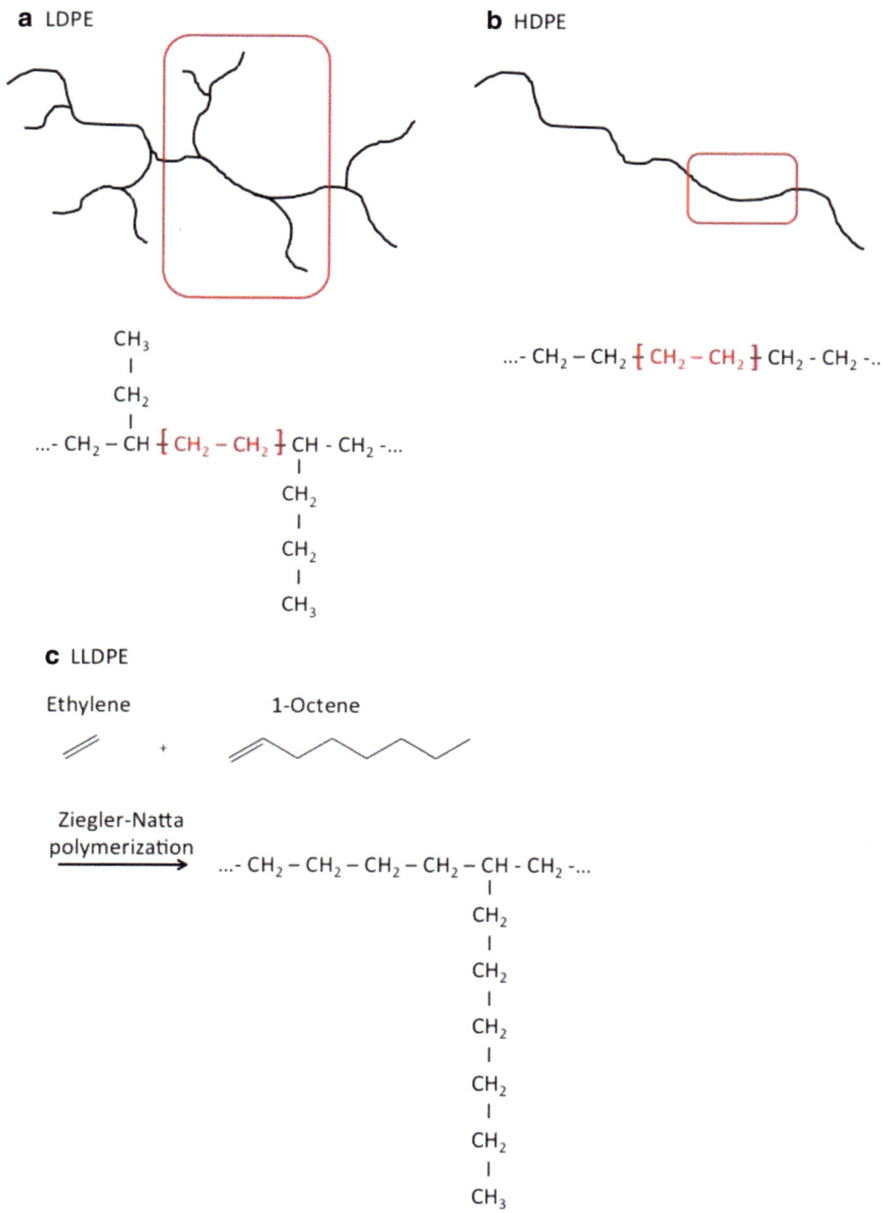

Fig. 7.2 General representations of various polyethylene variants. (**a**) LDPE; (**b**) HDPE; and (**c**) LLDPE for the case of an ethylene 1-octene copolymer

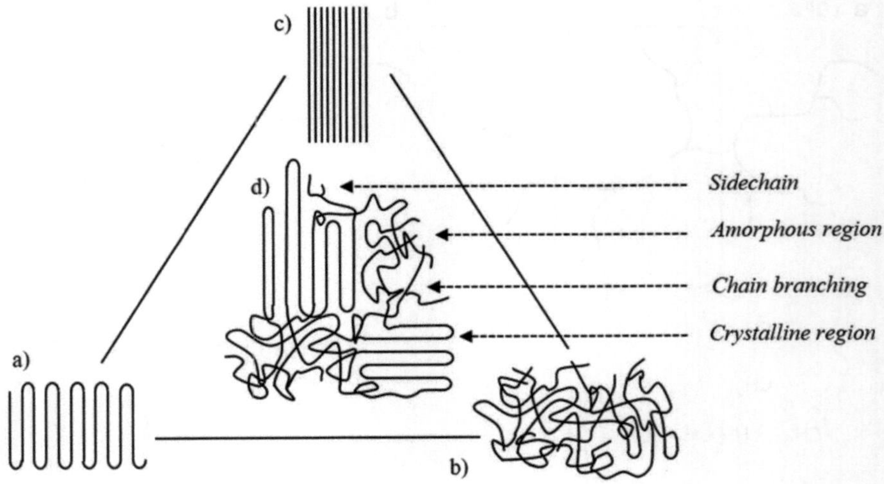

Fig. 7.3 (1) Different "folding" of a polymer chain: (**a**) lamella; (**b**) amorphous; (**c**) stretched; (**d**) combined structure – arrangement of a polymerchain [11, 12]

With increasing number of side chains on the polymer backbone, the order loosens up and amorphous areas (Fig. 7.3b) are formed. The macromolecular structure influences the material properties as it depends upon chain length, respectively molecular mass.

7.3 Catalysts for the Synthesis of PE

The production of LDPE follows a free-radical mechanism at high pressures, that is at pressures above 100 MPa and temperatures above 160 °C. These high pressures are necessary to achieve a reasonable concentration of monomer and to enable the dissolution of the produced polyethylene in unreacted ethylene. For the production of HDPE and LLDPEs transition metal catalysts of different types are used. The most important groups are described in brief in the following section.

7.3.1 Ziegler–Natta Catalysts

Ziegler–Natta catalysts are basically metal–organic mixed-catalysts and are based on the discovery that titanium chloride in presence of triethylaluminum polymerizes ethylene at ambient temperature and standard pressure with high crystallinity [9, 13]. Early Ziegler–Natta catalysts are based on β-$TiCl_3$, obtained as brown

$$TiCl_4 \xrightarrow[- AlEt_2Cl]{+ AlEt_3} TiEtCl_3 \longrightarrow TiCl_3 + 0.5\,(C_2H_4 + C_2H_6)$$

Fig. 7.4 Ziegler-Natta catalyst—first generation

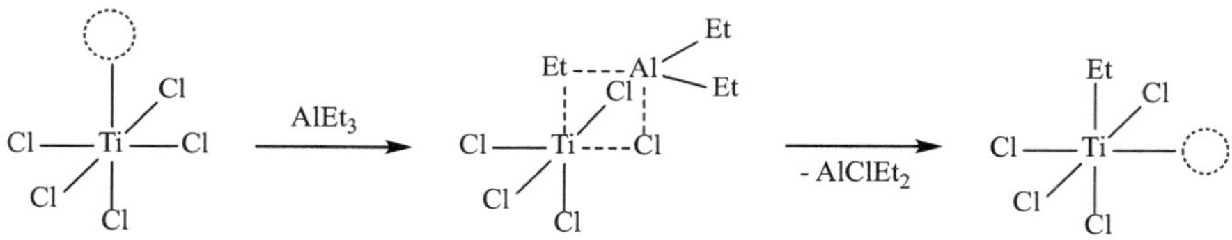

Fig. 7.5 Activation and initiation of the ZN-catalyst (the active site is marked by the circle)

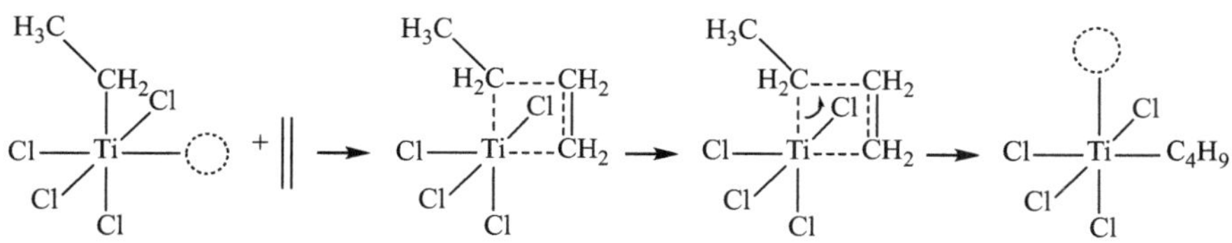

Fig. 7.6 Chain propagation due to insertion

needles, by reduction of $TiCl_4$ with $AlEt_2Cl$ and $AlEt_3$ (TEA) as co-catalyst (Fig. 7.4) [3].

Modern Ziegler–Natta catalysts are heterogenized on an anhydrous δ-$MgCl_2$ support, which has a somewhat similar crystal structure than δ-$TiCl_3$ [10]. An advantage of these systems is that most $TiCl_3$ molecules are available on the surface and can form active sites. Due to the adjustable grain size of the support the morphology as well as the molecular weight distribution [MWD] of the polymer can be controlled [3].

Cossee and Arlmann [14–16] described a comprehensive mechanism for the activation and polymerization of olefins with Ziegler–Natta systems, and pictured below (Fig. 7.5).

Chain propagation commences with an alkene coordinating to the active site, creating a π-complex. The olefin is then inserted between the metal–carbon-bond, switching the coordination place on titanium, leaving another active site for reaction. The growing polymer chain and the coordinated alkene are therefore continuously switching place during propagation (Fig. 7.6). The chain growth is by far (1000-fold) faster than termination, resulting in high molecular mass of the polymer. Chain transfer to hydrogen can terminate the growing chain and this reaction path is used to control the chain length (molecular weight) of the polymer. This termination reaction results in methyl terminated polymers. However, termination can also take place through a β hydrogen elimination, which includes two variants: the hydrogen may either be transferred to the coordinated olefin monomer or it is transferred to the

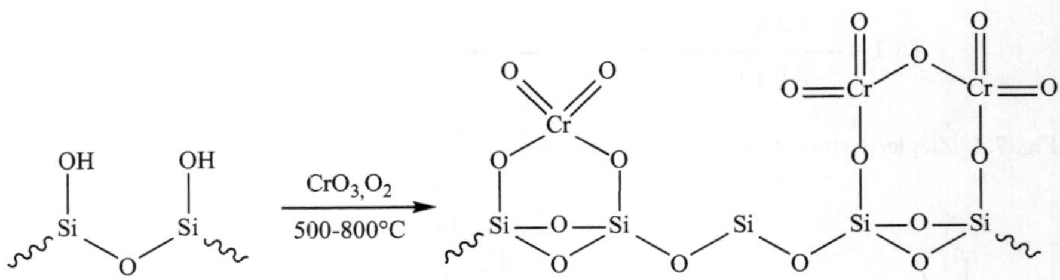

Fig. 7.7 Scheme for preparing a Phillips catalyst [3, 13]

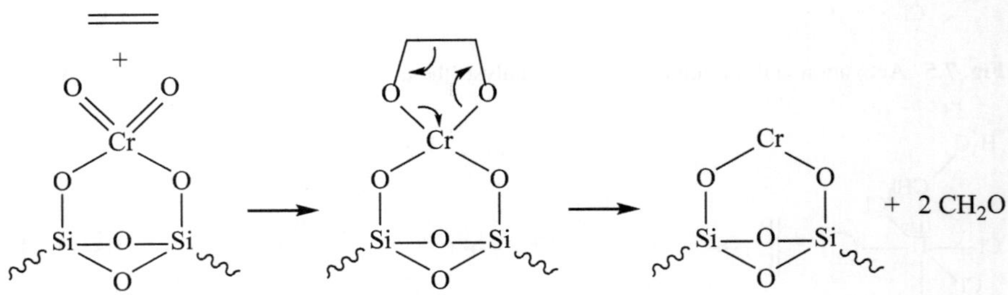

Fig. 7.8 Activation of the Phillips catalyst with ethylene [3]

active site of titanium; both resulting in an unsaturated polymer with vinyl groups. The catalyst is not deactivated and may grow new polymer chains. The fourth termination variant is chain transfer to the aluminum alkyl.

7.3.2 Phillips Catalysts

The so-called Phillips catalysts were developed by R. Banks and P. Hogan at the Phillips Petroleum Company and basically are made by treating silica with an aqueous CrO_3 solution. The chromium reacts with the silanol groups resulting in either silyl-monochromate or silyl-dichromate (Fig. 7.7) [3, 13]. After calcination in an oxidizing atmosphere at about 800 °C the catalyst is activated with CO or by ethylene (Fig. 7.8) [3].

Chromium based catalysts are mainly used for the production of linear high molecular weight PE (HDPE). Control of molecular weight is more difficult than for Ziegler–Natta catalysts as this is done through temperature. Hydrogen can't be used as a chain-transfer-agent, since water is formed, which acts as catalyst poison. Chain propagation is stopped by β elimination with hydride transfer to either chromium or another monomer unit. This results in a higher number of unsaturated polymer chains, which can crosslink at elevated temperatures during further processing.

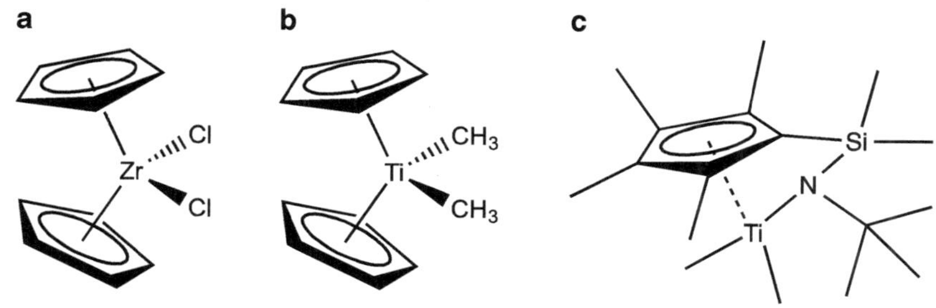

Fig. 7.9 Simple metallocene catalyst structures [3], (**a**) Zirconocene dichloride; (**b**) dimethyl titanocene; (**c**) constrained geometry catalyst (CGC) [17] for solution polymerization of polyethylene

7.3.3 Single Site Catalysts/Metallocenes

The polymerization activity of metal organic compounds (metallocenes) has already been described by Natta [4]; some basic structures are shown in Fig. 7.9. The discovery of Sinn and Kaminsky [5] that MAO boosts the activity of these systems triggered an enormous interest in academic research as well as industrial development.

In contrast to relatively ill-defined Ziegler–Natta catalysts, metallocene structures are compounds of high purity. They normally possess one active catalytic center, hence their name single site catalyst. This leads to polymers with narrow molecular weight distribution and an even co-monomer distribution in the case of copolymerization. Their uniformity helped determining the polymerization mechanism of Ziegler–Natta catalysts, especially explaining insertion of ethylene, as the mechanism is pretty similar for both [9, 10]. The early single site catalysts are built upon a metallocene complex, composed of a zirconium, titanium, or hafnium, with two ligands either Cl or CH_3 between two cyclopentadienyl rings (Fig. 7.9).

Due to the controlled geometry of later catalyst generations "constrained geometry catalysts" (CGC—see Fig. 7.9) it is possible to determine the tacticity, molecular weight distribution, and chain length of polymers.

Single site catalysts, not derived from metallocenes, but resting upon multidentate chelate complexes were introduced in the 1990s. Homogenous catalyst systems, composed of late-transition metals (Fe, Ni, Pd), are currently in the focus of olefin polymerization catalysis research [3, 13].

7.4 Processes for the Production of PE

Historically the first commercial PE production process was developed by ICI in 1933 following an accidental discovery [1]. However, the big success of PE material leading to the number one polymeric material used worldwide was mainly based on the low-pressure processes building on Ziegler–Natta and Phillips catalysts. In the coming paragraphs the main low-pressure processes are described in brief. It is important to mention that catalyst, process, and polymer design are closely interlinked and essential to tailor end properties of materials.

7.4.1 Suspension (Slurry) Process

Slurry or suspension processes use an inert hydrocarbon, as diluent and autoclaves or loop reactors are basic rector designs. Suspension processes are widely used for the production of HDPE and LLDPE. However, higher concentrations of co-monomer can create problems as a significant portion of the produced polymer can dissolve in the hydrocarbon diluent. Typical diluents are hexane, isobutane, or propane. Low-boiling diluents have the advantage to be easier removed from the polymer but require more care in the operation of the plant.

Autoclave Process

The typical representative for an autoclave process is the Hoechst process [10]. The pressure is between 0.5 and 1.0 MPa, and typical temperatures are in the range of 80–90 °C. The diluent is hexane and large reactors (ca. 100 m^3) are used. For more complex polymer design (bimodal materials) two reactors in series are employed. Slurry concentrations vary between 15 and 45 wt% dependent on the diluent, particle size and shape. Limiting factors are the heat transfer of the reaction heat to the cooling jacket and the viscosity of the slurry.

Loop Process

Phillips introduced the loop reactor for the production of PE. Compared to autoclaves it has the advantage of a higher surface-to-volume ratio, facilitating better hear removal and allowing short residence times [10]. Typical conditions are temperatures of 100 °C and pressures of 3–4 MPa, for Phillips catalysts and isobutane as diluent. However, conditions have to be tuned for actual used catalyst system and diluent.

7.4.2 Gas-Phase Process

A fluidized-bed reactor (FBR) for the production of HDPE was developed by Union Carbide in the late 1960s [10]. The gas-phase technology for PE production rapidly increased following the development of LLDPEs. Here the fluidized-bed process has the advantage of lower capital and operating costs compared with competing solution and high pressure processes. Another advantage is the ability to produce either LLDPE or HDPE in the same plant ("swing plant") according to demand. The gas-phase process can produce a wide range of molecular weights and densities since it has no constraints related to viscosity (solution process) and to solubility (slurry process) of the co-monomer or the hydrogen. To limit particle carryover into the recycle streams the fluidized-bed reactors have a characteristic design with an expanded section on top of the reactor.

7.4.3 *Solution Process*

In a solution process the ethylene is dissolved in a diluent such as cyclohexane and fed into the polymerization reactor at high pressures (ca. 10 MPa) and high temperatures (200–300 °C). Residence times are short (ca. 2 min) and soluble catalysts are used. The advantages of the solution process are that it can handle a broad range of co-monomers, handle very low densities and that it tends to produce more narrow molecular weight distributions. It is limited in the viscosity of the solution like the high-pressure process [10].

7.4.4 *Multistage Process*

As described in the next section an important tool to adjust polymer properties in PE is to create multimodal molecular weight and/or co-monomer distributions. Multi-stage processes bring substantial flexibility in the tailoring of polymer properties. Two-stage polymerizations are typically [18, 19] and basically all known polymerization technologies (slurry, gas phase, solution) can be operated in a series of reactors in order to achieve multimodal polyethylenes (PEs). Examples are Hostalen (LyondellBasell) [20] for a combination of slurry reactors and Unipol II (Dow) for gas-phase technology. But there are also combinations of different technologies such as Borstar (Borealis), a combination of slurry and gas phase [21]. Figure 7.10 [8] pictures the Borstar process just one example, also indicating the basic design of a loop and a gas-phase reactor. With all these technologies bimodal molecular weight distributions (MWDs) can be produced, Fig. 7.11 [22].

7.5 Multimodality: Tailoring the Polymer Properties

While average co-monomer content and distribution, and average molecular weight are influencing polyethylene features that are often observed as primary descriptors of the material, the actual physical performance of polyethylene depends mostly on

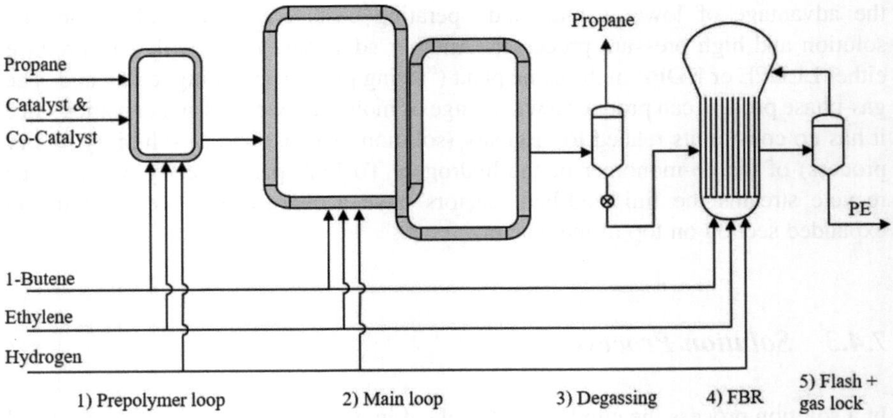

Fig. 7.10 Borstar polyethylene process with pre-polymerization loop, a main loop reactor and one fluidized-bed reactor (FBR) [8]

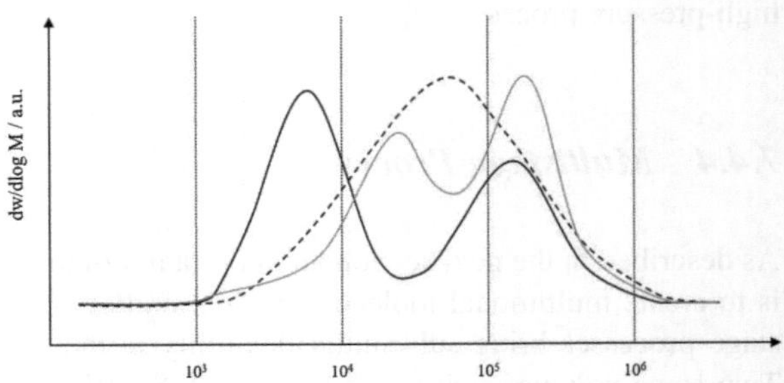

Fig. 7.11 Illustration of a typical unimodal (dashed) and two different bimodal molecular weight distributions (MWD) [22]

the composition of the bulk polymer. That means the relative amounts of polymer chains with different lengths and/or content of copolymer. The composition is often described as molecular weight distribution, and it is often combined with the distribution of co-monomer across the molecular weight.

Furthermore, fractions of polymer chains with certain length contribute to specific physical performance attributes. So is for example mechanical strengths, environmental stress crack resistance and melt strength of the material are dependent on the amount and length of longer chains, and shorter polymer chains contribute to better processability of the material.

Depending on the composition of bulk polyethylene, we often discuss modality of the material. When most chains constituting bulk polyethylene are of a similar length and co-monomer content, we describe the polymer as unimodal and narrow. Narrowness is a reflection of the molecular weight distribution (MWD). Typical example of a polyethylene that is narrow and unimodal is single site polyethylene, polyethylene made with a single site catalyst.

When polymer chains do differ in length, with longest chains much longer than the shortest one, and where polymer chains of all lengths between the two are also represented, we describe the polymer as unimodal and broad. Products made with Ziegler–Natta catalysts are typically broader materials and if a Chromium catalyst is used the materials are extremely broad, however still unimodal.

When it is possible to recognize at least two distinct groups of polymer chains with different lengths, we describe the polymer as multi- or bi-modal.

Less often discussed multimodality refers to differences in co-monomer content only. If a bulk polyethylene material contains two types of polymer chains that are mostly similar in length but differ in the number of branches, we discuss so-called density multimodality. A generic curve is shown in Fig. 7.11, exaggerating for emphasis.

7.5.1 Benefits of Multimodality for PE

Material performance requirements for most applications where polyethylene is being used are rather complex and not always achievable with a unimodal bulk polymer. So for example, if mechanical strength is needed for an application, the polymer will need to contain high molecular weight chains. Unimodal polyethylene with high average molecular weight has high viscosity and it is very difficult to process, it requires a lot of energy and throughput in conversion is low [18, 23]. It is therefore beneficial to combine long polymer chains with shorter ones. That way, the polyethylene is made broad, sometimes bimodal, and its viscosity will be reduced when melted and under shear (Fig. 7.12). The effect is known as shear thinning [24].

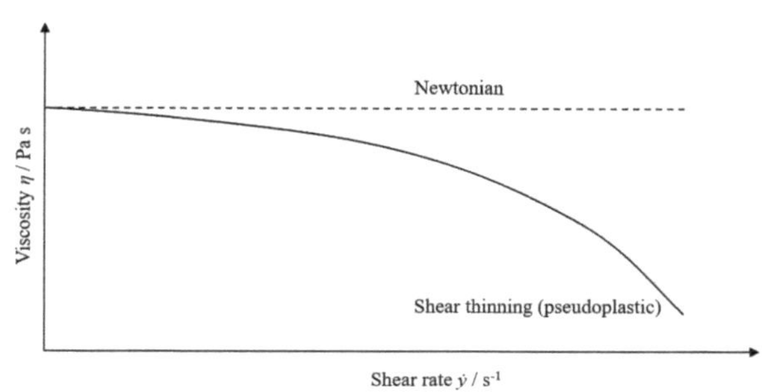

Fig. 7.12 Newtonian and shear thinning viscosity behavior

The application that is practically fully dependent on bimodality and presence of at least two types of polymer chains in the bulk material is polyethylene pipe. Polyethylene pipes are nowadays used in the construction of municipal building for transportation of fluids under pressure. They need to sustain high pressure over a period of at least 50 years or longer, as long as the building is in use. The pipes are being made from pure polyethylene and it is obvious that the material performance requirement is very high.

Polyethylene molecules, polymer chains, need to form a good barrier for the fluids in order to sustain the pressure. Therefore the chains need to crystallize leaving little room for molecules of the fluid to pass between the crystalline lamellae. This is the reason why polyethylene used for pipe manufacturing has very high density and little co-monomer is being used in its manufacturing. At the same time, and in order to further foster better crystallization, the polymer chains that are expected to form the barrier need to be of limited length, somewhat lower molecular weight.

Crystalline lamellae are however not sufficient. While material containing only high-density low molecular weight polyethylene chains organized in crystalline lamellae can sustain high pressure instantaneously, it will fail over time. Under stress, lamellae will start slipping and the pipe will crack along its length. In order to sustain pressure over time, the lamellae need to be connected and kept tied together. That is why the polyethylene material has to contain the so-called tie molecules. These are long polymer chains that are mostly amorphous but can and do partially co-crystallize in the primary crystallite blocks generating connection between them (Fig. 7.13).

Polyethylene that is used for manufacturing high performance pipes obviously needs to contain at least two types of molecules: short homo-polymer chains that crystallize and build lamellae, and long chains that will stay amorphous and therefore contain some co-monomer. With other words, the polyethylene needs to be bimodal [25].

Basically two ways reaching this target are possible, (a) multi-reactor and (b) multi-catalyst (dual catalyst) designs. In the first option a reactor cascade is

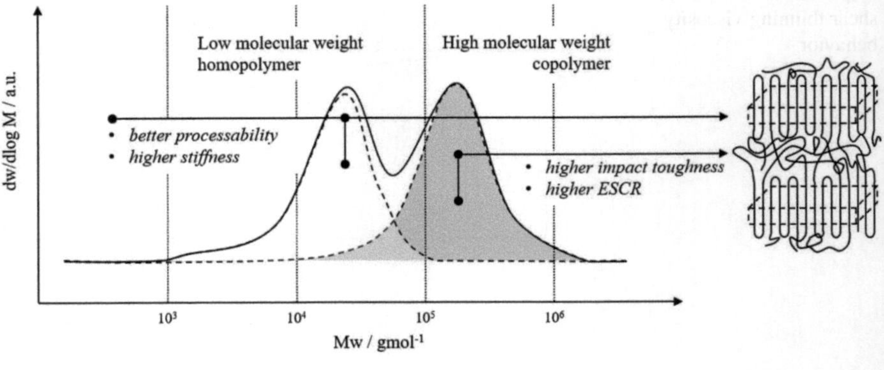

Fig. 7.13 Polyethylene pipe morphological schematics [25]

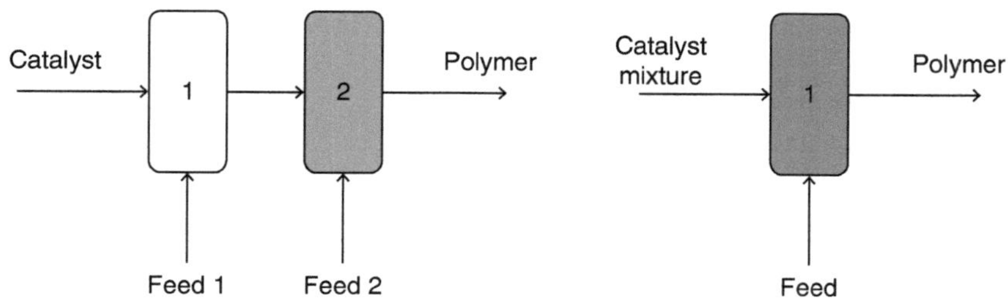

Fig. 7.14 Schematic drawing showing the different routes in obtaining bimodal PE resins

employed using only one catalyst–co-catalyst system. And for the second route a catalyst mixture is introduced in a single reactor. A schematic overview is given in Fig. 7.14.

7.5.2 *Multi-Reactor Design*

Pursuing the need to further develop polyethylene materials by securing additional degrees of freedom that would enable tailoring and fine-tuning the performance of the polymer, many companies have developed and implemented polymerization processes that include more than one polymerization reactor, most often operated in series. For a multi-reactor setup (see Fig. 7.10 as an example) this means, that during the polymerization process always the same catalyst chemistry is active. The advantage of this route is the high flexibility in developing new resins or optimizing existing grades. However, the employed catalyst system has to fulfill the following criteria:

- Producing a rather narrow molecular weight distribution.
- Possibility to vary the average molecular mass over a wide range with hydrogen.
- High activity over a long time, meeting the residence time requirements of the reactor train.
- Co-monomer incorporation in the required amounts.

Modern $MgCl_2$ supported Ti catalysts fulfill these demands in an excellent way. The average molecular weight can be adjusted over a wide range without substantial loss in activity (see Fig. 7.15) [26].

In addition the catalyst has to show acceptable copolymerization behavior with higher α-olefins. Because of the favorable incorporation of ethylene by the catalyst into the growing polymer chain, the reactors must be heavily loaded with the co-monomer. However, this can be handled because the co-monomer is recycled with the diluent stream.

Polymerization conditions in these reactors are different so that the catalyst produces different polymers depending on the location in the process train. The

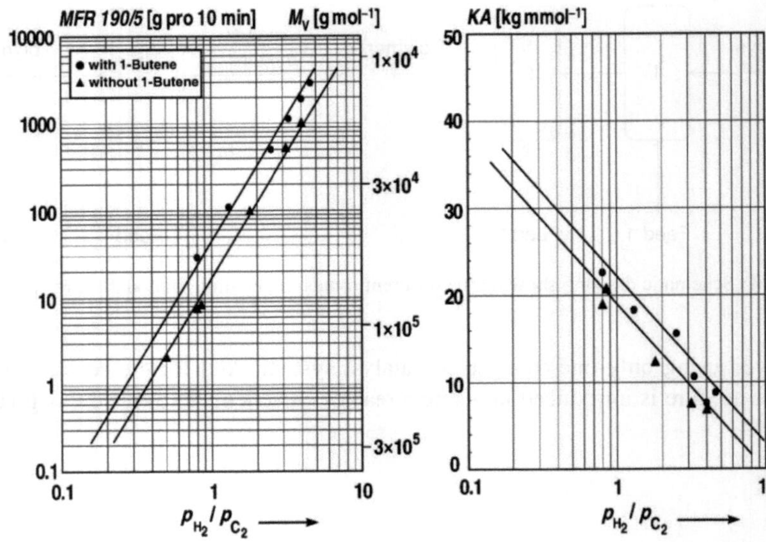

Fig. 7.15 Control of the melt flow rate MFR 190/5 with hydrogen (melt flow rate is a simple method to evaluate the average molecular weight) and influencing the catalyst productivity KA; $T = 85$ °C [26]

catalyst is being injected to the first reactor. In a particle forming process, polymer particle formation starts there. Polyethylene "grows" in and around the catalyst particle. The polymer chain characteristics, length and branching, of the polymer produced instantaneously depend on the polymerization conditions in the reactor where the particle is found.

In a typical cascade process (see Fig. 7.10), in the first reactor (in the example this is a loop reactor) high amounts of hydrogen are fed together with the ethylene in order to afford a low molecular mass PE. The second reactor (gas-phase reactor in the example of Fig. 7.10) is loaded with much less hydrogen to form a high- or very high molecular weight polymer. With this already two independent parameters, average molecular mass of the produced species in each reactor and their relative weight fraction, to tailor the molecular weight distribution of the PE resin (Fig. 7.11) are available.

In addition the cascade process allows a further tool for product modification: the incorporation of co-monomer in different amounts and quality in the different reactors. In this way a polymer blend is created comprising of a low molecular weight homopolymer and a high molecular weight copolymer (e.g., ethylene/1-butene copolymer). These materials build up a polymer blend, which in the solid state consists of crystalline regions (crystal lamella) and amorphous regions in between. The low molecular weight fraction mainly contributes to the crystalline regions whereas the high molecular weight copolymer forms the amorphous parts and acts as tie molecules connecting the crystalline areas.

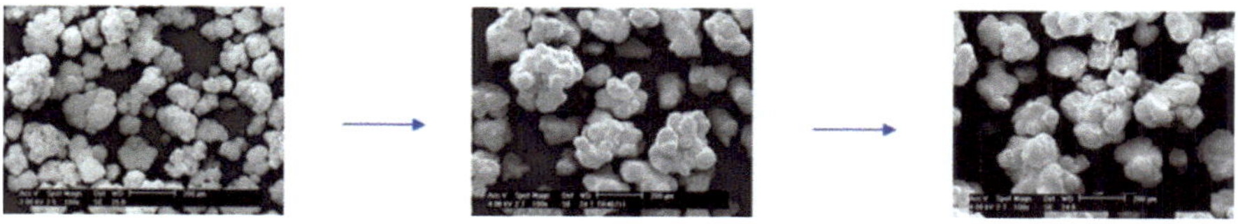

Fig. 7.16 Growth of polyethylene particles throughout the Borstar process, left picture shows particles from the pre-polymerization, the center picture material from the loop reactor and the right picture shows particles leaving the final gas-phase reactor [27]

The final result, polyethylene that exits the process, are polymer particles that each contain all fractions made along the process, in all reactors (Fig. 7.16). The morphology of the particles depends on the catalyst kinetics and early stages of the polymerization, but it more often described as the so-called onion structure. That means that polymer chains made in the first reactor are closer to the center of the particle. Chains made in the following reactor generate a layer covering the nucleus and so on. With a multi-reactor polymerization process operated in series, one generates multimodal polyethylene that is already mixed to a degree that enables good homogenization of the material.

Also a non-particle-forming process, solution polymerization process where polymerization occurs at high temperature and molten polymer is diluted-dissolved in a solvent, can produce multimodal polymer. The process that contains typically two reactors can be operated in series or parallel. Since particles are not being formed and the polymer fractions are mixed in the process, the significance of homogenization is less pronounced in the solution polymerization process.

7.5.3 One-Reactor Process

Earlier developed, simpler, and typically economically more efficient polymerization processes are based on one polymerization reactor operated in continues mode. Polymerization conditions in such a reactor are maintained homogeneous across the volume of the vessel and the reactor is described as a continuous stirred tank reactor (CSTR).

In a polymerization process that includes only one polymerization CSTR reactor compositional structure of polyethylene produced depends exclusively on the catalyst, that is, on the number of types of sites active in enabling the polymerization. This is quite obvious since the conditions are uniform and molecular weight and branching control is influencing only the average molecular weight and branching.

On the other hand, every type of active sites that can be found in the catalyst is characterized by its inherent ability to incorporate co-monomer and to propagate the polymer chain under the given polymerization conditions. That means, the composite result of the polymerization will be a mixture of polymer chains with different lengths and number of branches. Consequently, the so-called single site catalysts when used in a one-reactor polymerization process will produce narrow and unimodal polyethylene. In addition, the material will have the so-called uniform co-monomer distribution, all polymer chains regardless their length will contain branches distributed at same frequency, number of branches per carbon atoms of the back bone polymer. All metallocene and some post-metallocene catalysts belong to the category of single site catalysts.

Typically, narrow and unimodal polyethylene made with single site catalysts exhibits good mechanical properties in terms of toughness but it is very difficult to process requiring more energy for converting. Other typically used catalysts for polymerization of ethylene, Ziegler–Natta and chromium-based ones, contain many types of active sites that respond differently to the polymerization conditions. Therefore, when used in a one-reactor process they produce broader but still unimodal polyethylene. Unimodal polyethylene produced with Ziegler–Natta catalysts is typically better processible than the unimodal single site material. Even better processible is polyethylene made with chromium catalysts that contain large number of types of active sites. These broad polymers that contain little or no branches, HDPE, are due to high processability used for injection molding applications where high flow of a low viscosity material is needed.

Broad and multimodal, typically bimodal polyethylene can be produced in a one-reactor polymerization process when two different catalysts are being used. In contrast to the multi-reactor design described in the previous paragraph the multi-catalyst design uses only a single reactor to achieve multimodal resins. The concept of utilizing one bi-centric catalyst and/or two different catalysts that can produce bimodal polymer in one reactor is initially attractive since it appears less cost and capital intensive to modify the catalyst as oppose to modifying the process [28]. In practice, there are a few challenges that need to be addressed when adopting the strategy. Concepts in the catalyst development include hybrid systems loaded on a support [18, 29–34], and mixed catalysts. Mixed catalysts are a physical mixture of independent catalysts that are introduced together into the reactor [18, 29, 30].

A bi-centric catalyst would need to be designed for each polyethylene grade since it would not be possible to fine tune responses of such combination to changes in polymerization conditions. If two catalysts are used in a particle forming process, two types of polymer particles are formed keeping two polyethylene fractions physically separated. Mixing and homogenizing the two fractions can present a significant challenge that will be discussed later in the chapter.

The basic idea of the dual catalyst concept is that different active sites are present in the reactor at the same time able to produce different predominant molecular weights and densities at the given reactor condition (hydrogen content, co-monomer concentration). The two catalysts need to have different inherent ability to propagate polymer chain with different responses to the chain transfer agent. Operating under the same polymerization conditions, the two catalysts produce two types of polyethylene that combined presents a bimodal polyethylene. An example of the technology where a multimodal catalyst is used to produce bimodal polyethylene in one reactor is PRODIGY® Technology [35] developed by Univation.

Most described examples are based the combination of a heterogeneous Ziegler–Natta system [29, 30, 32, 34, 36] or Phillips catalyst [37–44] with a single-site component. In these examples the Ziegler–Natta and the Phillips catalyst systems provide the high molecular weight fraction giving the PE resin the desired toughness. The single site component accounts for the low molecular weight or lubricating part easing the processing of the bimodal PE material. However, also examples for the combination of different metallocene catalysts have been described [31, 37, 45].

7.5.4 *Properties of Bimodal PE Resins*

Bimodal PE resins show an excellent property combination originating form the bimodal molecular weight distribution and the heterogeneous co-monomer distribution. This can be best illustrated by the PE products used for pipe applications (drinking water-, gas-, and sewage pipes) [46, 47]. Pipe products made with this resins have a high stiffness (elastic modulus) due to the high-density homopolymer in the blend and a high resistance against slow crack growth (SCG) under stress [48–50]. Slow crack growth of pipe materials can be evaluated in laboratory scale for example by the full notch creep test [51, 52]. Today, bimodal PE resins used in pipe applications are classified as PE100 materials. This means the pipe must withstand a hoop stress of 10 MPa for up to 50 years at 20 °C (ISO 12162). Beside the resistance against SCG also the resistance against rapid crack propagation is an important feature. Rapid crack propagation can occur when a pipe is damaged, a crack can grow very quickly over a long distance driven by the energy stored in the pipe. This property is measured with the S4 test (small scale steady state, ISO 13477). Also here bimodal PE resins show an improved behavior compared with older generations of PE materials.

However, this class of materials is also of interest for many other applications. For packaging solutions, the weight of containers can be reduced due to the higher stiffness paired with high stress-crack-resistance also described as environmental stress crack resistance (ESCR, DIN 55457). Also film grades are based on bimodal PE resins accounting for a reduction of the film thickness (thin films below 10 μm).

This special property combination is the main driver for the success of PE materials in the abovementioned applications.

7.5.5 *Homogeneity*

Homogeneity of polyethylene is a feature that is needed for all applications. It is desirable that different polymer chains in a multimodal polyethylene are well mixed throughout the particles, pellets, and bulk material. The requirement extends of course only up to a certain level where the functionality of different types of polymer chains depends on the agglomeration of the same chains, that is, crystalline lamellae in pipe material. Still, at the macro and meso level, segregation of two or more types of polymer is lowering overall performance of the material.

By definition, multimodal polyethylene contains two or more types of polymer chains that are different in length, molecular weight. Consequently, it contains two or more types of polymer that when melted exhibit very different viscosities. Mixing of the materials with different viscosities is a challenge that sometimes leads to inhomogeneity of the bulk material. As described above, an answer to the challenge is in pre-mixing the polyethylene fractions already in the polymerization phase, during the particle formation.

In the process of polymerizing ethylene, a (supported) catalyst particle fractures to enable growth of polymer around the active sites. Depending on the quality of polymer generated in the first moment of polymerization, the particle can fracture more or less creating cavities open for further polymer growth or squeezing softer polymer out of the pores [53].

Depending on the very initial stages of the polymerization, polymer particle can be formed to enable good pre-mixing of different fractions or not [54].

There are obviously other possibilities to improve homogeneity of bimodal polyethylene. An example can be advances in the extrusion technology that contribute to improved products as well.

7.6 Summary and Conclusions

Polyethylene materials have been developed over the past 60 years into the largest polymeric material used worldwide serving many different applications. This success story is based on the ability to tailor the material properties of PE in a wide range and by the excellent price performance relationship for this class of polymers. The raw material, ethylene, is a large volume industrial chemical available at reasonable costs. The industrial processes are mature and working at low costs.

The development routes rest on the availability of optimized catalyst–co-catalyst systems, the ongoing tuning of the polymerization technology and a sound knowledge of physical and chemical processes on all scales. Understanding structure–property relationships along the whole value chain is essential as well. Especially bimodal PE materials with their ample possibilities to fine tune product properties will further expand the existing product windows.

Last but not least, PE is also shows an advantage regarding its reuse. After its lifetime it can easily be used for energy production or more preferably be reused like other thermoplastic materials.

PE is due to its versatility and relatively low price the material of choice also for the future. It also will help to solve one of our society biggest problems, the supply with fresh water. Only by the use of plastic pipes it will be possible to install water lines at an acceptable level of capital expenditure.

References

1. Plastics Europe: The Facts 2014. Retrieved November 9, 2018, from https://www.plasticseurope.org/download_file/view/479/179.
2. Roberts, R. M. (1989). *Serendipity: Accidental discoveries in science*. New York: Wiley.
3. Mülhaupt, R. (2004). Hermann Staudinger und der Ursprung der Makromolekularen Chemie. *Angewandte Chemie, 116*, 1072–1080.
4. Malpass, D. (2010). *Introduction to industrial polyethylene: Properties, catalysts, and processes*. Salem, MA: Wiley-Scrivener.
5. Galli, P., & Vecellio, G. (2001). Technology: Driving force behind innovation and growth of polyolefins. *Progress in Polymer Science, 26*, 1287–1336.
6. Shamiri, A., Chakrabarti, M. H., Jahan, S., Hussain, M. A., Kaminsky, W., Aravind, P. V., et al. (2014). The influence of Ziegler-Natta and metallocene catalysts on polyolefin structure, properties, and processing ability. *Materials (Basel), 7*, 5069–5108.
7. Kissin, Y. V. (2013). *Polyethylene end-use properties and their physical meaning*. Munich: Hanser Verlag.
8. Soares, J. B. P., & McKenna, T. F. L. (2012). *Polyolefin reaction engineering*. Weinheim: Wiley-VCH.
9. https://www.ptonline.com/articles/no-6%2D%2D-lldpe.
10. Whiteley, K. S., Heggs, T. G., Koch, H., Mawer, R. L., & Immel, W. (2005). *Polyolefins, Ulmann's Encyclopedia of industrial chemistry*. Weinheim: Wiley-VCH.
11. van Krevelen, D. W., & Nijenhuis, K. (2009). *Properties of polymers: Their correlation with chemical structure; their numerical estimation and prediction from additive group contributions* (4th ed.). Amsterdam, Boston: Elsevier.
12. Smallman, R. E., & Ngan, A. H. W. (2007). *Physical metallurgy and advanced materials* (7th ed.). Amsterdam, Boston: Butterworth Heinemann.
13. Steinborn, D. (2010). *Grundlagen der metallorganischen Komplexkatalyse*. Wiesbaden: Vieweg & Teubner.
14. Cossee, P. (1964). Ziegler-Natta catalysis I. Mechanism of polymerization of α-olefins with Ziegler-Natta catalysts. *Journal of Catalysis, 3*, 80–88.
15. Arlman, E. J. (1964). Ziegler-Natta catalysis II. Surface structure of layer-lattice transition metal chlorides. *Journal of Catalysis, 3*, 89–98.
16. Cossee, P., & Arlman, E. J. (1964). Ziegler-Natta catalysis III. Stereospecific polymerization of propene with the catalyst system TiCl-AlEt. *Journal of Catalysis, 3*, 99–104.
17. Hightower, J. W. (1996, June 30–July 5). *11th International congress on catalysis*, Baltimore, USA. (*Studies in surface science and catalysis*, vol. 101), Amsterdam, NY: Elsevier (1996).
18. Alt, F., Böhm, L. L., & Enderle, H. (2001). *Macromolecules, 163*, 135–143.
19. Abedi, S., & Hassanpour, N. (2006). Preparation of bimodal polypropylene in two-step polymerization. *Journal of Applied Polymer Science, 101*, 1456–1462.
20. Daftaribesheli, M. (2009). Ph.D. thesis, University Twente.
21. Fernandes, F. A. N., & Lona, L. M. F. (2004). Multizone circulating reactor modeling for gas-phase polymerization. I. Reactor modeling. *Journal of Applied Polymer Science, 93*, 1042–1052.
22. Ruff, M., & Paulik, C. (2012). Controlling polyolefin properties by in-reactor blending, 1—Polymerization process, precise kinetics, and molecular properties of UHMW-PE polymers. *Macromolecular Reaction Engineering, 6*, 302–317.
23. Böhm, L. L., Enderle, H. F., & Fleibner, M. (1992). High density polyethylene pipe resins. *Advanced Materials, 4*(3), 234–238.
24. Mezger, T. G. (2006). *The rheology handbook: For users of rotational and oscillatory rheometers* (2nd ed.). Hannover: Vincentz Network.
25. PE Pipe-design and Installation, American Water Works Association, 12 January 2011.
26. Böhm, L. L. (2003). The ethylene polymerization with Ziegler catalysts: Fifty Years after the discovery. *Angewandte Chemie, International Edition, 42*, 5010–5030.

27. Fredriksen, S. B. (2005, June). ECOREP III.
28. Liu, H. T., Davey, C. R., & Shirodkar, P. P. (2003). Bimodal polyethylene products from UNIPOLTM single gas phase reactor using engineered catalysts. *Macromolecular Symposia, 195*, 309–316.
29. Cho, H. S., Chung, J. S., & Lee, W. Y. (2000). Control of molecular weight distribution for polyethylene catalyzed over Ziegler-Natta/metallocene hybrid and mixed catalysts. *Journal of Molecular Catalysis A: Chemical, 159*, 203–213.
30. Cho, H. S., Choi, Y. H., & Lee, W. Y. (2000). Characteristics of ethylene polymerization over Ziegler-Natta/metallocene catalysts comparison between hybrid and mixed catalysts. *Catalysis Today, 63*, 523–530.
31. Rodrigues, S., Silveira, F., dos Santos, J. H. Z., & Ferreira, M. L. (2004). An explanation for experimental behavior of hybrid metallocene silica supported catalyst for ethylene polymerization. *Journal of Molecular Catalysis A: Chemical, 216*, 19–27.
32. UNIVATION. Retrieved from http://www.univation.com/catalysts.prodigy.php
33. Forte, M. M. C., da Cunha, F. O. V., dos Santos, J. H. Z., & Zacca, J. J. (2003). Ethylene and 1-butene copolymerization catalyzed by a Ziegler–Natta/metallocene hybrid catalyst through a 2 factorial experimental design. *Polymer, 44*, 1377–1384.
34. Hong, S. C., Mihan, S., Lilge, D., Delux, L., & Rief, U. (2007). Immobilized Me Si(C Me)(N-tBu)TiCl/(nBuCp) ZrCl hybrid metallocene catalyst system for the production of poly(ethylene-co-hexene) with pseudo-bimodal molecular weight and inverse comonomer distribution. *Polymer Engineering and Science, 47*, 131–139.
35. Ahmadi, M., Jamjah, R., Nekoomanesh, M., Zohuri, G. H., & Arabi, H. (2007). Ziegler-Natta/metallocene hybrid catalyst for ethylene polymerization. *Macromolecular Reaction Engineering, 1*, 604–610.
36. Lopez-Linares, F., Diaz-Barrios, A., Ortega, H., Matos, J. O., Joskowicz, P., & Agrifoglio, G. (2000). Toward the bimodality of polyethylene, initiated with a mixture of a Ziegler–Natta and a metallocene/MAO catalyst system. *Journal of Molecular Catalysis A: Chemical, 159*, 269–272.
37. Lhost, O., & Zandona, N. (2000, January 11). US Patent 6-013-595, Solvay.
38. Follestad, A., Almquist, V., Ommundsen, E., & Dreng, T. (2003, April 1). Borealis Technology Oy, US Patent 6-541-581-B1.
39. Follestad, A., Jens, K. J., Blom, R., & Dahl, I. M. (2004, September 21). Borealis Technology Oy, US Patent 6-794-326-B1.
40. Almquist, V., Dreng, T., Follestad, A., & Ommundsen, E. (1999, August 12). Borealis Technology Oy, Patent WO 99/40131.
41. Follestad, A., Almquist, V., Palmqvist, U., & Hokkanen, H. (2004, August 24). Borealis Technology Oy, US Patent 6-780-809-B1.
42. Yamamoto, K., Ishihama, Y., & Sakata, K. (2010). Preparation of bimodal HDPEs with metallocene on Cr-montmorillonite support. *Journal of Polymer Science, Part A: Polymer Chemistry, 48*, 3722–3728.
43. Moreno, J., van Grieken, R., Carrero, A., & Paredes, B. (2011). Development of novel chromium oxide/metallocene hybrid catalysts for bimodal polyethylene. *Polymer, 52*, 1891–1899.
44. Paredes, B., van Grieken, R., Carrero, A., Moreno, J., & Moral, A. (2012). Chromium oxide/metallocene binary catalysts for bimodal polyethylene: Hydrogen effects. *Chemical Engineering Journal, 213*, 62–69.
45. Chu, K. J., Soares, J. B. P., & Penlidis, A. (2000). Effect of hydrogen on ethylene polymerization using in-situ supported metallocene catalysts. *Macromolecular Chemistry and Physics, 201*, 340–348.
46. Scheirs, J., Böhm, L. L., Boot, J. C., & Leevers, P. S. (1996). PE100 resins for pipe applications: Continuing the development into the 21 century. *Trends in Polymer Science, 4*, 408–415.
47. Lüker, H., & Schulte, U. (1995). PE-Rohre – Kunststoffe mit verbessertem Leistungsniveau. *Kunststoffe, 85*, 1127–1128.

48. Richard, K., & Diedrich, G. (1956). Rohre aus Niederdruckpolyäthylen – Eigenschaften und Erprobung in Labor und Praxis. *Kunststoffe, 46*, 183–190.
49. Richard, K., Gaube, E., & Diedrich, G. (1959). Trinkwasserrohre aus Niederdruckpolyäthylen. *Kunststoffe, 49*, 516–525.
50. Gaube, E., Diedrich, G., & Müller, W. (1976). Rohre aus thermoplastischen Kunststoffen - Erfahrungen aus 20 Jahren Zeitstandprüfung. *Kunststoffe, 66*, 2–8.
51. Fleißner, M. (1987). Langsames Rißwachstum und Zeitstandfestigkeit von Rohren aus Polyethylen. *Kunststoffe, 77*, 45–50.
52. Fleißner, M. (1998). Experience with a full notch creep test in determining the stress crack performance of polyethylenes. *Polymer Engineering and Science, 38*, 330–340.
53. McKenna, T. F. L., Di Martino, A., Weickert, G., & Soares, J. B. P. (2010). Particle growth during the polymerisation of olefins on supported catalysts, 1—Nascent polymer structures. *Macromolecular Reaction Engineering, 4*, 40–64.
54. Ruff, M., & Paulik, C. (2012). Controlling polyolefin properties by in-reactor blending: 2. Particle design. *Macromolecular Reaction Engineering, 7*, 71–83.

Summary and Perspectives

The industrial development of polymers is an example of extraordinary example of multidisciplinary cooperation involving experts from different fields. In less than a century from the discovery of these novel synthetic polymers, which largely influenced the technology and society over the last decades, the number of application fields has impressively broadened. The production process has been developed and implemented in such a way that costs and performance are largely optimized. This has been possible thanks to a constant evolution within all areas of the business, with several breakthrough innovations in the fields of catalyst, process technology, and material design.

The revolutionary demand has led to the need of developments in parallel in the different areas in order to fulfill the fast growing request. Therefore, many scientific details of a production process, that in practice work very well, still need to be properly understood and connected, and research is still very active in many fields related to polymerization. Significant changes and improvements can still be expected in the industrial process. In this book, an overview of the polyolefin production and the state of the art in many of the specific areas is provided, merging academic research and industrial knowledge to fill the gap between scientific understanding and practical processes.

The core of the polymerization process is the catalyst. The catalyst provides the active sites for the polymerization reaction and represents the actual reactor; in the heterogeneous processes it provides also the initial 3D structure developing into polymer particles during the polymerization process and that can be easily transported, recovered, and processed.

In this book it is clearly addressed how any kind of innovation development, past, present, or future, needs to take into account the proper understanding, management, and handling of the catalyst system in the given process technology and in consideration of the targeted polymer. The understanding of catalyst structure, performance, and its proper use during the production spans over many different length scales. In fact, it covers from the order of Ångströms, for describing the basic insertion process, through the size of the growing particles with characteristic

A. R. Albunia et al. (eds.), *Multimodal Polymers with Supported Catalysts*,
https://doi.org/10.1007/978-3-030-03476-4

diameters in the order of 10^{-6}–10^{-3} m, up to its operability in reactors with volumes in the order of several tens of cubic meters. Most of the single process can nowadays be described and modelled, still it is highlighted that presently the main challenge is represented by the capability to connect the different scale, as well as to link different competences, from the catalyst chemistry, through the reaction engineering, to the material science.

Heterogeneous Ziegler–Natta catalysts are largely used by the industry. They offer such robust performance and are beyond comparison in performance and cost; in many locations $TiCl_3$ catalyst, developed by Solvay in the 1970s, still survive to this day. Nevertheless, the development of new catalyst systems is still very active. Many factors are contributing to this continuous development. On the one hand, there is the need of having a better control over the industrial process and the aim to produce new materials. On the other hand, the updates in the legislation, the new studies about possible dangerous substances and the consequent concerns, and the following regulatory restriction, also will pull continuously for revising and adjusting the chemical recipes and synthetic processes. It has been shown, for example, how this has driven the development of catalyst with increased productivity, in order to reduce the catalytic residuals, or the change from external to internal donors, and more recently the replacement of phthalates as donor and the development of the sixth generation of ZN catalysts.

The better control over the industrial process was the driving force for the development of suitable support for heterogeneous catalysts as well as of different reactors and process technologies. $MgCl_2$ and SiO_2 are the two most common inorganic supports in the polyolefin industry. They are in use are since long time due to a very reliable performance profile in heterophasic polymerization process. However, the success achieved with these systems, and at such low production cost, sometimes ended up to be a constraint for the research on alternative design of the support materials. This is further complicated by the number and complexity of the factors to be considered and fixed, such as chemical composition of the support, its surface characteristics, morphology (surface area, particle size, particle size distribution, and porosity), and mechanical strength. All these characteristics can be partly tailored by modifying the synthetic preparation process of the support. The target is defined by the performance of the catalyst in a desired process, namely the balance of productivity and operability, the kinetic profile, and polymer morphology.

Although we have seen that catalysts can be characterized quite well in their main properties, what happens to the particles injected into the reactor during the polymerization is still a topic of research. The development of the particle morphology is as important as it is difficult to define. When it comes to the practical application, often changes and continuous innovation on the process side are performed for the fine tuning of the overall production, based on the available catalyst and needed polymer design. Through this fine tuning, the control of catalyst fragmentation and of the final particle morphology and replication phenomena is possible. However, in such continuous evolution of the process, the gap existing in understanding and being able to model the particle growth was not easy to close, implying an essentially empirical use of the catalyst. The bases for the operative conditions are provided by the impact of process conditions on the catalyst performance, the heat and mass transfer limitations during the polymerization reaction, and the uses of prepolymerization stage. Process changes and developments in reactor design are trying to mitigate a rapid particle expansion, ensuring that the energy is properly dissipated, reducing the probability of particles disintegrating, or growing with poorly controlled morphology, or with low or high bulk densities (porosities). In many cases, all the process parameters need to be adjusted depending on the initial form of the catalyst particles and the physical properties of the polymer itself.

The aim of producing new materials is still pushing the research. In spite of the basic structural elements constituting polyolefins, consisting of simple alkenes, the technology has evolved into creating rather complex structures, made up of a population of macro-chains exhibiting a wide range of intramolecular and intermolecular distributions. The development of product property is still very much progressing. Polymer design is tailored by knowing the structure–property relationship and targeting to specialty materials for specific applications. Thus, the continuous update on polymer design and properties continues to redefine the technology requirements in the polyolefin industry.

The diversity of commercial polyolefin grades is the result of advances in both the development of novel catalytic systems and the use of new reactor configurations. The need of tailoring as much as possible the molecular structure is at the base of the most recent research in supported molecular catalysts. For polyethylene this is becoming already somehow commoditized, although there is still a reasonable year-on-year growth.

The opportunity of producing polyolefins with desired molecular profile in cascade-reactor technologies, also combining gas- and slurry-phase processes, offered a great tool of really tailoring the material properties, but increased the complexity of the process to be handled, as catalytic performance and morphological particle properties need to be robust and controllable throughout the process.

Multimodal processes and designs have been developed for both polyethylene and polypropylene. From material performance point of view, multimodality offers the possibility to prepare in the reactor mixture of materials with complete different properties. By having different fractions, the number of different combinations largely increases and the range of property profile that can be reached is also dramatically extended. The wide variety of possible microstructures within the final polymeric material affects the mechanical, rheological, and optical properties to the extent that it is impossible to talk about the properties of a single polyolefinic material, but rather about the properties window of a class of polyolefins.

Multimodality represents a unique way to maximize the blending at molecular level in a powerful and cost-effective way. Up to four or five different discrete fractions can be produced in one single formulation. The technology, which defines the kind, number, and size of reactors, provides some of the constraints on the structure and amount of each fraction which can be produced, and on the sequence of the polymerization cascade. Nevertheless, it offers a strategic differentiation for companies competing for a leadership position.

In PP the continuous phase (matrix) and/or the dispersed phase of monophasic and/or heterophasic PP can be tailored to get unique property profiles. Multiple fractions allow for the production of soft materials still maintaining good thermal stability, or retarding critical failure by enriching the high molecular phase (random copolymers); or achieving an excellent balance of stiffness and impact while reaching optimal results in surface aesthetics, organoleptics, and transparency (bimodal rubbers). Development of a new material is still at the center of the industrial competition, trying to reach out new properties, for example combining a homopolymer fraction and random fraction in a continuous matrix, or heterophasic PP/EPR (ethylene–propylene rubber) with a trimodal matrix or a trimodal discrete phase.

In PE multimodality offers the opportunity to broaden the molecular weight distribution, with the lower molecular weight fraction maximizing the processability and the higher molecular weight fraction the mechanical strength. Multimodality is also in terms of density, so the higher density fraction provides stiffness and the lower density fraction long-term properties.

The catalysts needed to make PP and PE in a multistage process need to fulfill all the typical requirements already listed for standard unimodal polymerization process and maintain the performance over longer residence times a at different polymerization conditions. The kinetic profile and the lifetime of the active sites have to last over a wide polymerization time. Qualitative and quantitative incorporation of comonomers, (stereo)selectivity, and hydrogen response need to cover broad operational window. The particles need to growth in a homogeneous going from one reactor to the other, so the initial shape and porosity of the catalyst, as well as the control of the process parameters is even more important. The catalyst fragmentation pattern and the particle forming process are key for reaching multimodal materials where the fractions are properly and statistically mixed, although this is further complicated by the effect of controlling and keeping narrow the residence time distribution in multiple reactors. The in-reactor homogeneity of the multimodal polymer might have a direct influence on the final material performance, although further post-reactor processing of the material can still influence it. The balance of all these effects and requirements is driving the development, and for some classes of material it is defining the success of a technology over others.

Multimodal polymers have not yet revealed their whole potential, and all the elements needed to have a successful technology still remain a topic of research and development. Companies, focused on understanding structure–property relationships along the whole value chain and in the commercialization of new multimodal technologies at low production costs, are the drivers for this continuous development. A sound knowledge of physical and chemical processes on all scales, where there are still many open gaps, and the research in catalyst and process technology will remain attractive research fields for the next decades.

Index

A. R. Albunia et al. (eds.), *Multimodal Polymers with Supported Catalysts*,
https://doi.org/10.1007/978-3-030-03476-4

MIX
Papier aus verantwortungsvollen Quellen
Paper from responsible sources
FSC® C105338

If you have any concerns about our products,
you can contact us on
ProductSafety@springernature.com

In case Publisher is established outside the EU,
the EU authorized representative is:
Springer Nature Customer Service Center GmbH
Europaplatz 3, 69115 Heidelberg, Germany

Printed by Libri Plureos GmbH
in Hamburg, Germany